AF358807

Physicochemical, Sensory and Nutritional Properties of Foods Affected by Processing and Storage

Series II

Physicochemical, Sensory and Nutritional Properties of Foods Affected by Processing and Storage Series II

Editors

Sidonia Martinez
Javier Carballo

Basel • Beijing • Wuhan • Barcelona • Belgrade • Novi Sad • Cluj • Manchester

Editors
Sidonia Martinez
Food Technology Area,
Faculty of Science
University of Vigo
Vigo
Spain

Javier Carballo
Food Technology Area,
Faculty of Science
University of Vigo
Vigo
Spain

Editorial Office
MDPI
St. Alban-Anlage 66
4052 Basel, Switzerland

This is a reprint of articles from the Special Issue published online in the open access journal *Foods* (ISSN 2304-8158) (available at: https://www.mdpi.com/journal/foods/special_issues/Foods_Affected_Processing_Storage_Series_II).

For citation purposes, cite each article independently as indicated on the article page online and as indicated below:

Lastname, A.A.; Lastname, B.B. Article Title. *Journal Name* **Year**, *Volume Number*, Page Range.

ISBN 978-3-7258-0233-3 (Hbk)
ISBN 978-3-7258-0234-0 (PDF)
doi.org/10.3390/books978-3-7258-0234-0

Cover image courtesy of Sidonia Martinez

Contents

About the Editors

Sidonia Martinez

Sidonia Martinez is an Associate Professor of Food Technology at the Faculty of Sciences of the University of Vigo, Spain, where she has worked since November 2007. She obtained her B.S. (1996) and Ph.D. (2002) degrees in Veterinary Medicine from the University of León (Spain).

In recent years, she has studied the improvement of traditional foods (meat, meat products; milk; dairy products; starter cultures, fish fishery products, vegetables, antioxidant components). She has worked with different Brassica spp. and fish species and developed different products using them. In this scientific field, she is the co-author of 50 articles published in different journals indexed in the JCR, most of them in prestigious international journals (Food Microbiology, Food Control, Food Chemistry, Meat Science, etc.), the co-author of six research articles in prestigious national journals not indexed in the JCR with reviewers, and the co-author of two teaching publications. She is also the co-author of one book and five book chapters, as well as 30 long communications and 80 short communications presented at international and national conferences/events. She acts as a Guest Editor for the journal Foods. She acts regularly as a reviewer for the most reputed journals in Food Science and Technology. She has reviewed more than 100 articles published by different international journals. She has collaborated in 18 projects funded by public institutions/administrations (Principal Investigator in two of them) and has signed 8 contracts with companies (Principal Investigator in five of them). She is focused on the development of new food products, with the present project being an important contribution and reinforcement of this current work.

Javier Carballo

Javier Carballo is a Full Professor of Food Technology at the Faculty of Sciences of the University of Vigo, Spain. He obtained his B.S. (1984) and Ph.D. (1989) degrees in Veterinary Medicine from the University of León (Spain), both with Special End of Degree Awards. Later, he completed research at the Station de Recherches Laitières, Jouy-en-Josas, France.

In the last 25 years, he has studied and improvement of traditional foods, meat quality and the effects of the animal's diet, and the oxidation of edible fats and the use of natural antioxidants as an alternative to chemical additives. As the former President (from 2008 to 2016) of the Food Microbiology Section at the Spanish Society of Microbiology, he is a member of the Scientific Committee of many national and international congresses, and he is part of the Editorial Board of several reputed international journals. He is currently an Associate Editor of the journals Frontiers in Microbiology and Food Microbiology, Guest Editor of the journal Foods, and acts regularly as a reviewer for the most reputed journals in the Food Science and Technology field, having reviewed a total of 423 manuscripts. He has also acted assiduously as a translator of scientific books in the field of Food Science and Technology.

Preface

The nature of foods, particularly their perishability and/or seasonality, as well as the new demands of consumers who are increasingly exigent rearding the organoleptic and nutritional quality of food and more concerned about the relationship between food and health make food processing and storage an activity of growing scientific interest and commercial importance.

Knowledge and understanding of the chemical, physical and microbiological effects of the treatments applied to raw materials, as well as the incorporation of new ingredients with different purposes, are vital to obtaining increasingly safer, nutritious, and healthy foods that are capable of surprising and fulfilling the needs and expectations of the most diverse consumers.

In recent years, research efforts in this domain have multiplied, and they have led to advances that have made it possible to obtain increasingly satisfactory foods with increasingly advantageous quality–price ratios.

Much progress has been made with regard to generating knowledge in this field. However, the diverse nature and behaviour of food components, the complexity of the interactions and effects that occur in such components as a consequence of the environmental conditions established during the different technological processes and during storage, and new consumer trends and preferences mean that research in this field should not cease; this is so as to achieve increasingly efficient economical and sustainable processes.

As a continuation of a previous book that collected the results of 19 research works carried out on this same topic, this volume aims to gather some of the latest advances in this field. We hope that this book serves as a useful tool for students, researchers, and professionals working in this area.

Sidonia Martinez and Javier Carballo
Editors

 *foods*

Editorial

Physicochemical, Sensory, and Nutritional Properties of Foods Affected by Processing and Storage Series II

Sidonia Martínez * and Javier Carballo

Food Technology Area, Faculty of Sciences, University of Vigo, 32004 Ourense, Spain; carbatec@uvigo.es
* Correspondence: sidonia@uvigo.es

Citation: Martínez, S.; Carballo, J. Physicochemical, Sensory, and Nutritional Properties of Foods Affected by Processing and Storage Series II. *Foods* **2024**, *13*, 156. https://doi.org/10.3390/foods13010156

Received: 20 December 2023
Accepted: 23 December 2023
Published: 2 January 2024

Food processing has several different purposes. First, it aims to increase the shelf life of foods by protecting them from physical, chemical, and biological agents of deterioration or to sanitize them by destroying the pathogens that can be present in raw materials and compromise consumer health. However, nowadays, consumers are increasingly demanding and better informed, and they are concerned with the quality of the foods they consume and the consequent effects on their health; they also demand new, sensorially surprising foods with new aromas, flavors, and textures. This new challenge facing food technology to incorporate new ingredients and additives to obtain more nutritive, safe, and surprising foods occupies a good part of the efforts of the current food industry. On the other hand, food storage is a necessary practice due to the seasonal nature of the production of some foods or the need to ensure their supply in other cases.

Food processing and storage are therefore normal and necessary activities to obtain and enjoy the wide range of foods that we know, retaining their safety and diverse sensory and nutritional profiles. However, sometimes these processes have negative effects that result in a decrease in nutritional value, in an alteration of sensory properties, or even in a generation of compounds that can be harmful to the health of the consumer. To avoid undesirable effects and obtain increasingly nutritious foods that meet the needs and expectations of consumers, it is necessary to constantly generate knowledge about the effects of technological treatments and storage on food components and properties to optimize these processes and minimize their negative effects.

In a previous Special Issue devoted to this same topic [1], 19 contributions were collected, including different studies on food canning, dehydration, fermentation, irradiation, marinating, and cooking, as well as on the use of preservatives and the effect of storage. This new Special Issue is a continuation of the previous one and presents new advances in and approaches to food processing and storage, the optimization of processes, and the incorporation of new components that allow the valorization of materials and the production new foods, therefore adding value to the food industry.

Heat treatment is one of the oldest and most effective procedures used to reduce the activity of microorganisms and enzymes and prolong the shelf life of foods. The effectiveness of heat treatment and its effects on the organoleptic characteristics of foods are conditioned by the environment in which it is applied. Daei et al. studied the effects of two mild thermal treatments (63 °C or 40 °C for 3 min) and in two different packaging media (brine with different NaCl concentrations or a vinegar solution) on some of the physicochemical (weight loss, phenolic compounds, ascorbic acid, and firmness) and microbiological (microbial load) characteristics of truffles (*Terfezia claveryi*) during 160 days of storage. A control of unheated truffles was also studied. Rthe authors showed that both methods were effective in reducing the microbial load. The 63 °C treatment in vinegar reduced the weight loss and microbial spoilage and increased the firmness of truffles during storage. However, the contents of antioxidant compounds (phenolics and ascorbic acid) decreased as the intensity of heat treatment increased. Despite the loss of antioxidant activity, treatment at 63 °C for 3 min in a vinegar solution proved effective in increasing the shelf life of truffles without appreciable damage in quality attributes.

Canning (heat treatment after packaging in hermetic containers) is a classical process used for food preservation. After the initial study by Appert [2], researchers have not stopped trying to optimize the treatments, packaging materials, and filling media. Marine foods are typically canned, resulting in high quality fish and seafood, which is appreciated by consumers. The chemical components and sensory attributes of marine foods are differently affected by canning; accordingly, widely varying effects of canning have been reported. In this Special Issue, two manuscripts offer valuable contributions to this topic. Gómez-Limia et al. evaluated the effect of a filling medium (sunflower oil, olive, or spicy olive oil) on the color (CieLab parameters) and sensory characteristics of European eel (*Anguilla anguilla*) for different steps of the canning process (after frying, canning, and 2 and 12 months of storage). The authors found significant color changes during the process both in the fish and in the filling medium They reported that the color changed the most in eel packed in olive oil during sterilization, and spicy olive oil was the filling medium that experienced the largest color changes, probably because of the migration of some spice components into the oil during canning and storage. The sensory characteristics of the eel were also influenced by the filling medium, with the eels packed in sunflower oil receiving the highest consumer scores. Biogenic amines are important in canned seafoods as quality indicators and because of their harmful effects on health. Qu et al. studied the effect of the storage (temperature and time) on the biogenic amine content of five canned seafood species (mud carp, sardine, mantis shrimp, scallop, and oyster). The levels of nine biogenic amines (histamine (HIS), phenylethylamine (PHE), tyramine (TYM), putrescine (PUT), cadaverine (CAD), tryptamine (TRY), spermine (SPM), spermidine (SPD), and octopamine (OCT)) were analyzed in these canned seafoods along 12 months of storage at different temperatures (4, 10, 25, and 30 °C). The authors detected HIS, PHE, PUT, and TYM in 100% of the samples; CAD, SPM, and SPD were detected in 60%; and OCT was detected in 40% of the samples. Additionally, they observed that the recommended maximum limits of HYS an TYR were exceeded in the canned mud carp and scallop when stored at 30 °C. They concluded that low storage temperatures inhibit the production of biogenic amine in canned seafood.

Pulsed electric field (PEF) is an emerging and promising technology used for destroying microorganisms at low temperatures in a continuous flow regime [3]. Extensive research efforts have recently been devoted to the use of this technology to replace others that are more harmful to the organoleptic characteristics of foods or to their safety. Delso et al. [4] studied the use of PEF to decontaminate red wine, avoiding the use more problematic classical techniques such as the addition of sulfur dioxide or the sterilizing filtration. The authors evaluated the suitable of PEF for the decontamination of red wine after alcoholic (elimination of *Saccharomyces cerevisiae*) and malolactic (elimination of *Oenococcus oeni*) fermentation. Their results showed that the populations of these two microbial species could be reduced up to 4.0 log10 cycles by applying PEF, and sublethal damage and synergistic effects with SO_2 were observed. After 4 months, wine treated with this technology after alcoholic fermentation lacked viable yeasts, whereas that treated after malolactic fermentation and 20 ppm of SO_2 contained less than 100 CFU/mL of *O. oeni*. No detrimental effects on the oenological or sensory parameters of treated wines were observed after storage.

Cooking is possibly the oldest of the procedures used for preparing food for consumption. Freeze-drying is an almost perfect dehydration method because the elimination of water in situ via sublimation maintains the structure of the food, favoring its rehydration and the conservation of textures, flavors, and aromas. Cambero et al. studied the textural properties, sensory quality, and rehydration kinetics of three Spanish chickpea varieties to identify the most suitable cooking conditions to obtain high-quality freeze-dried chickpeas for use in the preparation of a traditional Spanish dish called *cocido*. The authors also evaluated the post-rehydration sensory quality of various vegetables and meat portions (the other constituents of this dish), which were cooked under different conditions and freeze-dried. The results show that by selecting the appropriate cooking and freeze-drying conditions for the ingredients. the sensory quality of this traditional Spanish dish could be reproduced after rehydration of the freeze-dried components with water, heating in a microwave oven for 5 min, and letting them rest for 10 min.

Food storage is, in most cases, a necessary operation because some foods are seasonally produced; in other cases, adapting the pattern of production to those of consumption is difficult. Some food preservation procedures (i.e., canning, dehydration, and irradiation) destroy or inhibit the biological agents that cause spoilage and allow the long-term storage of food. However, such preservation procedures modify the organoleptic properties of foods. Some foods should therefore be stored in the fresh state, and other strategies to minimize their alteration during storage must be adopted. Such is the case of pomegranate, a fruit that has recently received attention due to its healthy components. The shelf life of this fruit ranges from 3–4 months when stored in air to 4–6 months when stored in a controlled atmosphere. One of the alterations that can appear during storage and that limit the shelf life of this fruit is husk scald (HS). This is a superficial brown discoloration that is restricted to the husk and does not affect the interior of the fruit but substantially limits its acceptance by consumers. Maghoumi et al. reviewed the factors involved in the development of HS, the modes of action of these factors, and their association with postharvest treatments. The authors discussed a hypothesis regarding the etiology and mechanism of development of HS in view of research on this subject; they also considered the postharvest treatments proposed for its control and the possible targets of these treatments.

Tomato is another fruit which is preferably consumed fresh and that is how it should be kept. It is a climacteric fruit that typically has as shelf life of 7–11 days when stored at room temperature. To extend the shelf life and maintain the quality it is crucial to control the processes of respiration and transpiration, as well as microbial contamination. To achieve these objectives, the application of edible coatings has emerged as an alternative in recent years. In this special issue, Flores-López et al. evaluated the effect of an alginate/chitosan nanomultilayer coating without (NM) and with *Aloe vera* liquid fraction (NM+Av) on the postharvest quality of tomato fruit stored at 20 °C and 85% relative humidity. Authors reported that both nanomultilayer coatings had comparable effects on firmness and pH values. However, the NM+Av coating significantly reduced weight loss and molds and yeasts counts compared to uncoated fruit. Moreover, it notably lowered O_2 consumption and CO_2 production, also inhibiting ethylene synthesis. The usefulness of the NM+Av coating was confirmed by visual evaluation and instrumental color measurement. In view of their results, authors concluded that the NM+Av coating is a potential alternative to improve tomato preservation.

Freezing followed by frozen storage is the most successful procedure for long-term storage, producing after-thawing products that differ minimally from the fresh products. However, for optimal results, the freezing, storage, and thawing conditions for each product must be correctly adjusted. Wang et al. studied the effects of two different storage temperatures (−18 °C and −55 °C) for 180 days of storage on three different parts (naked body, big belly, and middle belly) of bluefin tuna. The effects were evaluated through determining the tissues' water-holding capacity, malondialdehyde content, color differences, salt-soluble protein content, free amino acid contents, endogenous fluorescent protein content, and water distribution and migration. The authors found minimal quality changes during short-term storage, but storage at −55 °C significantly improved the tuna quality parameters compared with those achieved with storage at −18 °C. The results showed that the water-holding capacity, salt-soluble protein content, and water content of tuna decreased as the storage time increased. The authors concluded that because the tuna quality changed little during short-term frozen storage, a storage temperature of −18 °C is suitable for short-term storage prior to sale.

As indicated at the beginning of this Editorial, consumers are becoming increasingly concerned about the quality of the foods they consume and their effects on health. At the same time, they are also demanding and attracted to new foods with different and surprising organoleptic properties. This demand has forced the food industry to strengthen its efforts to incorporate new ingredients and additives and to develop new manufacturing processes capable of generating different aromas, flavors, and textures. To overcome such challenges, in-depth research is necessary to clarify the effects of new ingredients, additives, and processes and of their interactions, as well as to develop analytical methods and quality control procedures to monitor these new products. The last five articles contained in this special issue are devoted to this research topic. The textural attributes of cooked rice are the dominant indicators of eating quality. Their appropriate assessment is therefore critical to obtain a satisfactory product. Kaewsorn et al. evaluated the precision and sensitivity of a back-extrusion (BE) test to assess the texture of cooked germinated brown rice (GBR) during the production process. They studied the effects of different soaking and incubation durations on the hardness, toughness, and stickiness of Khao Dawk Mali 105 GBR, noting the high precision of the BE test in the measurement of these properties, as indicated by the repeatability and reproducibility results. The sensitivity was also satisfactory, as indicated by the high coefficient of variation of the texture properties.

Rice is also widely used as source of starch, having several different destinations in the food industry. Starch gels have many uses, such as for the development of gluten-free starch-based products. To improve the gel structure and stability of the rice starch (RS) used for this purpose, Xu et al. studied the effect of the addition of guar gum and locust bean gum to rice starch gel on its pasting, rheological properties, and freeze–thaw stability. Analyses were performed with a rapid viscosity analyzer, rheometer, and texture analyzer. The authors found that both gums can modify the pasting properties, as indicated by increases in the peak, trough, and final viscosities, thereby preventing the short-term tendency of the retrogradation shown by RS. The measurements of dynamic viscoelasticity also showed that the starch–gum system exhibited improved viscoelastic properties (higher storage modulus (G′) values) compared with those of starch alone. Compared with the control system, the hysteresis loop area of the systems containing guar gum or locust bean gum was reduced by 37.7% and 24.2%, respectively, indicating that the addition of gum could enhance the shear resistance and structure recovery properties. The thermodynamic properties showed that both gums retarded the short- and long-term retrogradations of the RS gels. The addition of galactomannans significantly improved the textural properties and freeze–thaw stability of the RS gel, with guar gum being more effective than locust bean gum, which could be attributed to the differences in the mannose-to-galactose ratio. The results of this study provide alternatives for gluten-free recipes with improved freeze–thaw stability and texture properties.

During legume processing, by-products with applications in the food industry are generated. Among these is aquafaba, the liquid obtained after cooking legumes. Fuentes Choya et al. studied the composition (total solid, protein, fat, and carbohydrate contents) and culinary properties (foaming and emulsifying capacities, and foam and emulsion stabilities) of *Pedrosillano* (a traditional Spanish variety) chickpea aquafaba produced with different cooking liquids (water, vegetable broth, meat broth, and the filling liquid from canned chickpeas). Then, using these aquafaba types as ingredients, they prepared and studied French egg-free baked meringues, using meringues containing egg white as control. The sensory characteristics of the meringues were analyzed using instrumental and panel-tester techniques. The authors found that the nature of the cooking liquid and the intensity of the heat treatment affected the composition and culinary properties of the aquafaba, which showed appropriate foaming and intermediate emulsifying capacities in all cases, with the aquafaba prepared using the filling medium of commercial canned chickpeas producing the meringue the most similar to that produced with egg white in terms of performance. Regarding the meringue characteristics, the aquafaba-based products had fewer alveoli, higher hardness and fracturability, and minimal color changes after baking compared with those manufactured using egg white. Those produced with aquafaba prepared using meat and vegetable broths were rated the lowest by a panel of testers.

Veganism is increasing among people, and new products are constantly being developed to satisfy vegan needs and preferences [4]. Mihaylova et al. reported the development of healthy vegan bonbons enriched with lyophilized peach powder. They prepared three different formulations containing 10%, 20%, and 30% lyophilized peach power (LPP). The bonbons immediately after manufacturing were analyzed in terms of physical (size, weight, moisture, ash, color, water activity, texture, and microscopic imagery), microbiological (total aerobic mesophiles, and mold and yeast counts), and nutritional (contents in protein, fat, sugars, fiber, monounsaturated fats, omega-3 fatty acids, energy) characteristics, as well as in terms of parameters indicating antioxidant properties (antioxidant activity, and flavonoid and phenolic contents). Analyses of texture and water activity were repeated after 5 days, and microbiological analyses were repeated after 3 and 5 days of storage. All the bonbons had the same weight, size, and ash content. The addition of LPP, however, decreased the protein, sugar, fat, omega-3 fatty acid, energy, and total phenolic contents, but increased the monounsaturated fatty acid and total flavonoid contents. The addition of LPP reduced the antioxidant activity and increased the hardness, fracturability, and adhesiveness of the bonbons, whereas color was not affected. Regarding the microbial load, the addition of LPPC inhibited the microbial growth during storage. The results of this study show the possibility of using lyophilized peach powder as an ingredient for the manufacture of heathy raw vegan bonbons.

Plant-based bioactive compounds (BCs) and their incorporation into processed foods have attracted attention from researchers and the food industry due being perceived as safer and healthier by consumers. In the last paper in this Special Issue, Vilas-Boas et al. compared the composition, antioxidant activity, and antimicrobial properties of polyphenol-rich extracts obtained from *Lavandula pedunculata* (LP), a native Portuguese species, using maceration (CE), or microwave-assisted (MAE) or ultrasound-assisted (UAE) extraction. The authors found that rosmarinic acid (58.68–48.27 mg/g dry extract) and salvianolic acid B (43.19–40.09 mg/g DE) were the representative phenolic compounds in the extracts. They also reported that the three methods allowed them to obtain extracts with high antioxidant activity, highlighting the ORAC results (1306.0 to 1765.5 mg Trolox equivalents (TE)/g DE). The extracts obtained using MAE and CE showed outstanding growth inhibition of *Bacillus cereus*, *Staphylococcus aureus*, *Escherichia coli*, *Salmonella enterica*, and *Pseudomonas aeruginosa* (>50%, at 10 mg/mL). The MAE extract had the lowest IC_{50} (0.98 mg DE/mL) for the inhibition of angiotensin-converting enzyme and the highest IC_{50} for the inhibition of α-glucosidase and tyrosinase (87 and 73% of inhibition, respectively, at 5 mg/mL). Moreover, these extracts were safe for caco-2 intestinal cells, where no mutagenic effects were detected. UAE was less efficient in obtaining LP polyphenol-rich extracts. The efficiency of MAE equaled that of CE, saving time and energy. From these results, the authors concluded that LP shows potential as a sustainable raw material, allowing diverse extraction methods to be used to safely obtain health-promoting food and nutraceutical ingredients.

Author Contributions: S.M. and J.C. Conceptualization, data curation, writing—original draft preparation, writing—review and editing; visualization, supervision, project administration. This manuscript has not been submitted to, nor is under review at, another journal or other publishing venue. All authors have read and agreed to the published version of the manuscript.

Conflicts of Interest: The authors declare no conflict of interest.

List of Contributions:

1. Daei, B.; Azadmard-Damirchi, S.; Javadi, A.; Torbati, M. Effects of mild thermal processing and storage conditions on the quality attributes and shelf life of truffles (*Terfezia claveryi*). *Foods* **2023**, *12*, 2212. https://doi.org/10.3390/foods12112212.
2. Gómez-Limia, L.; Carballo, J.; Rodríguez-González, M.; Martínez, S. Impact of the filling medium on the colour and sensory characteristics of canned European eels (*Anguilla anguilla* L.). *Foods* **2022**, *11*, 1115. https://doi.org/10.3390/foods11081115.
3. Qu, Y.; Wang, J.; Liu, Z.; Wang, X.; Zhou, H. Effect of storage temperature and time on biogenic amines in canned seafood. *Foods* **2022**, *11*, 2743. https://doi.org/10.3390/foods11182743.

4. Delso, C.; Berzosa, A.; Sanz, J.; Álvarez, I.; Raso, J. Microbial decontamination of red wine by pulsed electric fields (PEF) after alcoholic and malolactic fermentation: Effect on *Saccharomyces cerevisiae*, *Oenococcus oeni*, and oenological parameters during storage. *Foods* **2023**, *12*, 278. https://doi.org/10.3390/foods12020278.

5. Cambero, M.I.; García de Fernando, G.D.; Romero de Ávila, M.D.; Remiro, V.; Capelo, L.; Segura, J. Freeze-dried cooked chickpeas: Considering a suitable alternative to prepare tasty reconstituted dishes. *Foods* **2023**, *12*, 2339. https://doi.org/10.3390/foods12122339.

6. Maghoumi, M.; Amodio, M.L.; Fatchurrahman, D.; Cisneros-Zevallos, L.; Colelli, G. Pomegranate husk scald browning during storage: A review on factors involved, their modes of action, and its association to postharvest treatments. *Foods* **2022**, *11*, 3365. https://doi.org/10.3390/foods11213365.

7. Flores-López, M.L.; Vieira, J.M.; Rocha, C.M.R.; Lagarón, J.M.; Cerqueira, M.A.; Jasso de Rodríguez, D.; Vicente, A.A. Postharvest quality improvement of tomato (*Solanum lycopersicum* L.) fruit using a nanomultilayer coating containing *Aloe vera*. *Foods* **2023**, *13*, 83. https://doi.org/10.3390/foods13010083.

8. Wang, J.; Zhang, H.; Xie, J.; Yu, W.; Sun, Y. Effects of frozen storage temperature on water-holding capacity and physicochemical properties of muscles in different parts of Bluefin tuna. *Foods* **2022**, *11*, 2315. https://doi.org/10.3390/foods11152315.

9. Kaewsorn, K.; Maichoon, P.; Pornchaloempong, P.; Krusong, W.; Sirisomboon, P.; Tanaka, M.; Kojima, T. Evaluation of precision and sensitivity of back extrusion test for measuring textural qualities of cooked germinated brown rice in production process. *Foods* **2023**, *12*, 3090. https://doi.org/10.3390/foods12163090.

10. Xu, X.; Ye, S.; Zuo, X.; Fang, S. Impact of guar gum and locust bean gum addition on the pasting, rheological properties, and freeze-thaw stability of rice starch gel. *Foods* **2022**, *11*, 2508. https://doi.org/10.3390/foods11162508.

11. Fuentes Choya, P.; Combarros-Fuertes, P.; Abarquero Camino, D.; Renes Bañuelos, E.; Prieto Gutiérrez, B.; Tornadijo Rodríguez, M.E.; Fresno Baro, J.M. Study of the technological properties of *Pedrosillano* chickpea aquafaba and its application in the production of egg-free baked meringues. *Foods* **2023**, *12*, 902. https://doi.org/10.3390/foods12040902.

12. Mihaylova, D.; Popova, A.; Goranova, Z.; Doykina, P. Development of healthy vegan bonbons enriched with lyophilized peach power. *Foods* **2022**, *11*, 1580. https://doi.org/10.3390/foods11111580.

13. Vilas-Boas, A.A.; Gómez-García, R.; Machado, M.; Nunes, C.; Ribeiro, S.; Nunes, J.; Oliveira, A.L.S.; Pintado, M. *Lavandula pedunculata* polyphenol-rich extracts obtained by conventional, MAE and UAE methods: Exploring the bioactive potential and safety for use a medicine plant as food and nutraceutical ingredient. *Foods* **2023**, *12*, 4462. https://doi.org/10.3390/foods12244462.

References

1. Martínez, S.; Carballo, J. (Eds.) *Physicochemical, Sensory and Nutritional Properties of Foods Affected by Processing and Storage*; MDPI: Basel, Switzerland, 2021; 316p, ISBN 978-3-0365-2732-1.

2. Appert, N. *L'Art de Conserver, Pendant Plusieurs Années, Toutes les Substances Animales et Végétales*; Patris et Cie. Imprimeurs-Libraires: Paris, France, 1810; 116p.

3. Toepfl, S.; Siemer, C.; Saldaña-Navarro, G.; Heinz, V. Overview of Pulsed Electric Fields Processing for Food. In *Emerging Technologies for Food Processing*; Elsevier: Amsterdam, The Netherlands, 2014; pp. 93–114.

4. Alcorta, A.; Porta, A.; Tárrega, A.; Alvarez, M.D.; Vaquero, M.P. Foods for Plant-Based Diets: Challenges and Innovations. *Foods* **2021**, *10*, 293. [CrossRef] [PubMed]

 *foods*

Article

Effects of Mild Thermal Processing and Storage Conditions on the Quality Attributes and Shelf Life of Truffles (*Terfezia claveryi*)

Bahareh Daei [1], Sodeif Azadmard-Damirchi [2,*], Afshin Javadi [3] and Mohammadali Torbati [4]

[1] Department of Food Science and Technology, Mamaghan Branch, Islamic Azad University, Mamaghan 5375113135, Iran

[2] Department of Food Science and Technology, Faculty of Agriculture, University of Tabriz, Tabriz 5166616471, Iran

[3] Department of Food Hygiene, Faculty of Veterinary, Tabriz Medical Science, Islamic Azad University, Tabriz 5157944533, Iran

[4] Department of Food Science and Technology, Faculty of Nutrition and Food Sciences, Tabriz University of Medical Sciences, Tabriz 1561661885, Iran

* Correspondence: sodeifazadmard@yahoo.com

Abstract: This study investigated the effects of two mild thermal processing (MTP) (63 °C, 40 °C, 3 min) methods, in a brine storage medium (7–16% (w/v) NaCl) and a vinegar solution (5% vinegar, 1% salt, and 0.5% sugar), on some physicochemical properties of truffles (*Terfezia claveryi*). Weight loss, phenolic compounds, firmness, ascorbic acid and microbial loads were evaluated during 160 days of storage. It was demonstrated that a 5% vinegar treatment with 63 °C MTP was effective to reduce the weight loss, microbial spoilage and increased firmness and of truffles during storage. However, phenolic compounds and ascorbic acid content were decreased by heating. Both MTPs inhibited the microbial load, but the 63 °C, 3 min MTP was most effective and resulted in an immediate (3.05–3.2 log CFU/g) reduction in the total aerobic bacteria (TAB) and remained at an acceptable level during storage, while the 40 °C, 3 min MTP reduced (1.12–2 log CFU/g) of the TAB. The results of this study suggest that the 63 °C MTP and immersion in 5% vinegar increased the shelf life of the truffles without perceptible losses in quality attributes.

Keywords: brine; mild thermal; preservation; quality; vinegar

Citation: Daei, B.; Azadmard-Damirchi, S.; Javadi, A.; Torbati, M. Effects of Mild Thermal Processing and Storage Conditions on the Quality Attributes and Shelf Life of Truffles (*Terfezia claveryi*). *Foods* **2023**, *12*, 2212. https://doi.org/10.3390/foods12112212

Academic Editors: Sidonia Martinez and Javier Carballo

Received: 1 November 2022
Accepted: 26 November 2022
Published: 31 May 2023

1. Introduction

Truffles are polyphyletic soil fungi that belong to Ascomycota, Basidiomycota, and Zygomycota [1]. Desert truffles are a subclass of mycorrhiza ascomycetes fungi, mainly including species of the genus *Terfezia, Tirmania, Delastria, Tuber* and *Picoa* [2].

Currently, about 30 species of truffles are traded, and due to their unique aroma and potential health benefits, they are one of the most expensive and valuable products in the market [3]. Truffles have been a good source of natural antioxidants such as phenolic and flavonoids compounds, protein, essential amino acids, fatty acids, terpenoids, polysaccharides, minerals, carbohydrates and ergosterols, which have beneficial properties for human health [4]. Some of the potential health benefits of truffles are known, such as anti-tumor, antioxidant, antibacterial, anti-inflammatory, anti-mutagenic and hepatoprotective activities [5].

Truffles are highly perishable and have short postharvest life because of their high metabolic activity and also harvesting at the end of fruit ripening from natural soils, with high pest and microbial load that lead to a short shelf life (7–10 days) [6].

Due to their high nutritional value as well as their popularity and high cost, various methods are used to increase the storage time and shelf life. Various preservation techniques have been studied to preserve the physicochemical properties of truffles [7]. These methods cause major changes in the properties of truffles. Truffles used in the food industry require the least amount of heat treatment to preserve their organoleptic properties [8]. Thus, mild heat treatment is required to inactivate enzymes and stabilize the quality with minimal tissue damage and organoleptic properties [9].

Pickling is a common method for preservation of perishable and seasonal leafy vegetables [10]. It has been used for many years to preserve food and extend shelf life. This method involves preserving foods in high acid concentration to maintain texture and flavor during storage [11].

Therefore, the aim of this research aimed to evaluate the effect of different preservation methods including mild thermal processing, vinegar-pickling and immersion in different concentrations of brine on the physicochemical properties and microbial quality of truffles.

2. Materials and Methods

2.1. Materials

In this experimental research, fresh truffles were collected at maturity from Khoy (Western Azerbaijan Province, Iran) area, during April 2020 and randomly divided into three sets of 3 kg truffle each. Iodized food-grade salt with 99.2% purity and sugar were obtained from a local market (Tabriz, Iran). Commercial grape vinegar (5%) was purchased from Varda Company (Tehran, Iran). All chemicals were analytical grade and purchased from Merck KGaA (Gernsheim, Germany). The culture media of Plate Count Agar was purchased from (PCA, Scharlau Spain) and Potato Dextrose Agar from Sigma-Aldrich Co. (St. Louis, MO, USA).

2.2. Methods

Treatments and Storage Conditions

Fresh truffles were gently brushed under running tap water to remove soil and debris and then rinsed with sterile water. In order to prepare the brine, sodium chloride and distilled water were used. To prepare samples, cleaned truffles were immersed in different concentrations of brine (7, 10, 13, and 16%) and also in a vinegar solution (5% vinegar, 1% salt, and 0.5% sugar) in jars and sealed. A constant-temperature water bath was used to prepare the MTP treatments. The truffles jars were filled with hot brine or/and hot vinegar solution and then were heated in hot water pot at (63 and 40 °C) 3 min and immediately cooled in an ice-water bucket [12].

2.3. Proximate Composition of Fresh Truffles

Proximate composition of truffle samples was determined according to the AOAC methods [13]. In brief, oven-drying was used to determine dry matter of truffles, which were dried at 105 °C for 72 h until constant weight. Crude proteins were measured by the Kjeldahl method by calculating the total nitrogen $\times$ 4.38 (nitrogen factor). The Fehling method was used to determine the carbohydrate content. The ash content was also determined by combustion of the samples in a muffle furnace (Nabertherm Muffle Furnace, L 1/12-LT 40/12, Germany) for 4 h at 550 °C. Crude fat was extracted by the Soxhlet method using hexane as an extraction solvent [14].

2.4. Determination of the Phenolic Composition of T. claveryi

Phenolic compounds were analyzed according to the previously published literature [15]. Briefly, the samples were ground in a mixture of acetic acid and methanol (15:85) for 24 h and extraction was then centrifuged. The supernatant was mixed with *n*-hexane and vigorously mixed. Then the bottom solution was separated and 20 μL was injected into the HPLC after filtration. An England CECIL model of high-performance liquid instrument (Cecil CE-4900, Cambridge, England),, equipped with dual pump (Cecil, Cambridge, England)air bubble remover (Cecil) and ultraviolet detector (Cecil, 4201 UV/Vis) with C18 column reversed phase and pore diameter of 5 μm, was used. The phenolic compound peaks were determined by comparing their retention times with that of their reference standards.

2.5. Analysis of Ascorbic Acid

Vitamin C content was determined by iodometric titration at the first day (production day) and days 40, 80, 120, and 160 of storage [16] The truffle sample (10 g) was crushed and mixed with 150 mL of distilled water. The mixture was placed in a conical flask (wrapped with aluminum foil) and filtered through a Whatman filter paper No. 4 to obtain a clear extract. After adding 1 mL of starch solution (1%), it was titrated by the iodine solution until the appearance of a blue-black color. The content of vitamin C was calculated by multiplying the volume of iodine solution used for titration by 0.88 mg.

2.6. Weight Loss Percentage

The truffles were weighed before pickling and at days 1, 40, 80, 120, and 160 of storage. Weight loss (WL) was determined according to the standard method of AOAC [13] as described below and defined as the percentage of loss of weight of samples with respect to the initial weight:

$$WL\ (\%) = (IW - FW)/IW \times 100$$

(IW): initial weight, (FW): final weights

2.7. Firmness Analysis

A penetration test was performed to determine the tissue firmness of truffles using the previously described method [17]. Truffles were penetrated using a texture analyzer (CF1-250150–STM-1, SANTAM, Tehran. Iran) with a 5 mm cylindrical probe with a speed of 2.0 mm s^{-1}. Firmness was defined as the force recorded in a force-time curve obtained from the texture analyzer at a depth of penetration of 5 mm during the compression of the truffle by the cylinder probe. The results of the penetration test were expressed from the time vs. force curves in N mm^{-1}. Firmness was determined as the maximum force.

2.8. Microbial Analysis

Total aerobic plate counts (APCs) and yeast and molds (Y&M) were determined following the previously described method [18]. The number of total aerobic bacteria was enumerated after homogenization of 5 gr truffle from each pickle jar in 45 mL 1% peptone water with a stomacher at high speed for 2 min and then serially diluted by taking 1.0 mL in 9 mL of peptone water (10^{-1}–10^{-9}) and pour-plated by using the Plate Count Agar medium (PCA, Scharlau, Spain). The plates were incubated at 35 °C for 48 h and enumerated every 40 days up to 160 days. The results were expressed as log CFU/g. Yeasts and molds (Y&M) were estimated at the same preparation method by using the Potato Dextrose Agar medium (PDA, Sigma-Aldrich). The plates were incubated at 25 °C for 5–7 days.

2.9. Statistical Analysis

Statistical analyses were performed using SPSS 18.0 (SPSS Inc., Chicago, IL, USA). Significant differences were analyzed using ANOVA and Duncan's multiple-range test at a significance level of 0.05. All the experiments were carried out in triplicates.

3. Results and Discussion

3.1. Physicochemical Composition of T. claveryi

T. claveryi is a dark brown, oval and potato-shaped fruit body truffle with ivory interior and thin veins with an approximate diameter of 4.5 cm and a length of about 6.5 cm. The collected truffle samples had different weights and volumes. The average weight and volume of *Terfezia claveryi* were 71.5 g and 75.5 cm^3, respectively (Table 1).

Table 1. Appearance characteristics and proximate composition of truffles (*T. claveryi*).

Factor	Value
Weight (g)	71.5 ± 0.5
Volume (cm^3)	75.5 ± 0.4
Length (cm)	6.5 ± 0.3
Diameter (cm)	4.5 ± 0.2
Moisture (% in DW)	74.3 ± 0.5
Protein (% in DW)	11.7 ± 0.5
Carbohydrates (mg 100 g^{-1} DW)	14.6 ± 0.5
Vitamin C (mg/100 g)	4.2 ± 0.8
Fat (% in DW)	1.4 ± 0.5

T. claveryi contain various nutrients and have been well explored for its chemical profile and nutritional content. The nutritional value of truffles differs from region to region [19]. The average moisture content of the fresh truffle was 74.3 g/100 g, which is in close agreement with the previously reported data for *T. claveryi* [20]. However, some other studies found higher moisture content as high as 79.2–81.6% for truffles of Middle Eastern countries and Arabian truffles [21]. This level of moisture content is also an indication that truffles are highly perishable products.

Crude fat, crude protein and carbohydrates of samples were 1.4, 11.7 and 14.6%, respectively. The content of crude fat was higher than that reported for *T. claveryi* (0.89 to 1.10%) [21]. A higher amount of crude fat content (3.9%) has been reported for *T. claveryi* of the Iraq region [22] and in another study, it was from 3.5 to 5.5% for same species of truffle harvested from the Northwest of Iran [5]. The protein content of the samples obtained was about 11.7%, which was in agreement with those reported for black truffles (11.9 to 15.95%) [21,23], and which is close to the results reported for *T. claveryi* (13–17%) [5]. However, a lower amount of protein content (3.35%) was also reported previously [24]. The carbohydrate content of the samples obtained was 14.6%, which was lower than the previously reported results (28%) [25]. These differences in nutrition values could be due to the variation of the locations, soil characteristics and climate conditions.

3.2. Influence of Treatments on the Weight Loss

Weight loss occurs due to the transpiration and respiration of fruits and also, due to the osmosis process as a result of the exchange with the storage solution. Truffles have a moisture content of about 74% and thus they are very susceptible to rapid weight loss accompanied by visible shriveling. After 160 days, a clear increase in weight loss was observed in the control and treated truffles. This was because the membrane permeability increases during the storage period and this led to a decrease in the strength of cell tissue [26]. The results indicated that the rate of weight loss was significantly ($p < 0.05$) slower in truffles processed at 63 °C compared to the control samples (Table 2). Weight loss of the control samples significantly increased ($p < 0.05$), reaching values of 10.2 to 33.84% for different concentration of immersion liquid at the end of storage, while this value was 6.36 to 10.24 in sample treatments at 63 °C. Moreover, the weight loss rate in the 40 °C treated samples was higher than the 63 °C samples. The weight losses of the samples increased in the brine stored samples with increasing the salt concentration. Increasing the concentration of the brine increases the membrane absorption and transfer of the interstitial water to the surrounding fluid due to the high ionic strength, and the reverse osmosis process increases the weight loss of truffles. Additionally, with increasing the salt concentration of the brine and during storage, the tissue strength decreases and the membrane permeability increases, which reduces the weight of the truffles. Relative humidity of the environment and storage temperature are also important because of their influence on the vapor pressure differences between fruit and the surrounding medium.

Table 2. Effect of different percentages of brine and type of processing and storage time on the weight loss of truffle samples.

Preservation Method	Brine (Salt%)	Storage Days (D)				
		1	40	80	120	160
Control	5% vinegar	0.01 [Eb]	3.09 [Dd]	5.92 [Cd]	8.29 [Bf]	10.2 [Af]
	7%	0.01 [Eb]	3.49 [Dcd]	7.01 [Cc]	10.14 [Be]	13.41 [Ae]
	10%	0.01 [Eb]	3.62 [Dcd]	7.84 [Cc]	12.27 [Bd]	15.79 [Ad]
	13%	0.01 [Eb]	4.29 [Dbc]	9.20 [Cb]	14.44 [Bc]	18.65 [Ac]
	16%	0.01 [Eb]	7.60 [Da]	15.85 [Ca]	24.49 [Ba]	33.84 [Aa]
MTP 40 °C-3 min	5% vinegar	0.008 [Ec]	2.19 [Dd]	4.04 [Cef]	6.39 [Bg]	8.19 [Ag]
	7%	0.008 [Ec]	2.53 [Dcd]	4.72 [Ce]	7.29 [Bf]	9.64 [Af]
	10%	0.009 [Ec]	2.60 [Dc]	5.15 [Cde]	8.14 [Bf]	10.41 [Af]
	13%	1.01 [Ea]	3.59 [Dc]	5.64 [Cd]	12.21 [Bd]	15.5 [Ad]
	16%	1.09 [Ea]	4.84 [Db]	10.34 [Cb]	19.84 [Bb]	26.77 [Ab]
MTP 63 °C-3 min	5% vinegar	0.003 [Ec]	1.46 [Dd]	3.57 [Cf]	5.19 [Bg]	6.37 [Ah]
	7%	0.005 [Ec]	1.62 [Dd]	3.68 [Cf]	5.44 [Bg]	6.74 [Ah]
	10%	0.005 [Ec]	2.09 [Dd]	3.98 [Cf]	6.27 [Bg]	7.14 [Ag]
	13%	0.008 [Ec]	2.53 [Dcd]	4.72 [Ce]	7.29 [Bf]	9.01 [Af]
	16%	0.009 [Ec]	4.69 [Db]	6.7 [Ccd]	8.14 [Bf]	10.25 [Af]

Different lowercase letters in each column and different uppercase letters in each row indicate a significant difference at the 5% level.

Among the samples preserved with different methods, the truffles treated at 63 °C and pickled in the 5% vinegar solution had the lowest values of weight loss compared to the 40 °C treated and control non-treated samples during the storage.

3.3. Influence of Treatments on Phenolic Compounds

Phenolics compositions are very important from nutritional and stability points of view. *T. claveryi* is rich in secondary metabolites and antioxidants such as phenolic compounds. Due to the mentioned properties, determination of phenolic compounds is an important factor for the evaluation of *T. claveryi* quality.

The results of the phenolic composition and content of the *Terfezia claveryi* samples are presented in Table 3. Ten phenolic compounds from *T. claveryi* were identified in the fresh truffle samples. Protocatechuic acid was the predominant phenolic acid (26.67 mg/g), followed by gentisic acid, chlorogenic acid, p-hydroxy benzoic acid and other minor phenolic compounds. It has been reported that the analysis of the phenolic contents in the acetone extract of *T. claveryi* has resulted in the phenolic compounds, namely homogentisic acid, protocatechuic acid, gentisic acid, 3,4-dihydroxybenzaldehyde, syringic acid, p-coumaric acid, vanillic acid and some compound with lower amounts [27]. Generally, the obtained results were in agreement with the previously published data [27].

Both methods of thermal processing resulted in a significant reduction ($p < 0.05$) in all phenolic compounds. These compounds decreased significantly with heating and an increase in the salt concentration of the brine. The reduction ranges at 40 °C and 63 °C were 31.3–81% and 34.2–100% during storage, respectively (Table 3). The phenolic content decreased during storage, but reduction was less in the non-heated truffles stored in the vinegar solution. Also, a decrease in the phenolic contents was reported for kidney beans, field peas and chickpeas during the thermal processing [28]. This decrease in the phenolic compounds might be due to the breakdown of phenolics during the heat treatment and leaching of phenolics into the medium solution.

Table 3. Effect of different percentages of brine and type of processing and storage time on the phenolic content (μg/g dry weight of *T. claveryi*).

Time	Treatment		Catechin	p-Coumaric Acid	Ferulic Acid	Chlorogenic Acid	Syringic Acid	p-Hydroxy Benzoic Acid	Rutin	Proto Catechuic Acid	Eugenol	Gentisic Acid
Day 1			8.29 [A**]	14.48 [A]	12.15 [A]	24.67 [A]	2.47 [A]	25.49 [A]	5.58 [A]	26.67 [A]	8.45 [A]	26.40 [A]
Day 160	5% vinegar	A*	5.70 [BC]	9.10 [CD]	8.05 [C]	19.25 [CD]	1.75 [D]	18.54 [D]	3.18 [CD]	19.87 [C]	6.11 [C]	16.70 [DE]
		B	5.45 [C]	8.29 [D]	7.10 [D]	17.48 [E]	1.00 [I]	16.20 [G]	2.88 [D]	16.45 [F]	5.03 [E]	14.55 [G]
		C	6.15 [B]	10.18 [B]	9.40 [B]	20.85 [B]	2.12 [B]	20.05 [B]	4.10 [B]	22.20 [B]	6.80 [B]	20.35 [B]
	7% Brine	A	4.64 [CD]	6.83 [F]	6.50 [DE]	16.20 [FG]	1.64 [DE]	17.64 [F]	2.30 [E]	18.30 [D]	5.85 [D]	14.78 [G]
		B	3.20 [E]	5.35 [I]	5.48 [EF]	13.10 [H]	0.96 [I]	14.18 [I]	1.96 [EF]	15.64 [G]	4.94 [E]	13.40 [H]
		C	5.65 [BC]	8.45 [D]	7.87 [CD]	20.18 [BC]	2.05 [C]	19.40 [C]	3.70 [BC]	21.80 [B]	6.54 [C]	18.65 [CD]
	10% Brine	A	3.75 [E]	6.10 [FG]	6.15 [E]	15.35 [G]	1.20 [H]	15.40 [H]	2.18 [E]	17.65 [E]	5.67 [DE]	14.25 [GH]
		B	2.48 [F]	5.18 [GH]	5.12 [F]	12.49 [I]	0.58 [J]	12.55 [K]	1.40 [F]	14.80 [GH]	4.48 [FG]	12.18 [I]
		C	5.25 [BC]	8.10 [DE]	7.68 [CD]	19.46 [C]	1.87 [CD]	19.18 [CD]	3.45 [C]	21.18 [BC]	6.30 [C]	18.45 [CD]
	13% Brine	A	3.64 [E]	5.84 [G]	5.45 [EF]	14.97 [GH]	1.10 [H]	13.45 [J]	2.06 [E]	16.30 [F]	5.10 [E]	12.40 [I]
		B	1.83 [G]	4.96 [I]	4.55 [FG]	11.87 [J]	0.35 [J]	11.80 [L]	1.26 [F]	11.28 [I]	4.12 [G]	12.28 [I]
		C	4.67 [CD]	8.03 [DE]	7.40 [D]	19.16 [CD]	1.63 [DE]	18.96 [CD]	3.28 [CD]	20.15 [C]	6.18 [C]	17.36 [D]
	16% Brine	A	2.15 [F]	5.15 [GH]	4.98 [FG]	14.40 [GH]	0.45 [J]	12.15 [KL]	1.43 [F]	14.84 [GH]	4.65 [F]	12.19 [I]
		B	1.10 [H]	4.80 [I]	4.10 [G]	11.25 [J]	ND	10.50 [M]	1.10 [FG]	10.15 [I]	3.30 [H]	11.85 [J]
		C	3.28 [E]	7.30 [E]	6.80 [DE]	18.38 [D]	1.50 [F]	18.80 [CD]	2.65 [DE]	19.43 [CD]	5.80 [D]	16.80 [DE]

*: (A) MTP at 40 °C-3 min, (B) MTP at 63 °C-3 min, (C) Unheated control samples. **: Different capital letters in column indicate a significant difference in the probability level of 5%.

3.4. Influence of Treatments on the Texture

Texture is a vital indicator to evaluate the quality of pickled products that is extremely important in overall consumer acceptance. Retention of firmness can be explained by the delay in the breakdown of insoluble protopectin's into pectic acid and more soluble pectic. The firmness of all the samples decreased as the storage time increased. However, the heated treatment samples remained firmer than the control samples. The firmness of the control samples decreased by 77.8 to 96.6 % at 160 days. As can be seen from Table 4, the firmness of all samples also decreased with increasing of the salt concentration. This is due to the effect of the osmotic dewatering process as a result of the mass transfer in the concentration gradient between the sample and the osmotic solution, which was in line with the previously reported literature [29].

Table 4. Effect of different percentages of brine and type of processing and storage time on the texture of truffle samples.

Preservation Method	Brine (Salt%)	Storage Days (D)				
		1	40	80	120	160
Control	5% vinegar	6.78 [Aa]	4.78 [Bbc]	3.69 [Bbc]	2.48 [Cd]	1.5 [Dd]
	7%	6.78 [Aa]	4.0 [Bc]	2.88 [Ccd]	2.29 [Dd]	1.19 [Ed]
	10%	6.78 [Aa]	3.79 [Bc]	2.63 [Ccd]	1.83 [Dde]	0.96 [Ed]
	13%	6.78 [Aa]	3.68 [Bc]	2.23 [Cd]	1.19 [De]	0.91 [Dd]
	16%	6.78 [Aa]	3.41 [Bd]	1.9 [Cd]	0.93 [De]	0.23 [De]
MTP 40 °C-3 min	5% vinegar	6.78 [Aa]	5.81 [Bab]	4.29 [Cb]	3.69 [Dbc]	3.35 [Dbc]
	7%	6.81 [Aa]	5.38 [Bb]	4.08 [Cb]	3.56 [Dbc]	3.18 [Dbc]
	10%	6.79 [A]	4.63 [Bb]	3.78 [Cbc]	2.43 [Dd]	2.0 [Dd]
	13%	6.73 [Aab]	4.13 [Bcd]	2.66 [Ccd]	2.23 [CDd]	1.83 [Dd]
	16%	6.76 [Aa]	4.03 [Bc]	2.19 [Cd]	1.78 [CDde]	1.41 [Dd]
MTP 63 °C-3 min	5% vinegar	6.76 [Aa]	6.21 [ABa]	5.8 [Ba]	5.33 [BCa]	4.92 [Ca]
	7%	6.73 [Aab]	5.78 [Bab]	5.38 [BCab]	4.96 [Cab]	4.74 [Ca]
	10%	6.76 [Aa]	5.7 [Bab]	5.08 [BCab]	4.31 [Cb]	3.71 [Db]
	13%	6.69 [Ab]	5.11 [Bb]	4.39 [Cb]	3.82 [Dbc]	3.22 [Dbc]
	16%	6.73 [Aab]	4.1 [Bc]	3.48 [Cc]	3.09 [Cc]	2.53 [Dc]

Different lowercase letters in each column and different uppercase letters in each row indicate a significant difference at the 5% level.

Among the samples preserved with the different methods, the truffles processed at 63 °C had a significant tissue score compared to those processed at 40 °C and the control samples. This could be due to the reduced or eliminated activity of the pectin methyl esterase and polygalacturonase enzymes breaking the hydrogen bonds of their three-dimensional structure by heat [30]. After 160 days of storage at 4 °C, the hardness of the vinegared truffles treated with 63 °C was 4.92 N, about 1.57 N higher ($p < 0.05$) than that using the 40 °C heat treatment (3.35 N). This group of treatments achieved the best effect with regard to hardness. Truffles firmness was better maintained during storage if the samples had been pre-treated at 63 °C. Concerning the preservation method, the texture of the samples kept in 5% vinegar and 7% NaCl were in good condition compared with the other samples during storage (Figure 1).

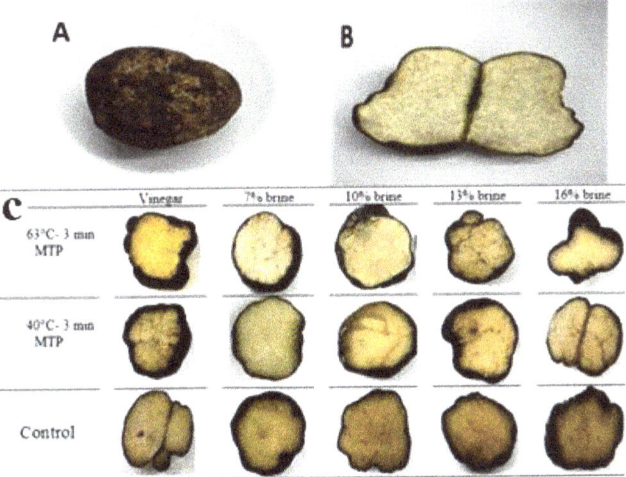

Figure 1. Macro features of *Terfezia claveryi* at first day: (**A**) mature ascocarps, (**B**) white-creamy texture and gleba of mature ascocarp, (**C**): at the end of storage.

Firmness of the unheated samples decreased almost 96.6% after 160 days of storage and up to 79.1% in the samples treated at 40 °C, while firmness reduction was up to 58% in the samples treated at 63 °C. These results were in agreement with previously published data [31], which reported that mild heating improved the cell structure and firmness with an increase in calcium amounts, leading to the crosslinking of pectin in the carrot samples.

3.5. Influence of Treatments on the Ascorbic Acid

The content of ascorbic acid in food products is an important nutritional factor that is usually evaluated and monitored during processing and storage. It was about 4.2 mg/100 g in the fresh truffles at the initial day, which reduced ($p < 0.05$) during the storage for all the preservation treatments (Table 5). However, the reduction rate of ascorbic acid was slower for the samples stored in the refrigerator in comparison to the ones stored at ambient temperature. Ascorbic acid is one of the most unstable vitamins and its content can be reduced during processing and storage due to degradation and Maillard reaction. Oxygen, heat, light, storage time, and storage temperature are factors affecting the ascorbic acid degradation rate [32].

Table 5. Effect of different percentages of brine and type of processing and storage time on the content of ascorbic acid of truffle samples.

Preservation Method	Brine (Salt%)	Storage Time (Day)				
		1	40	80	120	160
Control	5% vinegar	4.43 [Aa]	3.48 [Ba]	2.83 [Ca]	2.41 [CDa]	1.66 [Da]
	7%	4.41 [Aa]	3.5 [Ba]	2.68 [Ca]	2.24 [CDa]	1.51 [Da]
	10%	4.38 [Aa]	3.33 [Ba]	2.38 [Cab]	2.03 [Ca]	1.18 [Da]
	13%	4.41 [Aa]	3.08 [Bab]	2.13 [Cb]	1.88 [CDab]	1.1 [Dab]
	16%	4.38 [Aa]	3.03 [Bab]	2.01 [Cb]	1.78 [CDab]	1.03 [Da]
MTP 40 °C-3 min	5% vinegar	4.48 [Aa]	3.28 [Bab]	2.44 [Cab]	2.10 [Da]	1.38 [Ea]
	7%	4.48 [Aa]	3.13 [Bab]	2.28 [Cab]	1.98 [Dab]	1.23 [Ea]
	10%	4.48 [Aab]	3.10 [Bab]	2.13 [Cb]	1.76 [Dab]	1.04 [Ea]
	13%	4.48 [Aa]	3.08 [Bab]	2.00 [Cb]	1.61 [Db]	0.92 [Eb]
	16%	4.48 [Aa]	2.82 [Bb]	1.92 [Cbc]	1.48 [CDb]	0.81 [Db]
MTP 63 °C-3 min	5% vinegar	3.78 [Ab]	3.13 [Bab]	2.30 [Cab]	1.94 [CDab]	1.33 [Da]
	7%	3.38 [Ac]	3.03 [Aab]	2.16 [Bb]	1.72 [BCab]	0.98 [Cb]
	10%	3.34 [Ac]	2.78 [Bb]	2.02 [Cb]	1.33 [Db]	0.92 [Db]
	13%	3.36 [Ac]	2.43 [Bb]	1.81 [BCbc]	1.13 [Cb]	0.76 [Db]
	16%	3.40 [Ac]	2.18 [Bc]	1.68 [BCc]	0.96 [Cb]	0.63 [Cb]

Different lowercase letters in each column and different uppercase letters in each row indicate a significant difference at the 5% level.

The ascorbic acid content decreased significantly ($p < 0.05$) with the thermal processing and also with increasing the salt concentration in the filling medium. The loss of the ascorbic acid content with increasing the salt concentration could be explained by increasing the osmotic gradient, which causes the transfer of soluble material such as ascorbic acid to the brine [33]. As expected, the reduction in ascorbic acid in the MTP samples during storage is due to the heat stress [34]. This reduction was higher in the samples processed at 63 °C and 40 °C, respectively. Among the samples preserved with different methods, the truffles pickled in the vinegar solution with no heat treatment had the highest ascorbic acid content during the storage.

3.6. Microbiological Analysis

The effect of the salt concentration of the brine solutions and the change in aerobic bacteria counts during storage are shown in Table 6. Mild thermal processing is an effective method to eliminate microbial contaminants and extend shelf life. These gentle treatments consist of placing the products at a temperature of 50–90 °C for a period of 1–5 min [35]. The results showed that both thermal treatments of truffles in all groups were able to reduce the microbial flora in the fresh-made samples. However, from these results, 63 °C, 3 min were the most effective to reduce 3.05–3.2 log CFU/g of the total aerobic bacteria (TAB) when compared to the control, while 40 °C, 3 min caused a 1.12–2 log CFU/g reduction of the TAB. The microbial counts were increased in all treatment samples during storage, but the rate of increase in 63 °C MTP was slower than that over the two treated samples. These results indicated that MTP at the 40 °C treatment applied in this study could not have the protective and sufficient effects on the inhibition of microbial spoilage, which led to a further increase in microbial loads during storage.

Table 6. Effect of different percentages of brine and type of processing and storage time on the total count of truffles of truffle samples.

Preservation Method Temperature	Brine (Salt%)	Storage Days (D)				
		1	40	80	120	160
Control	5% vinegar	4.08 [Da]	4.16 [Db]	4.78 [Cbc]	5.28 [Bc]	7.53 [Ab]
	16%	4.13 [Da]	4.22 [Db]	5.29 [Cab]	5.98 [Bb]	8.83 [Ab]
	13%	4.11 [Ea]	4.26 [Dab]	5.43 [Cab]	6.18 [Bb]	8.65 [Ab]
	10%	4.16 [Ea]	4.78 [Da]	5.83 [Ca]	6.78 [Bab]	8.38 [Ab]
	7%	4.13 [Ea]	4.92 [Da]	6.13 [Ca]	7.33 [Ba]	8.23 [Aa]
MTP 40 °C	5% vinegar	2.08 [Ca]	2.53 [BCb]	3.05 [Bbc]	3.86 [ABc]	4.15 [Ad]
	16%	2.96 [Ca]	3.92 [Ba]	4.46 [Bab]	4.84 [Ab]	5.78 [Ac]
	13%	2.88 [Ca]	3.78 [BCa]	4.33 [Bab]	4.66 [ABbc]	5.78 [Ac]
	10%	2.76 [Ca]	3.63 [Bab]	4.18 [BCc]	4.58 [ABbc]	5.55 [Ac]
	7%	2.63 [Ba]	3.56 [Bab]	3.96 [Bc]	4.24 [ABbc]	5.51 [Ac]
MTP 63 °C	5% vinegar	0.88 [Cb]	1.52 [Bb]	2.23 [Bb]	2.86 [ABd]	3.03 [Ad]
	16%	1.26 [Cb]	2.73 [Ba]	3.50 [ABab]	3.73 [Abc]	3.83 [Acd]
	13%	1.24 [Cb]	2.46 [Bab]	3.26 [ABab]	3.68 [ABc]	3.78 [Acd]
	10%	1.16 [Cb]	2.33 [Bb]	3.15 [ABb]	3.58 [Ac]	3.73 [Ad]
	7%	1.13 [Cb]	2.18 [Bb]	3.03 [ABbc]	3.53 [Ac]	3.70 [Ad]

Different lowercase letters in each column and different uppercase letters in each row indicate a significant difference at the 5% level.

The effect of MTP and the salt concentrations of the brine on the yeast and molds of the samples during storage is shown in Table 7. Similar results occurred in the yeast and molds reduction. A temperature of 63° C was also able to reduce the yeast and molds load to an acceptable and safe level.

Table 7. Effect of different percentages of brine and type of processing and storage time on the yeast and mold count of truffles of truffle samples.

Preservation Method Temperature	Brine (Salt%)	Storage Days (D)				
		1	40	80	120	160
Control	5% vinegar	3.10 [Da]	3.66 [Db]	3.95 [Cbc]	4.00 [Bc]	4.15 [Ab]
	16%	3.13 [Ea]	3.98 [Da]	4.84 [Ca]	4.73 [Ba]	4.85 [Aa]
	13%	3.16 [Ea0]	3.98 [Da]	4.63 [Ca]	4.68 [Bab]	4.78 [Ab]
	10%	3.11 [Ea]	3.96 [Dab]	4.45 [Cab]	4.56 [Bb]	4.68 [Ab]
	7%	3.13 [Da]	3.92 [Db]	4.15 [Cab]	4.38 [Bb]	4.63 [Ab]
MTP 40 °C	5% vinegar	0.63 [Ca]	0.70 [BCb]	0.76 [Bbc]	0.86 [ABc]	0.92 [Ad]
	16%	0.95 [Ca]	0.99 [Ba]	1.16 [Bab]	1.24 [Ab]	1.28 [Ac]
	13%	0.88 [Ca]	0.95 [BCa]	0.98 [Bab]	1.06 [ABbc]	1.12 [Ac]
	10%	0.76 [Ca]	0.83 [Bab]	0.86 [BCc]	0.98 [ABbc]	1.06 [Ac]
	7%	0.68 [Ba]	0.76 [Bab]	0.85 [Bc]	0.91 [ABbc]	0.95 [Ac]
MTP 63 °C	5% vinegar	0.45 [Cb]	0.50 [Bb]	0.56 [Bb]	0.65 [ABd]	0.68 [Ad]
	16%	0.65 [Cb]	0.73 [Ba]	0.77 [ABab]	0.82 [Abc]	0.88 [Acd]
	13%	0.63 [Cb]	0.66 [Bab]	0.74 [ABab]	0.78 [ABc]	0.85 [Acd]
	10%	0.56 [Cb]	0.60 [Bb]	0.65 [ABb]	0.68 [Ac]	0.78 [Ad]
	7%	0.53 [Cb]	0.58 [Bb]	0.63 [ABbc]	0.67 [Ac]	0.75 [Ad]

Different lowercase letters in each column and different uppercase letters in each row indicate a significant difference at the 5% level.

The results showed that with increasing the salt concentration, the growth of the aerobic bacteria was delayed. After 160 days of storage, the aerobic bacteria count in the thermal processed samples was significantly lower, while the unheated samples showed a sharp increase rate, and the colonies were uncountable at the end of storage with visible spoilage (>8 log CFU/g).

The growth rate of the mesophilic population in the 5% vinegar treatment was lower compared to the brine treatments during storage. It can be due to the antimicrobial role of the high concentrations of vinegar and penetrating the cell wall and denaturing cell plasma proteins. These results are in line with a previous study that reported that acetic acid in vinegar can act as an antimicrobial agent and lead to bacterial cell death [36]. Based on the obtained results, the combined effect of the MTP treatments and immersion in the preservative solutions against microbial contaminations, combined the treatments at 63 °C MTP and immersion in 5% vinegar, appears to be an efficient and sustainable method to ensure microbial safety and to extend the refrigerated shelf life of fresh truffles.

4. Conclusions

As our main conclusion, mild thermal processing at 63 °C-3 min can be considered as an efficient and sustainable technique that preserves the quality and organoleptic properties of truffles. Moreover, due to the antimicrobial efficacy of the 5% vinegar and 1% salt solution for the growth inhibition of yeast and mold and enhancement of shelf life up to 160 days, this can be suggested as a safe preservation medium and method for truffles during storage. These observations suggest that mild thermal processing, pickling in 5% vinegar solution can substantially enhance the shelf life of truffles and can be suggested for the marketing of this precious product.

Author Contributions: B.D. performed the experiments and wrote the manuscript. S.A.-D., A.J. and M.T. were supervisors, designed the experiments, analyzed the data and revised the manuscript. All authors have read and agreed to the published version of the manuscript.

Funding: This research received no external funding.

Institutional Review Board Statement: Not applicable.

Informed Consent Statement: Not applicable.

Data Availability Statement: Data are contained within the article.

Acknowledgments: The authors appreciate the support of Islamic Azad University—Mamaghan branch to accomplish this research.

Conflicts of Interest: The authors declare no conflict of interest.

References

1. Benucci, G.M.N.; Bonito, G.M. The truffle microbiome: Species and geography effects on bacteria associated with fruiting bodies of hypogeous Pezizales. *Microb. Ecol.* **2016**, *72*, 4–8. [CrossRef] [PubMed]
2. Khabar, L. Mediterranean Basin: North Africa. In *Desert Truffles*; Springer: Berlin/Heidelberg, Germany, 2014; pp. 143–158. [CrossRef]
3. Vahdatzadeh, M.; Splivallo, R. Improving truffle mycelium flavour through strain selection targeting volatiles of the Ehrlich pathway. *Sci. Rep.* **2018**, *8*, 9304. [CrossRef] [PubMed]
4. Patel, S.; Rauf, A.; Khan, H.; Khalid, S.; Mubarak, M.S. Potential health benefits of natural products derived from truffles: A review. *Trends Food Sci. Technol.* **2017**, *70*, 1–8. [CrossRef]
5. Alirezalu, K.; Azadmard-Damirchi, S.; Achachlouei, B.F.; Hesari, J.; Emaratpardaz, J.; Tavakolian, R. Physicochemical Properties and Nutritional Composition of Black Truffles Grown in Iran. *Chem. Nat. Compd.* **2016**, *52*, 290–293. [CrossRef]
6. Campo, E.; Marco, P.; Oria, R.; Blanco, D.; Venturini, M.E. What is the best method for preserving the genuine black truffle (*Tuber melanosporum*) aroma? An olfactometric and sensory approach. *LWT* **2017**, *80*, 84–91. [CrossRef]
7. Nazzaro, F.; Fratianni, F.; Picariello, G.; Coppola, R.; Reale, A.; Di Luccia, A. Evaluation of gamma rays influence on some biochemical and microbiological aspects in black truffles. *Food Chem.* **2007**, *103*, 344–354. [CrossRef]
8. Tirillini, B.; Maggi, F.; Venanzoni, R.; Angelini, P. Enhanced duration of truffle sauce preservation due to addition of linoleic acid. *J. Food Qual.* **2019**, *2019*, 8421898. [CrossRef]
9. Halpin, B.; Lee, C. Effect of blanching on enzyme activity and quality changes in green peas. *J. Food Sci.* **1987**, *52*, 1002–1005. [CrossRef]
10. Ashaolu, T.J.; Reale, A. A holistic review on Euro-Asian lactic acid bacteria fermented cereals and vegetables. *Microorganisms* **2020**, *8*, 1176. [CrossRef]

11. Behera, S.S.; El Sheikha, A.F.; Hammami, R.; Kumar, A. Traditionally fermented pickles: How the microbial diversity associated with their nutritional and health benefits? *J. Func. Foods* **2020**, *70*, 103971. [CrossRef]
12. Mills-Gray, S. *Pickling Basics—In a Pickle. Nutrition and Health*; UME: Columbia, MO, USA, 2015; pp. 1–4.
13. AOAC. *Official Methods of Analysis*; Williams, S., Ed.; Association of Official Analytical Chemists: Arlington, VA, USA, 2000.
14. Bouatia, M.; Touré, H.; Cheikh, A.; Eljaoudi, R.; Rahali, Y.; Idrissi, O.M.B.; Khabar, L.; Draoui, M. Analysis of nutrient and antinutrient content of the truffle (*Tirmania pinoyi*) from Morocco. *Int. Food Res. J.* **2018**, *25*, 174–178.
15. Lister, C.E.; Lancaster, J.E.; Sutton, K.H.; Walker, J.R. Developmental changes in the concentration and composition of flavonoids in skin of a red and a green apple cultivar. *J. Sci. Food Agric.* **1994**, *64*, 155–161. [CrossRef]
16. Nurhidayati, U.A.; Ali, U.; Murwani, I. Yield and quality of cabbage (*Brassica oleracea* L. var. Capitata) under organic growing media using vermicompost and earthworm Pontoscolex corethrurus inoculation. *Agric. Agric. Sci.* **2016**, *11*, 5–13. [CrossRef]
17. Gao, M.; Feng, L.; Jiang, T. Browning inhibition and quality preservation of button mushroom (*Agaricus bisporus*) by essential oils fumigation treatment. *Food Chem.* **2014**, *149*, 107–113. [CrossRef] [PubMed]
18. Li, Y.; Padilla-Zakour, O.I. High Pressure Processing vs. Thermal Pasteurization of Whole Concord Grape Puree: Effect on Nutritional Value, Quality Parameters and Refrigerated Shelf Life. *Foods* **2021**, *10*, 2608. [CrossRef]
19. Bokhary, H.; Parvez, S. Chemical composition of desert truffles Terfezia claveryi. *J. Food Comp. Anal.* **1993**, *6*, 285–293. [CrossRef]
20. Wahiba, B.; Wafaà, T.; Asmaà, K.; Bouziane, A.; Mohammed, B. Nutritional and antioxidant profile of red truffles (*Terfezia claveryi*) and white truffle (*Tirmania nivea*) from southwestern of Algeria. *Der. Pharm. Lett.* **2016**, *8*, 134–141.
21. AL-RUQAIE, I.M. Effect of different treatment processes and preservation methods on the quality of truffles: I. Conventional methods (drying/freezing). *J. Food Proc. Pres.* **2006**, *30*, 335–351. [CrossRef]
22. Kagan-Zur, V.; Roth-Bejerano, N. Studying the brown desert truffles of Israel. *Israel J. Plant Sci.* **2008**, *56*, 309–314. [CrossRef]
23. Murcia, M.A.; Martínez-Tomé, M.; Vera, A.; Morte, A.; Gutierrez, A.; Honrubia, M.; Jiménez, A.M. Effect of industrial processing on desert truffles Terfezia claveryi Chatin and Picoa juniperi Vittadini): Proximate composition and fatty acids. *J. Sci. Food Agri.* **2003**, *83*, 535–541. [CrossRef]
24. Vahdani, M.; Rastegar, S.; Rahimizadeh, M.; Ahmadi, M.; Karmostaji, A. Physicochemical characteristics, phenolic profile, mineral and carbohydrate contents of two truffle species. *Mag. Int. Sci. Agric. Technol.* **2017**, *19*, 1091–1101.
25. Ammarellou, A. Protein profile analysis of desert truffle (*Terfezia claveryi* Chatin). *J. Food Agri. Env.* **2007**, *5*, 62.
26. Fan, K.; Zhang, M.; Fan, D.; Jiang, F. Effect of carbon dots with chitosan coating on microorganisms and storage quality of modified-atmosphere-packaged fresh-cut cucumber. *J. Sci. Food. Agri.* **2019**, *99*, 6032–6041. [CrossRef] [PubMed]
27. Kıvrak, İ. Analytical methods applied to assess chemical composition, nutritional value and in vitro bioactivities of Terfezia olbiensis and Terfezia claveryi from Turkey. *Food Anal. Methods* **2015**, *8*, 1279–1293. [CrossRef]
28. Parmar, N.; Singh, N.; Kaur, A.; Virdi, A.S.; Thakur, S. Effect of canning on color, protein and phenolic profile of grains from kidney bean, field pea and chickpea. *Food Res. Int.* **2016**, *89*, 526–532. [CrossRef]
29. Basiri, S.; Gheybi, F. The effect of osmotic dehydration and various methods of drying pumpkin on quality and color of final product. *J. Food Res.* **2019**, *29*, 141–153.
30. Tanaka, A.; Hoshino, E. Similarities between the thermal inactivation kinetics of Bacillus amyloliquefaciensα-amylase in an aqueous solution of sodium dodecyl sulphate and the kinetics in the solution of anionic-phospholipid vesicles. *Biotechnol. Appl. Biochem.* **2003**, *38*, 175–181. [CrossRef]
31. Imaizumi, T.; Tanaka, F.; Uchino, T. Effects of mild heating treatment on texture degradation and peroxidase inactivation of carrot under pasteurization conditions. *J. Food Eng.* **2019**, *257*, 19–25. [CrossRef]
32. Burdurlu, H.S.; Koca, N.; Karadeniz, F. Degradation of vitamin C in citrus juice concentrates during storage. *J. Food Eng.* **2006**, *74*, 211–216. [CrossRef]
33. Yalim, S.; Özdemır, Y. Effects of preparation procedures on ascorbic acid retention in pickled hot peppers. *Int. J. Food Sci. Nut.* **2003**, *54*, 291–296. [CrossRef]
34. Grzelakowska, A.; Cieślewicz, J.; Łudzińska, M. The dynamics of vitamin C content in fresh and processed cucumber (Cucumis sativus L.)/Dynamika zmian zawartości witaminy C w ogórku świeżym (*Cucumis sativus* L.) i jego przetworach. *Chem. Didact. Ecol. Metrol.* **2013**, *18*, 97–102. [CrossRef]
35. Orsat, V.; Gariepy, Y.; Raghavan, G.; Lyew, D. Radio-frequency treatment for ready-to-eat fresh carrots. *Int. Food Res. J.* **2001**, *34*, 527–536. [CrossRef]
36. Shweta, R.; Geeta, I.; Ganessin, A. Comparative evaluation of antimicrobial properties of piperidine and apple cider vinegar on Streptococcus mutansand in-silico demonstration of the mechanism of action. *RGUHS J. Dent. Sci.* **2021**, *13*, 42–50.

 *foods*

Article

Impact of the Filling Medium on the Colour and Sensory Characteristics of Canned European Eels (*Anguilla anguilla* L.)

Lucía Gómez-Limia, Javier Carballo, Miriam Rodríguez-González and Sidonia Martínez *

Food Technology, Faculty of Science, Campus As Lagoas s/n, University of Vigo, 32004 Ourense, Spain; lugomez@uvigo.es (L.G.-L.); carbatec@uvigo.es (J.C.); miriamrodrigzg@gmail.com (M.R.-G.)
* Correspondence: sidonia@uvigo.es; Tel.: +34-988-387-080

Abstract: The different vegetable oils used in canned fish as a filling medium have a preserving effect and contribute to the palatability of the product. In this study, the colour of European eels and the filling medium (sunflower oil, olive oil or spicy olive oil) was measured at different steps of the canning process. The sensorial characteristics of canned eels packed in the different oils were also evaluated. Colour scores (CieLab values) were higher in canned eels packed in sunflower and spicy olive oil than in canned eels packed in olive oil. The changes in colour parameters depended on the type of oil, the stage of the process and the storage time. Colour changes in canned eels packed in olive oil were highest during the sterilization process. Spicy olive oil was the filling medium in which the colour change was greatest, probably due to the migration of some of the spice components into the oil. Organoleptic properties were directly related to the type of oil used as the filling medium. The canned eels packed in sunflower oil were those awarded the highest scores in consumer tests, although the preferences varied depending on the age and gender of the consumers.

Keywords: olive oil; sunflower oil; spicy olive oil; canned eel; colour; sensory analysis

Citation: Gómez-Limia, L.; Carballo, J.; Rodríguez-González, M.; Martínez, S. Impact of the Filling Medium on the Colour and Sensory Characteristics of Canned European Eels (*Anguilla anguilla* L.). *Foods* **2022**, *11*, 1115. https://doi.org/10.3390/foods11081115

Academic Editor: Susana Casal

Received: 9 March 2022
Accepted: 11 April 2022
Published: 13 April 2022

1. Introduction

Sensory properties are one of the most important quality attributes of any product and directly determines consumer satisfaction or overall acceptance based on colour, taste, flavour and aroma. The level of acceptance by consumers plays a key role in the development of new food products. Consumer decisions about food products largely depend on their sensory profile [1].

The colour of food has an important influence on consumer opinion about the quality of the product. Colour provides one of the most important visual cues regarding the sensory properties (e.g., taste and flavour) [2]. Colour is correlated with other quality parameters such as sensory and nutritional characteristics, and it can help to indicate defects [3,4]. It can also be used as an indicator of the intensity of heat treatments and to predict the deterioration resulting from heat exposure [5]. Colour can be determined by visual inspection or through instrumental assessment. Visual inspection is rather subjective and varies greatly from one person to another. Robust results can only be obtained by using trained assessors. As the visual inspection method is time consuming and involves specially trained observers, the instrumental method is recommended for conducting colour analysis. Colour spaces and numerical values are usually used to construct models to represent colours in two- or three-dimensional spaces [4,6].

The sensory qualities of food are very important in determining whether consumers will accept or reject the product. The sensory quality of fish or a fish product can be established by evaluation of various sensory attributes such as appearance, colour, taste, texture, aroma and flavour. However, these attributes can undergo changes during processing.

Canning is one of the oldest methods of food processing. Canned fish are economically important products in many countries, supporting a high market demand due to convenience and affordability. Canned fish have a relatively long shelf life and are thus widely available at all times. The synergistic effects of heat treatment and the filling medium plays an important role in the canning process. The procedures used to prepare raw fish for canning, the pre-cooking process, type of heat treatment, filling medium and storage conditions are carefully selected to lengthen the shelf life, preserve the quality and achieve optimal sensory characteristics. The quality of canned fish can vary with different factors such as the quality of the raw product, processing conditions, packaging material and filling ingredients [7]. The sensory and nutritional characteristics are the result of complex interactions between the fish and filling medium. Colour and sensory properties are among the most important parameters for determining changes in food quality, and they have an important impact on the acceptability of canned fish. In the canning process, brine or oil are generally used to improve the taste, aroma and appearance of the fish and to yield a compact product with a soft, juicy texture. The vegetable oils in which the fish are packed have a preserving effect and contribute to the palatability of the product. On the other hand, the oils can be used as part of a commercial strategy of adding value to canned fish. The filling oil can dilute and/or partly extract some components, and lead to heat transfer in fish muscle [8]. Processing can also cause changes in the sensory and physical properties of the oils. Olive and sunflower oils are two of the most common filling media used in canning.

Eels are commercially valuable in Europe (mainly Spain, Portugal, Italy and the Netherlands) and Asia (mainly Japan, China, Korea and Taiwan). The eels can be processed in different ways such as freezing, smoking, canning and jelling. Although canned eel is a commercially important product in some countries, information on its quality parameters is scarce. Eel populations have declined greatly because of various factors threatening their survival, such as environmental impacts associated with the construction of diverse obstacles in rivers and an increase in pollution of human origin [9]. Canning enables eels to be consumed throughout the year, while respecting restrictions or quotas aimed at protecting the species. Larger eels are least valued for fresh consumption, and their exploitation, instead of smaller eels (the commercial weight is usually between 150 and 250 g, depending on the country of sale), could help to increase the reproductive success of the species.

Although different filling media are widely used for canning fish, available information on how different oils affect the sensory quality of fish is scarce, and the effects on the sensory properties of canned eels have not been investigated.

In this context, the aim of the present study was to evaluate how the colour of European eels (*Anguilla anguilla* L.) and different filling media (vegetable oils) were affected at different stages of the canning process by the type of oil used and by the storage time. The sensorial characteristics of canned eels were also evaluated.

2. Materials and Methods

2.1. Samples and Fish Preparation

2.1.1. Samples Preparation

Eels (weighing between 200 and 400 g) were purchased from a local market (Mariscos Vivos del Grove, Plaza de Abastos,) in Ourense (Galicia, NW Spain) after being captured in the Ulla River (Galicia, NW Spain). The eels were eviscerated and transferred to the laboratory. The fish was washed, packed in vacuum bags and stored at $-20\,^{\circ}$C until further processing.

2.1.2. Eel Canning Process

Before canning, frozen eels were thawed in 12% brine. The head and skin were removed, and the fish were sliced. The fish slices were fried at 190 $^{\circ}$C for 2 min in a conventional frying pan to eliminate the water present and to prevent the formation of a water-oil mixture in the cans. Two different frying media were used: refined sunflower and an olive oil mixture (refined olive oil + virgin olive oil). The fried eel slices were cooled (30 $^{\circ}$C) and placed in glass jars (6 or 7 slices in each). The hot filling medium was then added: refined sunflower oil (eel slices previously fried in this oil), olive oil (refined olive oil + virgin olive oil) or spiced olive oil (refined olive oil + virgin olive oil plus chilli and pepper) (eel slices previously fried in olive oil) (Figure 1). In the case of the spiced olive oil, approximately 0.20 g of dried chilli and 0.30 g of dried pepper were added inside the glass jar before sterilization.

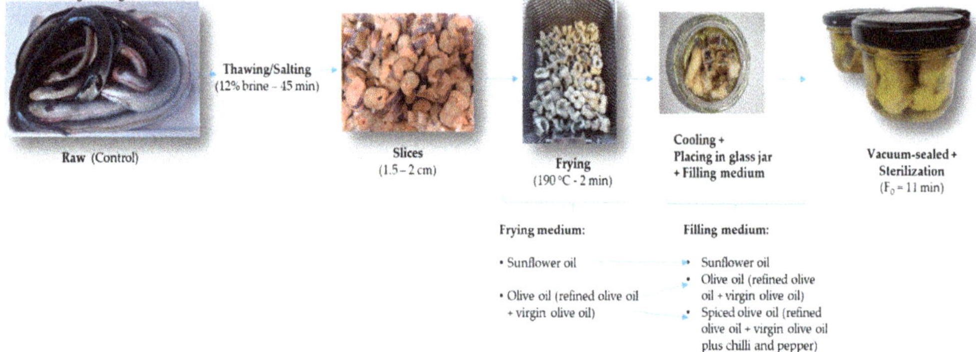

Figure 1. Flow chart of the canned eel production process.

The jars were vacuum-sealed and sterilized at 118 °C for 30 min, after which they were cooled and stored at room temperature in a dark room.

For colour determination, the eels and the different oils were sampled at different stages: before processing, after frying the fish and after the sterilization treatment.

2.1.3. Storage

The canned eels and oils were also sampled after two months and one year of storage at room temperature. The jars were opened, and the filling medium was carefully drained off through a 3 mm pore sieve. The sensory analysis of the canned eels was carried out after one year of storage.

2.2. Colour Determination

Colour measurements were made with a portable colorimeter (Chroma Meter CR−400, Konica Minolta Sensing, Inc., Osaka, Japan). The colour was expressed by the values of the coordinates in the CIEL*a*b* space: L^* (lightness), a^* (balance between red and green) and b^* (balance between yellow and blue).

The colorimetric parameters hue angle ($H°$) (colour of sample as defined by its location in a 360° axis; 0 or 360° = red, 90° = yellow, 180° = green and 270° = blue), chroma (C^*) (colour saturation, increasing from 0) and total colour change (ΔE) were calculated as follows:

$$H° = 180° + arctg\ (b^*/a^*),\ for\ a^* < 0\ and\ b^* > 0$$

$$C^* = \sqrt{(a)^2 + (b)^2}$$

$\Delta E = \sqrt{[(L_0 - L)^2 + (a_0 - a)^2 + (b_0 - b)^2]}$, colorimetric coordinates of raw eels indicated by "0".

All determinations were made at least in triplicate.

2.3. Sensorial Analyses

2.3.1. Descriptive Sensory Evaluation

The aim of the sensorial analysis was to identify the important sensory attributes of canned eels in different filling media. A panel of 11 trained tastes recruited from the Faculty of Science of Ourense (University of Vigo, Spain) was used to evaluate the intensity of each attribute.

The panellists were trained in sensory evaluation of fish according to International Organization for Standardization (ISO) standards [10] and they were familiarized with the questionnaire on the sensory parameters.

The different descriptive attributes were measured according to ISO [11,12] standards. The final descriptions selected were defined according to UNE-EN-ISO 5492:2010/A1:2017 [13]. Sensory descriptive terms (15 attributes) were generated and agreed on among the panel members. The descriptive attributes and their definitions are included in Table 1. The descriptive parameters of fish (bitterness, acid, sweet, salty, metallic, pungent, colour, gloss, appearance, aroma intensity, preference, hardness, adhesion, residual taste in the mouth and aftertaste) were evaluated by the panellists on a scale ranging from 1 (absence of sensation) to 10 (extremely intense sensation) (Figure 2). The sensory profile of the samples was determined from the scores awarded in the descriptive analysis.

Table 1. Descriptive attributes and their definitions adapted from UNE-EN-ISO 5492:2010.

Descriptor	Definition
External aspect	
Colour	Attribute of products inducing a colour sensation. Colour ranging from white to brownish
Glossiness	The amount of light reflected from the muscle, ranging from dull to glossy
Appearance	All the visible attributes of fish
Texture in mouth	
Hardness	Force required to achieve a given deformation, penetration, or breakage of the fish muscle.
Adhesiveness	Force required to remove the muscle with tongue that sticks to the mouth or to a substrate.
Residual taste	Sensation perceived whilst the product was in the mouth.
Aroma	
Preference	Preference between two or more samples.
Intensity	Perceived strength at differing concentrations of certain volatile compounds
Taste	
Bitter	The intensity of the taste associated with dilute aqueous solutions of various substances such as quinine or caffeine
Acid	Taste associated with aqueous solutions of most acid substances such as citric acid.
Sweet	Taste associated with aqueous solutions of sugar such as sucrose, or aspartame.
Salty	The intensity of the taste associated with salt solutions
Pungent	An irritating, sharp sensation, or piercing sensation in the buccal and nasal mucous membranes
Metallic	Taste associated with flavour reminiscent of metal, slightly oxidized metal, or salts.
Aftertaste	Sensation that occurs after elimination of the product

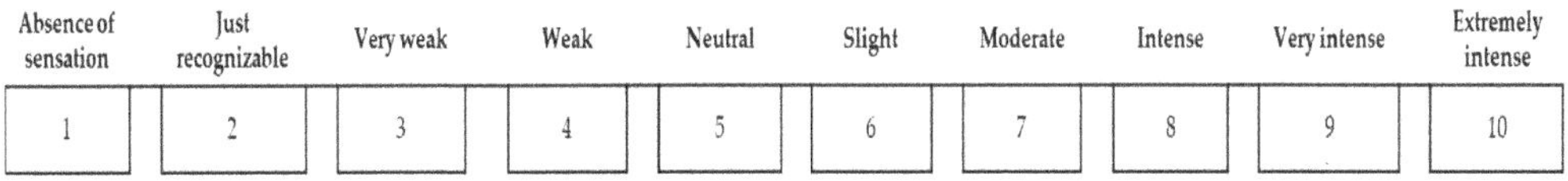

Absence of sensation	Just recognizable	Very weak	Weak	Neutral	Slight	Moderate	Intense	Very intense	Extremely intense
1	2	3	4	5	6	7	8	9	10

Figure 2. Scores of intensity of each descriptor.

A colour scale was created to enable the colour of the samples to be evaluated, taking into account the different shades that the fish acquired during treatment (Figure 3).

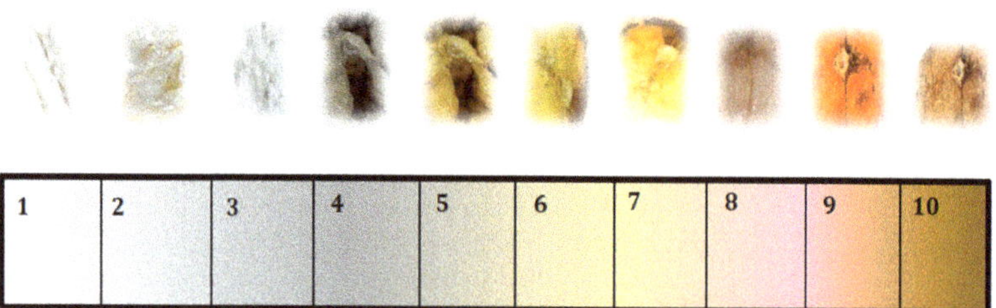

Figure 3. Colour scale prepared for rapid assessment of the colour of eel muscle (1= lighter and whitish to 10= darker and brownish).

2.3.2. Consumer Tests

Samples of canned eels packed in the different oils were also used for the consumer test. The test consumer group was formed by 97 volunteers from the teaching staff, graduate students and technicians of Faculty of Science (Ourense, Spain). The testers were of both sexes (63 women and 34 men) and different ages (31 < 20 years; 46 from 21 to 35 years; 20 > 35 years). Prior to consumer testing, a supplementary questionnaire was filled out to obtain more information about the age and gender of the testers and their weekly fish consumption. The panellists ate fish products often and they were notified about the fish that they had to taste.

The consumer study involved analysis to investigate the perceptions of consumers on the previously selected attribute group, and degree of acceptance. The testers were divided into two groups, with 5–6 individuals in each session. The testers evaluated three samples identified with a three-digit code, corresponding to the three different types of canned eels (packing in sunflower oil, olive oil or spiced olive oil) in each session. Testers ate pieces of bread and sipped water to neutralize the taste between sample testing. The samples were presented simultaneously to testers.

The parameters of flavour, aroma, texture, appearance and general appreciation were evaluated by the testers in a 10-point hedonic scale test, with a score of 1 indicating extreme dislike, and a score of 10 indicating excellent (Figure 4). The final scores reported were the average scores for the 97 testers.

Dislike extremely	Dislike very much	Disslike moderately	Dislike slightly	Neutral/ Neither like nor dislike	Like slightky	Like moderately	Like very much	Like extremely	Excellent
1	2	3	4	5	6	7	8	9	10

Figure 4. Hedonic 10-point scale.

2.4. Statistical Analysis

All analyses were conducted at least in triplicate. The data were examined by analysis of variance (ANOVA). The least significant difference (LSD) test was applied, with a 95% confidence interval ($p \leq 0.05$), for comparison of the mean values, using the statistical software Statistica version 8.1 (Statsoft© Inc., Tulsa, OK, USA).

The sensory data were examined by Friedman analysis of variance, to test for any differences between the overall preferences for canned eels. The significance of differences between samples was determined by the Fisher test ($\alpha = 0.05$), modified for non-parametric data.

3. Results and Discussion

3.1. Colour

Vision is usually the first sense used in making purchasing decisions. The colour of food products is a very important quality parameter for consumers.

The colour parameters measured in raw eels and in canned eels sampled at each stage of the canning process (frying, sterilization, and after room storage for 2 and 12 months), and packed in sunflower oil, olive oil or spiced olive oil (olive oil plus chilli and pepper) are shown in Table 2.

Table 2. Colour parameters of raw and canned European eels packed in sunflower oil, olive oil or spiced olive oil at the different steps of the canning process, and after 2 and 12 months of storage. Values are means ± standard deviations of three replicates.

		L^*	a^*	b^*	$H°$	C^*	ΔE
	Raw	74.05 ± 0.24 [a]	−4.95 ± 0.13 [ac]	14.60 ± 0.18 [a]	108.74 ± 0.55 [a]	15.15 ± 0.35 [a]	–
Frying	Sunflower oil	72.68 ± 1.63 [b]	−5.12 ± 0.20 [ab]	16.69 ± 0.36 [c]	108.08 ± 1.67 [a]	17.56 ± 0.21 [b]	10.78 ± 0.30 [ab]
	Olive oil	71.94 ± 1.89 [b]	−4.72 ± 0.51 [acd]	19.56 ± 0.34 [de]	103.56 ± 1.18 [b]	20.13 ± 0.45 [c]	11.17 ± 0.23 [a]
Sterilization	Sunflower oil	77.61 ± 0.70 [c]	−5.34 ± 0.11 [b]	18.88 ± 1.07 [de]	106.02 ± 0.86 [c]	19.64 ± 1.03 [c]	11.63 ± 0.06 [ac]
	Olive oil	76.14 ± 1.27 [c]	−4.68 ± 0.09 [cd]	21.78 ± 2.95 [d]	101.84 ± 1.02 [bd]	22.09 ± 3.22 [c]	13.31 ± 0.16 [d]
	Spiced olive oil	71.31 ± 1.61 [b]	−4.92 ± 0.40 [acd]	19.57 ± 1.48 [de]	103.74 ± 0.64 [b]	20.18 ± 1.54 [c]	11.46 ± 1.38 [ac]
2 months storage	Sunflower oil	76.30 ± 1.09 [c]	−4.69 ± 0.24 [cd]	19.00 ± 0.56 [de]	103.88 ± 1.10 [b]	19.58 ± 0.49 [c]	10.88 ± 0.23 [a]
	Olive oil	72.76 ± 0.73 [b]	−4.57 ± 0.15 [d]	17.61 ± 0.81 [ce]	104.37 ± 1.21 [b]	17.59 ± 0.41 [b]	9.92 ± 0.25 [b]
	Spiced olive oil	67.22 ± 1.58 [d]	−4.82 ± 0.16 [c]	18.50 ± 1.24 [de]	104.08 ± 0.99 [b]	19.12 ± 1.24 [c]	12.43 ± 0.51 [cd]
12 months storage	Sunflower oil	72.29 ± 1.63 [b]	−4.35 ± 0.14 [d]	19.25 ± 0.50 [de]	102.98 ± 0.23 [b]	19.75 ± 0.49 [c]	10.69 ± 0.01 [ab]
	Olive oil	67.22 ± 1.86 [d]	−3.59 ± 0.21 [e]	20.02 ± 0.85 [d]	100.20 ± 0.96 [d]	20.34 ± 0.81 [c]	12.31 ± 0.88 [cd]
	Spiced olive oil	70.93 ± 1.75 [bd]	−4.44 ± 0.04 [d]	19.25 ± 0.50 [de]	102.98 ± 0.23 [b]	19.75 ± 0.49 [c]	10.70 ± 0.38 [ab]

[a–e] Means in the same column with different letters differ significantly ($p < 0.05$).

The L^* coordinate represents the luminosity and varies between 0 (black) and 100 (white). The raw eel presented a mean luminosity value of 74.05. This is higher than values reported for other species. Huang et al. [14] reported an L^* value of 45 in Pacific tuna. Fuentes-Zaragoza et al. [15] reported a value of 47.96 in hake, and Sánchez-Zapata et al. [16] reported a value of 57.47 for this same species. L^* values are generally lower in fatty fish with high proportions of dark muscle [17]. Greater luminosity is usually related to greater whiteness, a parameter that is valued in some fish such as hake or cod. Lower luminosity is usually associated with older and/or stale products. In fish with low concentration of pigments, such as carotenoids or haemopigments, reflection of light is favoured, and the L^* value will therefore be higher.

The a^* coordinate represents the variation between red (>0) and green (<0). High a^* values are usually related to a higher concentration of haemopigments, mainly myoglobin, in the dark muscle and lower concentrations of carotenoids [18]. In the raw eel muscle, the low a^* values (mean −4.95) indicated a light colouration, with slightly greenish tones.

The b^* coordinate represents the variation between yellow (>0) and blue (<0). In the raw eels the mean b^* value of 14.60 indicated slightly yellowish tones. Variations in this value are usually due to the presence of a greater or lower amounts of red muscle [16].

Hue angle ($H°$) is used to describe the difference of a certain colour with reference to grey colour with the same lightness. It varies between 0 and 360°. In this case, 90° represents yellow, 180° represents blue-green and 270° represents blue [19]. In the raw eel, the mean $H°$ value of 108.74° indicated yellowish-greenish tones.

The C^* coordinate, or chroma, represents the purity of the colour, and is used to define the degree of difference of a hue in comparison to a grey colour with the same lightness. The C^* mean value in raw eels was 15.15. In previous studies, it has been pointed out that the chroma parameter generally performs in a similar way to a^* and b^*, i.e., the factors that modify these coordinates also affect the C^* values [20].

In food processing, colour serves as a cue for temperature and time control during cooking/processing and is correlated with changes in aroma and flavour.

Frying caused a decrease in brightness (L^*) relative to raw eels. This decrease may be due to chemical reactions that take place in the frying oil, such as hydrolysis, oxidation, polymerization and others [21].

Frying did not cause any significant changes in parameter a^*. However, the b^* and C^* values increased. This indicates a tendency towards yellowish colours. The b^* and C^* values were higher in the eels fried in olive oil than in eels fried in sunflower oil. Frying caused a decrease in hue angle ($H°$) in eels fried in olive oil, but not in those fried in sunflower oil. During frying, heat and mass transfer occurs between the fish and the oil, causing physicochemical changes that affect the colour of the fish and the oil. Caramelization and Maillard reactions also take place at the surface, and evaporation of water and absorption of oil modify the colour of the fried product. Frying often results in darkening of the muscle due to these reactions. The colour changes during frying are determined by the temperature of the oil, the type of oil and the shape and size of the food [22]. Pérez-Palacios et al. [23] reported more intense browning in hake fillets fried in olive oil than in those fried in sunflower oil.

The sterilization treatment caused a significant increase in the luminosity of the canned eels packed in sunflower oil and olive oil, but not in canned eels packed in spicy olive oil. The a^* coordinate did not vary significantly during the sterilization process, but the b^* and chroma values increased in the eel packed in sunflower oil. Hue angle decreased during sterilization in eels packed in sunflower oil. In the spiced eels, b^*, chroma and hue angle did not vary during sterilization. The denaturation of myoglobin and the oxidation of some pigments such as carotenoids cause changes in the colour of the muscle from red to paler pinkish colours, which leads to an increase in luminosity. The increase in L^* values could also be due to the leaching of fish pigments or of white connective tissue containing collagen between the segments of muscles [24]. In the canned eels packed in spicy olive oil, the spices appear to significantly affect the colour parameters, probably due to the presence of pigments or prevention or enhancement of certain reactions in the fish muscle.

In the canned eels packed in sunflower oil, the L^*, b^* and C^* values remained stable during the first two months of storage, but a^* approached positive values and $H°$ decreased. After storage for 12 months, the L^* value decreased in these canned eels. In the canned eels packed in olive oil, the L^*, b^* and C^* values decreased during the first two months of storage. In this case, L^* and $H°$ continued to decrease until twelve months, and the value of a^*, b^* and C^* values increased. In canned eels packed in spicy olive oil, a decrease in L^* was observed after two months of storage. These changes can be explained by denaturation of some pigments and haem-proteins inside the fish muscle [25]. The changes in colour in fish during storage are also associated with oxidation of pigments and the decomposition and polymerization of primary products of lipid oxidation [26]. Changes in filling medium also affect the colour of the fish.

Canned eels packed in olive oil underwent the greatest global colour change (ΔE) during the sterilization process (mean values of 13.31). The changes in all the colour parameters are considered indicative of the modification of the colour in the fish and the filling medium because of the heat treatments applied and processes (such as oxidation) occurring during storage.

The colour of oils is influenced by the pigment content and has been widely used as a quality index. The colour parameters measured in the raw sunflower oil and fresh olive oil and in sunflower oil, olive oil and spicy olive oil after each step of the canning process and after storage for 2 and 12 months are shown in Table 3.

Table 3. Colour parameters in the raw oils and after each stage of processing and storage. Values are means ± standard deviations of three replicates.

		L^*	a^*	b^*	$H°$	C^*	ΔE
Raw	Sunflower oil	75.53 ± 2.38 [ab]	−3.85 ± 0.07 [a]	4.05 ± 0.06 [a]	133.50 ± 0.13 [a]	5.59 ± 0.09 [a]	–
	Olive oil	77.41 ± 1.83 [b]	−10.49 ± 0.32 [d]	27.46 ± 1.87 [b]	110.95 ± 0.73 [b]	29.40 ± 1.86 [b]	–
Frying	Sunflower oil	75.84 ± 3.03 [ab]	−4.72 ± 0.04 [b]	7.17 ± 0.64 [c]	123.46 ± 2.11 [c]	8.59 ± 0.55 [c]	4.47 ± 0.35 [a]
	Olive oil	73.08 ± 1.11 [c]	−9.61 ± 0.06 [e]	27.78 ± 0.76 [b]	109.08 ± 0.37 [b]	29.40 ± 0.73 [b]	4.48 ± 1.09 [a]
Sterilization	Sunflower oil	78.99 ± 1.90 [bd]	−4.16 ± 0.08 [c]	4.61 ± 0.46 [a]	132.66 ± 0.40 [a]	6.17 ± 0.40 [a]	4.55 ± 0.63 [a]
	Olive oil	72.97 ± 0.88 [c]	−9.90 ± 0.36 [e]	27.04 ± 1.11 [b]	110.10 ± 0.25 [b]	29.07 ± 0.92 [b]	4.53 ± 0.95 [ab]
	Spiced olive oil	72.72 ± 2.13 [c]	−9.36 ± 0.20 [ef]	25.62 ± 1.62 [b]	110.22 ± 0.91 [b]	27.67 ± 1.46 [b]	6.08 ± 0.90 [b]
2 months storage	Sunflower oil	78.18 ± 1.91 [bd]	−4.13 ± 0.06 [c]	4.21 ± 0.43 [a]	134.58 ± 2.63 [a]	5.91 ± 1.33 [a]	4.66 ± 0.97 [a]
	Olive oil	78.10 ± 0.99 [bd]	−9.96 ± 0.19 [e]	25.04 ± 0.98 [b]	111.63 ± 0.41 [b]	27.04 ± 0.97 [b]	2.68 ± 0.56 [c]
	Spiced olive oil	82.78 ± 2.13 [d]	−8.87 ± 0.57 [f]	19.23 ± 2.66 [d]	114.63 ± 1.05 [d]	21.57 ± 1.24 [d]	10.81 ± 1.02 [d]
12 months storage	Sunflower oil	76.79 v ± 1.14 [b]	−4.89 ± 0.13 [b]	7.31 ± 0.65 [c]	124.12 ± 1.73 [c]	8.65 ± 0.60 [c]	4.49 ± 0.59 [a]
	Olive oil	76.30 ± 0.99 [b]	−10.77 ± 0.30 [d]	32.37 ± 2.27 [e]	108.46 ± 0.99 [e]	34.12 ± 2.67 [e]	2.16 ± 0.72 [c]
	Spiced olive oil	72.89 ± 0.41 [c]	−10.49 ± 0.17 [d]	36.25 ± 3.12 [e]	106.24 ± 1.58 [e]	37.75 ± 2.95 [e]	11.87 ± 0.71 [d]

[a–f] Means in the same column with different letters differ significantly ($p < 0.05$).

No significant differences were observed between raw sunflower oil and olive oil in regard to luminosity (L^*). In sunflower oil, the different steps during canning and storage did not cause significant changes ($p > 0.05$) in this parameter. However, in olive oil, the L^* value was significantly lower ($p < 0.05$) after the frying process. The decrease in L^* is associated with darkening of the oil, due to the formation of polymers and dissolution of non-polar compounds in the medium [20]. The loss of luminosity may also be associated with the formation of the Maillard products resulting from interactions between food and the frying oil [27]. Sterilization causes significant changes in L^* values, relative to the fresh oil, in olive and spiced olive oil. Storage for 2 months caused an increase in luminosity in olive oil. With regard to spicy olive oil, the luminosity increased during the first months of storage, but decreased after 12 months. These changes may be due to the exchange of substances between the food and filling oil. Sánchez-Gimeno et al. [20] observed an increase in luminosity of fish fried in olive oil, but a decrease in the same fish fried in sunflower oil.

Regarding the greenness/redness (a^* value), as expected the values were lower in olive oil (−10.49) than in sunflower oil (−3.85) due to the greenish hue of the former. In sunflower oil, the a^* values decreased after frying (−4.72) and after sterilization (−4.16). During storage, a^* initially remained stable (2 months), but then decreased after 12 months of storage. The a^* values in olive oil and spicy olive oil were higher (redder colour) after frying and sterilization and at the beginning of storage and decreased after 12 months of storage. Positive a^* values ($a^* > 0$, redness), which would indicate a less greenish colouration, were not observed in any of the samples. Reddish colouration is not acceptable from the point of view of oil quality as it is related to combined oxidized fatty acid and pyrolytic condensation products [28].

The b^* value was lower in sunflower oil (4.05) than in olive oil (27.46). The colour of oils is chiefly influenced by two groups of constituents: carotenoids and chlorophyll. A yellow colour is associated with the presence of carotenoids in oils, while a green colour is associated with the presence of chlorophylls. Carotenoids predominate in sunflower oil and are responsible for the yellow colour of the oil [29]. Olive oil contains more chlorophyll which gives the oil a greenish colour. In addition, the extraction technology and the maturity of the fruit from which the oil is extracted determine the intensity of the colour of the oil.

Regarding sunflower oil, an increase in the b^* value was observed after frying. Browning reactions, which are caused by oxidation products in the frying oil, can change the colour of the oil [27]. During the sterilization process and the first months of storage, the b^*

value remained stable and increased after 12 months. The increase in b^* values may be due to the oxidation of carotenoids and to the exchange of different components between the eel and the filling medium. Sánchez-Gimeno et al. [20] reported that the b^* values increased in olive oil and sunflower oil after frying.

The $H°$ values were higher in raw sunflower oil (133.50°) than in olive oil (110.95°). The frying treatment decreased this parameter in sunflower oil, but not in olive oil. There were no changes in $H°$ values in olive oil or spicy olive oil after the sterilization process; however, an increase in this parameter was observed in sunflower oil after this treatment. After 12 months of storage, $H°$ values decreased in the three types of filling media.

The changes in the C^* values (chroma) followed the same trend as in b^* values. The C^* value was lower in raw sunflower oil (5.59) than in raw olive oil (29.40). Frying produced an increase in C^* values in sunflower oil, while sterilization caused a decrease in these values. No significant changes were observed in the C^* values in olive oil after frying and sterilization. The C^* values increased after 12 months of storage.

ΔE represents the overall colour change. The greatest changes took place in the spicy olive oil and mainly during storage. The more pronounced changes in the spicy olive oil may be due to migration of some of the components of the pepper and/or chilli into the oil.

Vegetable oils are susceptible to oxidation. Oxidation is an important cause of deterioration in oil quality during processing and storage [30]. A significant difference ($p < 0.05$) in most colour parameters was observed in canned filling media relative to the control sample. The thermal stability of oils depends on their chemical structure. Reda [31] reported that saturated oils are more stable than unsaturated oils. Moreover, during storage, the colour of the oil changes substantially over time. These changes may be due to the dilution and partial extraction of some components [8]. Light, heat and oxygen cause the degradation of pigments such as chlorophylls and carotenoids that give the oil its colour. Chlorophylls and carotenoids play an essential role in the oxidative stability of oils [32]. The loss of colouration during heat treatments is related to the thermolability of the pigments present in the oils.

3.2. Sensorial Analysis

3.2.1. Sensory Description

The sensory quality of a food is the result of the interaction between the product and consumer. Acceptance or rejection of food by humans is based on the sensations it produces. The sensory quality of a food is evaluated by sensory analysis. Sensory analysis is an organized assessment of the colour, aroma, flavour, texture and appearance of a food. Different studies on fish consumption reveal that the principal factors responsible for the acceptance or rejection of a product are sensory characteristics [33].

The differences observed in the descriptive tests, in relation to the intensity of each of the attributes, can be observed in Figure 5, which shows the different sensory profiles of the canned eels, according to the perceptions of the panellists.

Prior to consumption, the external aspect influences decisions on the intention of purchase or consumption of a food by affecting expectations of palatability [34]. Colour significantly affects food acceptance. Colour scales are available for evaluation of some fish, such as salmon and tuna, and for the skin of some aquarium fish. However, colour scales are seldomly used for fresh or processed muscle of other fish. In this work, a colour scale was developed with the different colourations observed in the eel muscle after different treatments (Figure 3). The colour parameters revealed differences between canned eels due to the colour of each of the oils used as filling medium. Colour scores were higher for canned eels packed in sunflower and spicy olive oil than for canned eels packed in olive oil. These findings are consistent with the CieLab values obtained (Table 2), while the values for canned eel packed in sunflower oil and spicy olive oil after 12 months of storage were more similar to each other than those of olive oil. The canned eels packed in olive oil were yellower-greener and canned eels packed in sunflower and spicy olive were pinker. The olive oil used as filling medium, due to its characteristic organoleptic imprint, could had partially attenuated the typical eel colour. Colour scores were significantly negatively correlated with a^* (r = −0.99), b^* (r = −0.98), C^* (r = −0.97) and ΔE values (r = −0.97) and positively correlated with H° (r = 0.97).

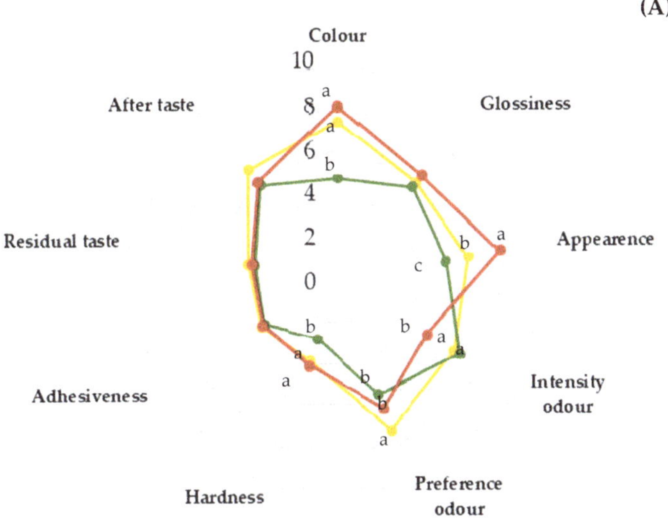

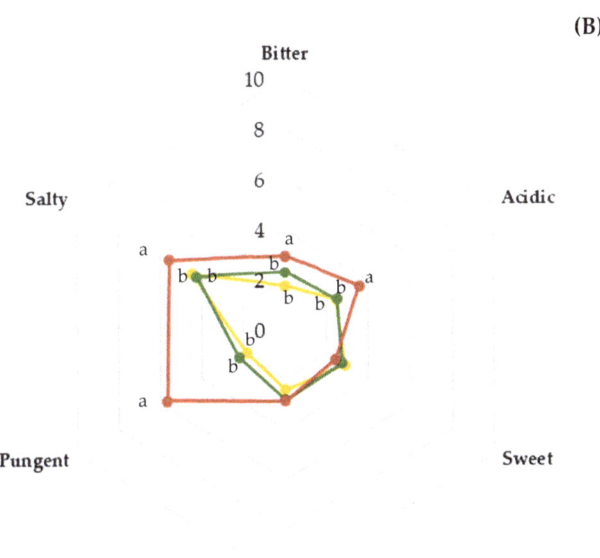

Figure 5. Graphic expression of the sensory profiles of canned eel packed with different filling media. (**A**): texture in mouth and aroma; (**B**): taste. Plotted values are the means of 11 observations. Different letters indicate significantly different ($p < 0.05$).

The highest scores for glossiness and appearance were also obtained for canned eels packed in spicy olive oil.

Texture is an important parameter of fish also associated with consumer acceptability [35]. The texture is strongly related to the amino acid composition and protein structure, which undergo important changes during processing and storage. However, hardness was lower in canned eels packed in olive oil than in canned eels packed in sunflower oil and spicy olive oil. This may be due to the different changes that proteins and amino acids undergo during the process of canning eels in different filling media [36]. No differences in aftertaste residual or adhesiveness were observed between the different canned eel products.

Aroma is an important sensory property of fish products, and it can provide information about any physical/chemical alterations that have occurred during handling and processing. Canned eels packed in spicy olive oil were characterized by lower scores for aroma intensity. Canned eels packed in sunflower oil were awarded higher scores for preferred aroma. The oils used as filling media provide characteristic aromas to the fish.

As expected, the panel of assessors perceived significant differences in the pungency of the canned eels packed in spicy olive oil and the other two canned eel products (Figure 5B). The pungent taste is due to the chilli and pepper contained in spicy canned eels. Spicy canned eels were also slightly more bitter and acidic. Capsaicinoids are responsible for the spiciness of chili and peppers and are directly related to the pungency. On the other hand, different studies have indicated that capsaicin can evoke a bitter taste [37]. Organic acids present in spices can also contribute to the acidity of the canned fish.

No differences in basic taste were observed between canned eels packed in sunflower oil and canned eels packed in olive oil.

These results revealed that the filling medium significantly affected the sensory properties of the canned eels.

The non-parametric Friedman's test is usually used to analyse the data obtained from a multiple-samples ranking test [38] and allows the establishment of the degree of probability with which it can be shown that panellists recognize differences between the samples tested. The F value was calculated according to the standard, yielding a value of 13.27. Taking into account this value and the numbers of samples and panellists it is confirmed that the participants in the sensory analysis were able to detect differences between the canned eels, as the observed F value was higher than the estimated F value (5.99 for $\alpha = 0.05$).

3.2.2. Consumer Preference

The average acceptance value scores awarded by 97 testers, expressed in %, are shown in Figure 6. Participants responded on a scale between 1 (extreme dislike) and 10 (excellent). A high proportion of the testers rated all canned eels with positive expressions on the hedonic scale. The test product with the greatest number of positive evaluations was the canned eels packed in sunflower oil, followed by the spicy canned eels.

In the case of the purchase intention, 45% of the participants would buy the canned eels packed in sunflower oil, 24% would buy canned eels in spicy packed olive oil and 31% would buy canned eels packed in olive oil.

The sense of taste may play an important role in food preferences and food choices. In consumers, different factors, such as demographics and social factors, can affect the overall scores regarding satisfaction with products. Numerous studies have attempted to clarify the influence of gender and/or age on sensory perception of and preference for different foods [39,40]. Age, gender, employment status and frequency of consumption of canned fish or shellfish are very important factors. In this tasting, 76% of the participants were regular consumers (once or twice a week) of this type of product. The average female and male consumer preferences for canned eel attributes are illustrated in Figure 7A. The results indicated that females preferred spicy canned eels packed in olive oil (41%) and canned eels packed in sunflower oil (37%), while men mostly preferred canned eels packed in sunflower oil (65%). Previous studies of the effects of gender on sensory function indicated that females have more sensitive and more accurate sensory perception than males [41].

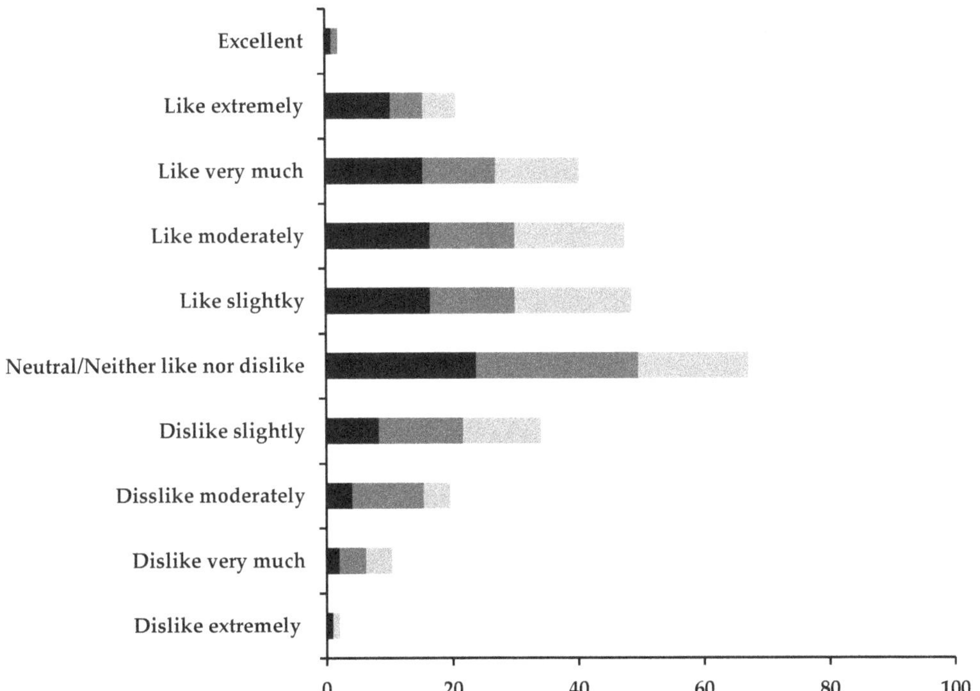

Figure 6. Distribution of consumer acceptance scores for three candy bars. Average acceptance values scores awarded by 97 testers expressed in %.

The age of the consumers varied as follows <20 years old (about 31.96%), between 21 and 35 (about 47.42%) and the remaining were over 35 years old (20.62%). It was found that 63% of tasters under 20 years of age preferred canned eels packed in sunflower oil, and only 20% preferred canned eels packed in olive oil. Those over 35 years of age also showed similar preferences (55% preferred canned eels packed in sunflower oil and 20% canned eels packed in olive oil). The middle group showed a similar preference for canned eels packed in sunflower oil and canned eels packed in spicy olive oil.

Mojet et al. [42] studied the effect of gender and age on the threshold sensitivity of the basic tastes in 22 young adults (11 males and 11 females) and 21 older adults (10 males and 11 females). They reported that age, but not gender, significant y affected the sensitivity.

The consumers generally considered that all of the canned eel products were acceptable, with moderately high scores awarded to all. However, canned eels packed in sunflower oil were ranked the best of the three canned eels in sensory acceptation.

The type of filling medium played an important role in the formation of volatile compounds during the preparation of canned fish. Some filling media can slightly alter the typical fish sensory characteristics.

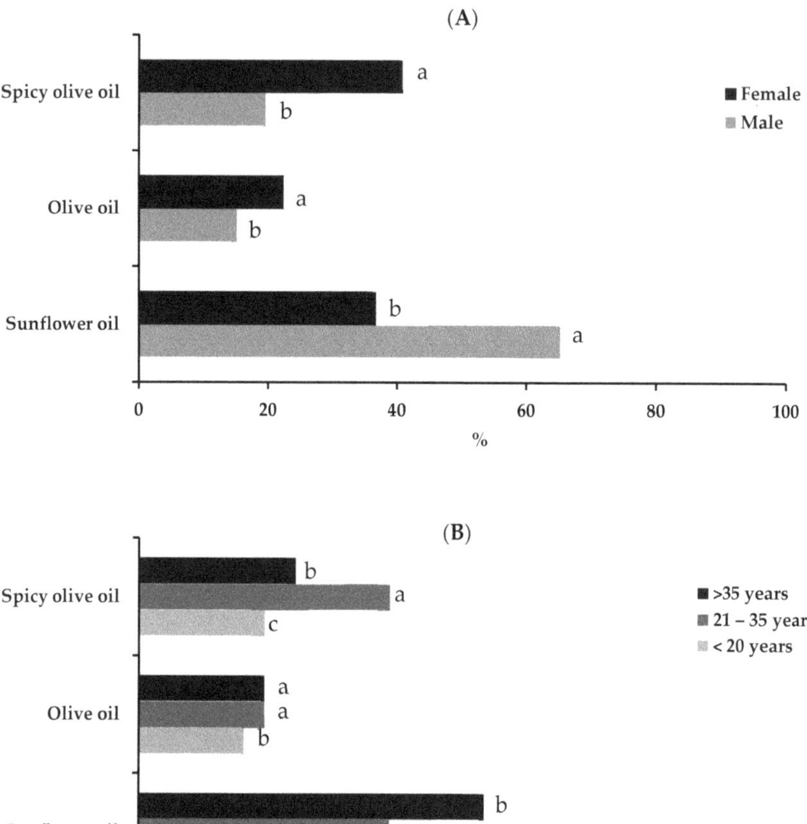

Figure 7. Mean score of preferences by gender (**A**) and age group (**B**). Different letters indicate significant differences ($p < 0.05$) in each type of canned eels.

4. Conclusions

Marked changes in the colour of canned eels and the filling medium were detected as a result of the canning process and subsequent storage. The changes that took place depended on the processing step, the filling medium and the storage time. Luminosity, b^* (variation between yellow and blue), hue angle and chroma were the parameters most affected. The addition of spices (pepper and chilli pepper) to the filling medium caused important changes in the colour of the medium. The changes in the spicy olive oil may be due to migration of some of the components of the pepper and/or chilli into the oil. The changes in the colour in the canned eels and filling medium during the manufacture of canned fish products have an important effect on the perceived quality.

A sensory analysis of canned eels was carried out after the cans had been stored at room temperature for one year. The canned eels packed in sunflower oil were awarded the highest scores by testers, who indicated that this was the product that they were most likely to purchase. Therefore, it is concluded that canning is a good option for larger eels, due to the high level of acceptance. However, the preferences varied depending on the age and gender of the consumers. Females preferred spicy canned eels packed in olive oil and canned eels packed in sunflower oil, while males mainly preferred canned eels packed in sunflower oil. The youngest and oldest consumers preferred the eels canned in sunflower oil and those of 21–35 years old equally preferred the eels canned in spicy olive oil and those canned in sunflower oil.

The present results on colour and sensory quality provide valuable information regarding consumer acceptance of canned fish and the effect on sensory properties that may be expected as a result of using different filling media in canned fish.

Author Contributions: Conceptualization, S.M.; formal analysis, L.G.-L. data curation, L.G.-L., M.R.-G., J.C. and S.M.; writing—original draft preparation, L.G.-L. and S.M.; writing—review and editing, J.C. and S.M.; supervision, S.M. All authors have read and agreed to the published version of the manuscript.

Funding: This research was funded by Xunta de Galicia, Spain (CITACA Strategic Partnership), grant number ED431E 2018/07.

Informed Consent Statement: Informed consent was obtained from all subjects involved in the study.

Data Availability Statement: Data is contained within the article.

Acknowledgments: Lucía Gómez-Limia acknowledges receiving financial support from the University of Vigo through a predoctoral scholarship. The authors would also like to thank GAIN (Axencia Galega de Innovación) for supporting this study (grant number IN607A2019/01).

Conflicts of Interest: The authors declare no conflict of interest.

References

1. Wilkinson, C.; Dijksterhuis, G.B.; Minekus, M. From food structure to texture. *Trends Food Sci. Technol.* **2000**, *11*, 442–450. [CrossRef]
2. Spence, C. On the psychological impact of food colour. *Flavour* **2015**, *4*, 21–47. [CrossRef]
3. Abdulla, M.Z.; Guan, L.C.; Lim, K.C.; Karim, A.A. The application of computer vision and tomographic radar imaging for assessing physical properties of food. *J. Food Eng.* **2004**, *61*, 125–135. [CrossRef]
4. Leon, K.; Mery, D.; Pedreschi, F.; Leon, J. Colour measurement in L*a*b* units from RGB digital images. *Food Res. Int.* **2006**, *39*, 1084–1091. [CrossRef]
5. Pathare, P.B.; Opara, U.L.; Al-Julanda Al-Said, F. Color Measurement and analysis in fresh and processed foods: A review. *Food Bioprocess Technol.* **2013**, *6*, 36–60. [CrossRef]
6. Trusell, H.J.; Saber, E.; Vrhel, M. Color image processing. *IEEE Signal Process. Mag.* **2005**, *22*, 14–22. [CrossRef]
7. Morsy, M.K. Quality enhancement of canned little tunny fish (*Euthynnus alletteratus*) by whitening solutions, pre-cooking time and filling medium. *J. Food Process. Technol.* **2016**, *7*, 632. [CrossRef]
8. Medina, I.; Sacchi, R.; Biondi, L.; Aubourg, S.P.; Paolillo, L. Effect of packing media on the oxidation of canned tuna lipids. Antioxidant effectiveness of extra virgin olive oil. *J. Agric. Food Chem.* **1998**, *46*, 1150–1157. [CrossRef]
9. Cobo, F.; Hernández, J.S.; Vieira, R.; Servia, M.J. Seasonal downstream movements of the European eel in a Southwestern Europe river (River Ulla, NW Spain). *Nova Acta Científica Compostel.* **2014**, *21*, 77–84.
10. *UNE-ISO 4121:2006*; Sensory Analysis—Guidelines for the Use of Quantitative Response Scales. ISO: Geneva, Switzerland, 2006.
11. *ISO 11035:1994*; Sensory Analysis—Identification and Selection of Descriptors for Establishing a Sensory Profile by a Multidimensional Approach. International Organization for Standardization (ISO): Geneva, Switzerland, 1994.
12. *UNE-ISO 6658:2005*; Sensory Analysis. Methodology. General Guidance. International Organization for Standardization (ISO): Geneva, Switzerland, 2008.
13. *UNE-EN-ISO 5492:2010/A1:2017*; Sensory Analysis—Vocabulary—Amendment 1. International Organization for Standardization (ISO): Geneva, Switzerland, 2017.
14. Huang, M.C.; Cho, S.C.; Ochiai, Y.; Watabe, S. Evaluation of biochemical properties of burnt and normal meat in pacific bluefin tuna (*Thunnus orientalis*). *Int. J. Nutr. Food Sci.* **2017**, *6*, 203–210. [CrossRef]
15. Fuentes-Zaragoza, E.; Pérez-Alvarez, J.A.; Sánchez-Zapata, E. Effect of ingredients concentration upon colour of heat treated hake (*Merluccius australis*) batters. *Ópt. Pura Apl.* **2009**, *324*, 9–21.
16. Sánchez-Zapata, E.J.; Pérez-Álvarez, J.A.; Fernández-López, J.; Barber-Valles, X. Descriptive study of reflectance spectra of Hake (*Merluccius australis*), Salmon (*Salmo salar*) and light and dark muscle from Tuna (*Thunnus thynnus*). *J. Food Qual.* **2010**, *33*, 391–403. [CrossRef]
17. Chaijan, M.; Benjakul, S.; Visessanguan, W.; Faustman, C. Changes of pigments and colour in sardine (*Sardinella gibbosa*) and mackerel (*Rastrelliger kanagurta*) muscle during iced storage. *Food Chem.* **2005**, *93*, 607–617. [CrossRef]

18. Sikorski, E.; Kolakowska, A.; Sun-Pan, B. Composición nutritiva de los principales grupos de organismos alimenticios marinos. In *Tecnología de los Productos del Mar: Recursos, Composición Nutritiva y Conservación*; Sikorski, E., Ed.; Acribia: Zaragoza, Spain, 1994; pp. 41–70.

19. Ersus, S.; Yurdagel, U. Microencapsulation of anthocyanin pigments of black carrot (*Daucus carota* L.) by spray drier. *J. Food Eng.* **2007**, *80*, 805–812. [CrossRef]

20. Sánchez-Gimeno, A.C.; Negueruela, A.I.; Benito, M.; Vercet, A.; Oria, R. Some physical changes in Bajo Aragón extra virgin olive oil during the frying process. *Food Chem.* **2008**, *110*, 654–658. [CrossRef]

21. Kusucharid, C.; Jangchud, A.; Thamakorn, P. Changes in characteristics of palm oil during vacuum and atmospheric frying conditions of sweet potato. *Kasetsart J. (Nat. Sci.)* **2009**, *43*, 298–304.

22. Krokida, M.K.; Maroulis, Z.B.; Saravacos, G.D. The effect of the method of drying on the colour of dehydrated products. *Int. J. Food Sci. Technol.* **2001**, *36*, 53–59. [CrossRef]

23. Perez-Palacios, T.; Casal, S.; Petisca, C.; Ferreira, I.M. Nutritional and sensory characteristics of bread-coated hake fillets as affected by cooking conditions. *J. Food Qual.* **2013**, *36*, 375–384. [CrossRef]

24. Haard, N.F. Biochemistry and chemistry of color and color changes in sea foods. In *Advances in Seafood Biochemistry: Composition and Quality*; Flick, G.J., Martin, R.E., Eds.; Technomic Publishing Company: Lancaster, PA, USA, 1992; pp. 305–360.

25. Gómez, B.; Munekata, P.E.S.; Gavahian, M.; Barba, F.J.; Martí-Quijal, F.J.; Bolumar, T.; Bastianello Campagnol, P.C.; Tomasevic, I.; Lorenzo, J.M. Application of pulsed electric fields in meat and fish processing industries: An overview. *Food Res. Int.* **2019**, *123*, 95–105. [CrossRef]

26. Cropotova, J.; Tappi, S.; Genovese, J.; Rocculi, P.; Rosa, M.D.; Rustad, T. The combined effect of pulsed electric field treatment and brine salting on changes in the oxidative stability of lipids and proteins and color characteristics of sea bass (*Dicentrarchus labrax*). *Heliyon* **2021**, *7*, e05947. [CrossRef]

27. Maskan, M. Change in colour and rheological behaviour of sunflower seed oil during frying and after adsorbent treatment of used oil. *Eur. Food Res. Technol.* **2003**, *218*, 20–25. [CrossRef]

28. Pedreschi, F.; Moyano, P.; Kaack, K.; Granby, K. Color changes and acrylamide formation in fried potato slices. *Food Res. Int.* **2005**, *38*, 1–9. [CrossRef]

29. Kaynak, G.; Ersoz, M.; Kara, H. Investigation of the properties of oil at the bleaching unit of an oil refinery. *J. Colloid Interface Sci.* **2004**, *280*, 131–138. [CrossRef] [PubMed]

30. Morello, J.R.; Motilva, M.J.; Tovar, M.J.; Romero, M.P. Changes in commercial virgin olive oil (cv. Arbequina) during storage, with special emphasis on the phenolic fraction. *Food Chem.* **2004**, *85*, 357–364. [CrossRef]

31. Reda, S.Y. Comparative Study of Vegetable Oils Subjected to the Thermal Stress. Master's Dissertation, Universidade Estadual de Ponta Grossa, Ponta Grossa, Brasil, 2004.

32. Criado, M.N.; Romero, M.P.; Casanovas, M.; Motilva, M.J. Pigment profile and colour of monovarietal virgin olive oils from Arbequina cultivar obtained during two consecutive crop seasons. *Food Chem.* **2008**, *110*, 873–880. [CrossRef]

33. Myrland, O.; Trondsen, T.; Johnson, R.S.; Lund, E. Determinants of seafood consumption in Norway: Lifestyle, revealed preferences and barriers to consumption. *Food Qual. Prefer.* **2000**, *11*, 169–188. [CrossRef]

34. Spence, C. What is so unappealing about blue food and drink? *Int. J. Gastronomy Food Sci.* **2018**, *14*, 1–8. [CrossRef]

35. Cheng, J.H.; Sun, D.W.; Han, Z.; Zeng, X.A. Texture and structure measurements and analyses for evaluation of fish and fillet freshness quality: A review. *Compr. Rev. Food Sci. Food Saf.* **2014**, *13*, 52–61. [CrossRef]

36. Gómez-Limia, L.; Franco, I.; Martínez-Suárez, S. Effects of processing step, filling medium and storage on amino acid profiles and protein quality in canned European eels. *J. Food Comp. Anal.* **2021**, *96*, 103710. [CrossRef]

37. Green, B.G.; Hayes, J.E. Capsaicin as a probe of the relationship between bitter taste and chemesthesis. *Physiol. Behav.* **2003**, *79*, 811–821. [CrossRef]

38. *ISO 8587:2006*; Sensory Analysis-Methodology-Ranking. International Organization for Standardization (ISO): Geneva, Switzerland, 2006.

39. Wansink, B.; Cheney, M.M.; Chan, N. Exploring comfort food preferences across age and gender. *Physiol. Behav.* **2003**, *79*, 739–747. [CrossRef]

40. Spence, C.; Velasco, C.; Knoeferle, K. A large sample study on the influence of the multisensory environment on the wine drinking experience. *Flavour* **2014**, *3*, 8. [CrossRef]

41. Da Silva, L.A.; Lin, S.M.; Teixeira, M.J.; de Siqueira, J.T.T.; Filho, W.J.; de Siqueira, S.R.D.T. Sensorial differences according to sex and ages. *Oral Dis.* **2014**, *20*, e103–e110. [CrossRef] [PubMed]

42. Mojet, J.; Christ-Hazelhof, E.; Heidema, J. Taste perception with age: Generic or specific losses in threshold sensitivity to the five basic tastes? *Chem. Senses* **2001**, *26*, 845–860. [CrossRef]

 *foods*

Article

Effect of Storage Temperature and Time on Biogenic Amines in Canned Seafood

Yinghong Qu [1], Jingyu Wang [1], Zhidong Liu [2,*], Xichang Wang [1] and Huimin Zhou [1]

[1] Shanghai Engineering Research Center of Aquatic Product Processing and Preservation,
College of Food Science and Technology, Shanghai Ocean University, Shanghai 200120, China

[2] Key Laboratory of Oceanic and Polar Fisheries, Ministry of Agriculture and Rural Affair,
East China Sea Fishery Research Institute, Chinese Academy of Fishery Sciences, Shanghai 200090, China

* Correspondence: zdliu1976@163.com

Abstract: Biogenic amines in canned seafood are associated with food quality and human health. In this study, a total of nine biogenic amines (histamine (HIS), phenylethylamine (PHE), tyramine (TYM), putrescine (PUT), cadaverine (CAD), tryptamine (TRY), spermine (SPM), spermidine (SPD), and octopamine (OCT)) were used as standards. The biogenic amines of five canned seafood species (canned mud carp, canned sardine, canned mantis shrimp, canned scallop, and canned oyster) were investigated every three months for 12 months at different storage temperatures (4, 10, 25, and 30 °C). The biogenic amine contents were determined by the ultrasound-assisted dispersive solid-phase extraction method combined with reversed-phase high-performance liquid chromatography-photodiode array detection (UADSPE-RPLC-PDA). These results showed a detection rate of 100, 60, and 40% for HIS, PHE, PUT, and TYM; CAD, SPM, and SPD; OCT in all the samples, respectively. The contents of histamine and tyramine exceeded the recommended maximum limits (50 and 100 mg kg^{-1}) in the canned mud carp and canned scallop when stored at 30 °C, indicating their potential health risks ($p < 0.05$). This result also indicates that low temperatures could inhibit the BAs content of canned seafood during storage. Overall, storage temperature and time can be used as the primary means to monitor and control the quality and safety of canned seafood.

Keywords: biogenic amines; canned seafood; storage time; storage temperature

Citation: Qu, Y.; Wang, J.; Liu, Z.; Wang, X.; Zhou, H. Effect of Storage Temperature and Time on Biogenic Amines in Canned Seafood. *Foods* **2022**, *11*, 2743. https://doi.org/10.3390/foods11182743

Academic Editors: Sidonia Martinez and Javier Carballo

Received: 9 August 2022
Accepted: 31 August 2022
Published: 7 September 2022

1. Introduction

Canned seafood is recommended due to its good nutritional value, ease of storing at room temperature, and high stability [1]. Canned seafood is generally considered safe, which is achieved through a heat treatment sufficient to kill the vegetative bacteria and bacterial spores and being stored in closed vessels. However, many studies have indicated that the storage environmental factors could significantly affect the quality and safety of canned seafood. Aquatic raw materials are rich in protein; therefore, the protein decomposition of canned aquatic products leads to the formation of biogenic amines (BAs) by microorganism-induced amino acid decarboxylase or the transamination of aldehydes or amino acid transaminases during processing and storage [2]. As such, a low BAs level is acceptable for human health. However, excessive BAs accumulation by the human body could be toxic, resulting in serious public health and food safety concerns [3,4]. The common BAs found in food mainly include HIS, TYM, PUT, and CAD. HIS and TYR pose severe acute effects on human health. PUT and CAD pose low toxicological properties that can act as a precursor of carcinogenic N-nitrosamines under the presence of nitrite [5,6]. The sum of certain BAs are the primary indicators of the quality and safety of aquatic products [6]. The formation of BAs in canned seafood might have significant effects on the nutritional and quality and safety of canned products after yielding, which is mediated by various factors, including the raw materials, microorganisms, and processing and storage conditions, etc. [7,8]. The BAs contents of canned seafood vary to a great extent due to the differences in species, sex, season, stage of maturity, feeding, living environment, and the style of raw materials. The USA Food and Drug Administration (FDA) stipulates that the intervention level of HIS in fish products in the United States is 50 mg kg^{-1} based on the toxicity of HIS at 500 mg kg^{-1} [9]. HIS is the only BA with regulatory limits. Hence, the production and accumulation of excessive BAs content in canned seafood should be limited during storage [10,11]. Thus, the storage temperature and time are of critical importance to controlling BAs contents in canned seafood during storage.

However, the effect of storage conditions (storage temperature range, storage time, etc.) on certain quality and safety characteristics on the quality and safety of canned seafood species is very scarce and unsystematic, especially BAs. Moreover, the profile and content of BAs vary from canned seafood species to species [12]. Although some studies have focused on this aspect, those are limited to a few species, such as tuna, mackerel, sardines, crustaceans, and mollusks [7,8]. This study aimed to investigate the changes in BAs content in various canned seafood species (canned mud carp, canned sardine, canned mantis shrimp, canned scallop, and canned oyster) stored under different conditions (4, 10, 25 and 30 °C) for 12 months and probed into the quality and safety changes during storage.

2. Materials and Methods

2.1. Samples

Commercially sterilized canned seafood (canned mud carp (*Cirrhinus molitorella*), canned sardine (*Sardinops melanostictus*), canned mantis shrimp (*Portunus trituberculatus*), canned scallop (*Patinopecten yessoensis*), and canned oyster (*Carnis Ostreae*) ($n = 300$)) without damage and rust were purchased from the canned seafood manufacturers (latest production batch) from the 1st to the 7th of August 2020. These are prevalent product in China market.

2.2. Determination of Biogenic Amines

A total of nine BAs (PUT, CAD, SPM, SPD, TYR, OCT, PHE, HIS, and TRM) were determined in the canned seafood by the ultrasound-assisted dispersive solid-phase extraction method combined with reversed-phase high-performance liquid chromatography-photodiode array detection (RP-HPLC-PDA) (Waters 2695 Separations Module, Waters, Milford, CT, USA) [13]. About 5 g of aliquot was weighed and placed in a 50 mL centrifuge tube. Later, 20 mL of TCA (5%, w/v) was added, and the mixtures were homogenized for 3 min (Ouhor, Shanghai, China), followed by ultrasonication for 30 min (KO5200E ultrasonic bath, Kun Shan Ultrasonic Instruments Co., Shanghai, China), and finally centrifuged at 5000 rpm for 10 min. The supernatant was transferred to a 50 mL brown volumetric flask, and the volume was fixed to the scale with 5% trichloroacetic acid. Afterward, 10 mL of n-hexane was added to the above sample extraction solution (10 mL) for complete fat removal, and the ultrasonic treatment was repeated for 10 min. After static stratification, the upper organic layer was discarded. About 5 mL of the degreased solution was transferred to a 50 mL centrifuge tube. The pH was alkalized up to 12 by adding NaOH (2 mol L^{-1}) (Mettler Toledo, Greifensee, Switzerland). About 5 mL of n-butanol/chloroform (1/1) mixed solution was added to purify the extracted solution. After being ultrasonicated for 10 min and centrifuged at 5000 rpm for 5 min, the upper organic layer was adjusted to 10 mL by adding the mixed solution of n-butanol/chloroform (1/1). Finally, 5 mL of the above purified extracted solution was taken into the test tube and blown to dryness by nitrogen in a 40 °C water bath. After being added to 1 mL of hydrochloric acid (0.1 mol L^{-1}) and ultrasonicated for 5 min, the extract could be used for derivatization and determination according to the method reported by Ishimaru, et al. and Gong, et al., respectively [14–16]. All standards/compounds were separated and identified by their retention times. The reagents were procured from Sigma-Aldrich (Goettingen, Germany) (amines and dansysl chloride) or Sigma-Aldrich (St. Louis, MO, USA) (acetonitrile, acetone, n-butanol, and n-hexane). The results are expressed in mg kg^{-1} wet weight.

The chromatograms containing BAs in the standard solutions showed a good peak resolution, with selectivity, sharpness, and symmetry. The correlation coefficients of the linear regression lines were better than 0.999 for all compounds. The LODs and LOQs of all BAs ranged between 0.08 and 0.25 mg kg^{-1} and 0.27 to 0.83 mg kg^{-1}, respectively. The precision, repeatability, reproducibility, recovery, and accuracy values of this method were assessed according to the previously reported method [9]. These results were consistent with the previous study results [9,17], demonstrating the feasibility of this method to detect BAs content in canned seafood.

2.3. Storage Experiments

Canned seafood (canned mud carp, canned sardine, canned mantis shrimp, canned scallop, and canned oyster, *n* = 300) was used as the samples for the storage experiment. These samples were stored at 4, 10, 25, and 30 °C and the BAs content was determined every three months during the storage for 12 months, respectively. Different canned seafood species were randomly collected and mixed with three samples from each species at the scheduling points to ensure the uniformity and representativeness of the samples.

2.4. Index Value

The synergism of various BAs induces toxicity in canned seafood, therefore, evaluating the biogenic amine index (BAI) is necessary to determine its quality. The interrelationship of HIS, PUT, CAD, SPD, and SPM was used to formulate a chemical index of decomposition for canned seafood [17]. This index was consistent with the sensory evaluation scores, confirming its adequacy for the quality evaluation of seafood products [18,19]. The relationship between the 5 amines was analyzed according to the 3 sensory quality classes, with the following generated formula:

$$\text{Chemical Index} = (\text{HIS} + \text{PUT} + \text{CAD})/(1 + \text{SPM} + \text{SPD}) \tag{1}$$

where, PUT, CAD, HIS, SPD, and SPM are the contents (mg kg^{-1}) of putrescine, cadaverine, histamine, methylamine spermidine, and spermine, respectively. A chemical index of BAs within 0–1 indicated the good quality of canned seafood, while a range higher than 10 would correspond to an unacceptable quality [20].

The chemical index level was not consistent for different species; the levels were even different for two differently packed products of the same type. Therefore, the chemical index should be studied along with other quality indicators. The amine index proposed by Duflos, et al., (1999) [21] was used to assess the spoilage:

$$\text{Amine Index} = (\text{HIS} + \text{PUT} + \text{CAD})/(\text{HIS} + \text{PUT} + \text{CAD} + \text{TRM} + \text{TYR} + \text{PHE} + \text{SPD} + \text{SPM}) \times 100 \tag{2}$$

where, PUT, CAD, HIS, TRM, TYR, PHE, SPD, and SPM are the contents (mg kg^{-1}) of putrescine, cadaverine, histamine, tryptamine, tyramine, phenylethylamine, spermidine, and spermine, respectively. According to the defined reference limits, the amine index values of below 25 indicate good quality, while the index values higher than 66 correspond to inedibility.

2.5. Statistical Analysis

Statistical analysis was performed using Origin (OriginLab Inc., Northampton, MA, USA) and SPSS 26.0 software (SPSS Inc., Chicago, IL, USA). Data were analyzed for the degree of variation by calculating the mean of at least three determinations and standard deviations (SDs) of the results. The significance of differences was evaluated using a one way analysis of variance (one-way ANOVA). A *p* value of less than 0.05 was considered statistically significant.

3. Results

3.1. Changes of BAs in Canned Mud Carp during Storage

The total BAs content in canned mud carp stored at 4, 10, 25, and 30 °C significantly increased (*p* < 0.05) with prolonged storage time. A total of even BAs (TRM, PHE, PUT,

HIS, TYR, SPD, and SPM) was detected in caned mud carp during storage, without the occurrence of CAD and OCT (Table 1). After being stored at 4, 10, 25, and 30 °C, the total BAs in canned mud carp increased from the initial values of 7.02 mg kg^{-1} to 40.56, 57.25, 79.64, 45.00, and 122.71 mg kg^{-1}, respectively. The sum of all seven amines was <300 mg kg^{-1} in all samples (maximum value: 122.71 mg kg^{-1}). The higher the storage temperature, the higher the total BAs in caned mud carp. This obtained value was much lower than the results reported by Kim, et al. (2009) [22]. When canned mud carp was stored at 4, 10, 25, and 30 °C for 12 months, the HIS content reached 15.32, 23.74, 27.36, and 52.81 mg kg^{-1}, respectively. Thus, based on the limit levels of HIS set by the USA FDA (50 mg kg^{-1}), canned mud carp is unsuitable for storage at 30 °C for 12 months. When canned mud carp was stored at 4, 10, 25, and 30 °C for 12 months, the SPM level reached 5.32, 8.74, 15.59, and 17.47 mg kg^{-1}, respectively. TRM and PUT levels showed a significantly increasing trend when stored at 4 and 10 °C, while SPD showed a significantly increasing trend at 25 and 30 °C ($p < 0.05$). The SPM level changed the most when canned mud carp was stored at 25 °C. This might have happened as the pH of canned seafood content increased with the increased storage temperature. Previous studies have also reported correlations between pH and BAs formation in seafood [23].

Table 1. Changes of BAs in canned mud carp during storage at different temperatures.

BA Contents/(mg kg^{-1})	Temperature (°C)	Time (Month)				
		0	3	6	9	12
TRM	4	1.57 ± 0.08 [c]	3.84 ± 0.56 [A,b]	3.99 ± 0.82 [A,b]	5.17 ± 0.52 [A,a]	5.38 ± 0.36 [B,a]
	10	1.57 ± 0.08 [c]	1.43 ± 0.32 [C,c]	1.69 ± 0.04 [B,c]	2.57 ± 0.35 [C,b]	5.82 ± 0.28 [B,a]
	25	1.57 ± 0.08 [e]	2.58 ± 0.26 [B,d]	3.66 ± 0.31 [A,c]	4.96 ± 0.27 [A,b]	6.73 ± 0.23 [A,a]
	30	1.57 ± 0.08 [d]	3.48 ± 0.24 [A,b]	3.45 ± 0.37 [A,b]	3.90 ± 0.42 [B,b]	6.92 ± 0.77 [A,a]
PHE	4	1.02 ± 0.09 [b]	1.33 ± 0.17 [B,b]	1.38 ± 0.04 [C,b]	2.43 ± 0.77 [B,a]	2.91 ± 0.09 [C,a]
	10	1.02 ± 0.09 [c]	1.94 ± 0.93 [B,bc]	2.54 ± 0.36 [B,b]	3.83 ± 0.56 [B,a]	4.63 ± 0.12 [B,a]
	25	1.02 ± 0.09 [d]	1.98 ± 0.17 [B,c]	2.76 ± 0.37 [B,c]	3.89 ± 0.34 [B,b]	4.99 ± 0.81 [B,a]
	30	1.02 ± 0.09 [e]	3.02 ± 0.41 [A,d]	5.13 ± 0.79 [A,c]	11.86 ± 1.37 [A,b]	13.48 ± 1.02 [A,a]
PUT	4	1.31 ± 0.03 [c]	1.48 ± 0.98 [c]	2.13 ± 0.04 [C,c]	5.02 ± 0.47 [B,b]	6.28 ± 0.93 [B,a]
	10	1.31 ± 0.03 [d]	1.64 ± 0.05 [d]	3.37 ± 0.83 [B,c]	5.18 ± 0.30 [B,b]	6.99 ± 0.61 [B,a]
	25	1.31 ± 0.03 [d]	2.30 ± 0.46 [cd]	2.58 ± 0.37 [BC,c]	5.39 ± 0.92 [B,b]	7.12 ± 0.91 [B,a]
	30	1.31 ± 0.03 [e]	2.39 ± 0.63 [d]	4.29 ± 0.21 [A,c]	7.03 ± 0.29 [A,b]	9.74 ± 1.05 [A,a]
HIS	4	ND [d]	3.75 ± 0.34 [C,c]	10.96 ± 1.99 [B,b]	12.38 ± 1.65 [C,b]	15.32 ± 2.05 [C,a]
	10	ND [d]	2.43 ± 0.49 [D,d]	15.03 ± 1.42 [AB,c]	18.45 ± 2.42 [C,b]	23.74 ± 1.28 [B,a]
	25	ND [c]	15.78 ± 0.29 [A,b]	17.47 ± 1.09 [A,b]	26.02 ± 3.09 [B,a]	27.36 ± 2.57 [B,a]
	30	ND [e]	8.39 ± 0.12 [B,d]	20.58 ± 5.99 [A,c]	38.20 ± 6.20 [A,b]	52.81 ± 4.02 [A,a]
TYR	4	0.92 ± 0.09 [bc]	0.47 ± 0.05 [C,c]	0.36 ± 0.08 [C,d]	1.16 ± 0.63 [C,b]	2.21 ± 0.13 [C,a]
	10	0.92 ± 0.09 [d]	1.23 ± 0.40 [B,cd]	1.58 ± 0.32 [B,c]	2.48 ± 0.29 [BC,b]	3.10 ± 0.22 [C,a]
	25	0.92 ± 0.09 [d]	1.35 ± 0.49 [B,cd]	2.30 ± 0.63 [B,c]	6.43 ± 1.02 [A,a]	6.91 ± 0.37 [B,a]
	30	0.92 ± 0.09 [d]	2.04 ± 0.12 [A,cd]	3.54 ± 0.88 [A,b]	3.06 ± 0.93 [B,bc]	8.31 ± 0.95 [A,a]
SPD	4	1.51 ± 0.07 [d]	1.66 ± 0.14 [A,d]	2.38 ± 0.08 [B,c]	2.82 ± 0.21 [B,b]	3.12 ± 0.17 [C,a]
	10	1.51 ± 0.07 [c]	1.58 ± 0.09 [A,c]	1.85 ± 0.12 [B,b]	1.94 ± 0.12 [B,b]	4.28 ± 0.24 [C,a]
	25	1.51 ± 0.07 [d]	2.07 ± 0.18 [A,d]	3.54 ± 0.32 [A,c]	8.09 ± 1.52 [A,b]	10.75 ± 0.32 [B,a]
	30	1.51 ± 0.07 [d]	0.93 ± 0.48 [B,d]	4.05 ± 0.51 [A,c]	9.21 ± 1.62 [A,b]	13.98 ± 1.61 [A,a]

Table 1. *Cont.*

BA Contents/(mg kg^{-1})	Temperature (°C)	Time (Month)				
		0	3	6	9	12
SPM	4	0.69 ± 0.48 [c]	0.75 ± 0.51 [c]	2.74 ± 0.93 [C,b]	4.35 ± 0.61 [B,a]	5.32 ± 0.39 [C,a]
	10	0.69 ± 0.48 [c]	2.39 ± 1.93 [c]	5.85 ± 0.23 [B,b]	8.39 ± 2.01 [AB,a]	8.74 ± 0.63 [B,a]
	25	0.69 ± 0.48 [d]	1.59 ± 0.20 [d]	5.06 ± 0.09 [B,c]	12.49 ± 3.03 [A,b]	15.59 ± 0.37 [A,a]
	30	0.69 ± 0.48 [c]	1.49 ± 0.18 [c]	8.08 ± 0.37 [A,b]	9.30 ± 3.21 [A,b]	17.47 ± 2.19 [A,a]

ND: not detected. Within each column and for each storage time of each amine, different capital letters (A–D) indicate significant differences ($p < 0.05$); within each row and for each storage temperature, different lowercase letters (a–e) indicate significant differences ($p < 0.05$). The absence of a letter indicates that no significant differences were found ($p > 0.05$).

3.2. Changes of BAs in Canned Sardine during Storage

Fish is widely consumed among protein-rich foods [24]. The total BAs in the canned sardine significantly increased with the storage temperature increasing for 12 months ($p < 0.05$). A total of eight BAs (TRM, PHE, PUT, HIS, OCT, TYR, SPD, and SPM) were detected in the canned sardine during storage without the occurrence of CAD, HIS, and OCT (Table 2). After storage at 4, 10, 25, and 30 °C, the total BAs in canned sardine increased from the initial values of 13.29 mg kg^{-1} to 71.21, 91.61, 146.00, and 158.52 mg kg^{-1}, respectively. Meanwhile, the content increased at a faster rate at 30 than at 4 °C. The sum of all eight amines was <300 mg kg^{-1} in all samples (maximum value: 158.52 mg kg^{-1}). HIS significantly increased in all samples during storage and decreased slowly after 6 months of storage. Similarly, SPD and SPM significantly increased ($p < 0.05$) when stored at 4 °C; TRM and OCT significantly increased ($p < 0.05$) when stored at 10 and 25 °C; PHE, SPD, and OCT significantly increased ($p < 0.05$) when stored at 30 °C. However, there was no significant difference in TYR and CAD ($p > 0.05$) when stored at 4, 10, and 30 °C. Notably, CAD was detected, which significantly increased ($p < 0.05$) when stored at 25 °C. This might have happened because CAD could not be decomposed at room temperature. These eight BAs showed an increasing tendency with increased storage temperature (Table 2), which was consistent with the results of Gómez-Limia, et al. (2020) [7]. Bilgin and Gençcelep also detected TYM in canned tuna, chunk canned tuna, marinated anchovies, canned mackerel, and canned sardines at levels ranging between ND and 48.63 mg kg^{-1} [23]. Gómez-Limia, et al. (2020) reported that the hydrolysis reactions and the interactions of the components in canned seafood continued when stored at room temperature [7].

Table 2. Changes of BAs in canned sardine during storage at different temperatures.

BA Contents/(mg kg^{-1})	Temperature (°C)	Time (Month)				
		0	3	6	9	12
TRM	4	5.14 ± 0.34 [b]	5.60 ± 0.64 [ab]	6.48 ± 1.37 [A,ab]	7.03 ± 1.09 [C,a]	7.21 ± 0.96 [C,a]
	10	5.14 ± 0.34 [c]	5.58 ± 0.66 [c]	5.90 ± 0.79 [AB,c]	9.45 ± 1.07 [B,b]	12.49 ± 0.18 [B,a]
	25	5.14 ± 0.34 [d]	5.26 ± 0.45 [d]	6.23 ± 1.18 [AB,c]	13.06 ± 0.19 [A,b]	16.48 ± 0.85 [A,a]
	30	5.14 ± 0.34 [c]	5.14 ± 1.04 [c]	5.87 ± 0.45 [B,c]	8.39 ± 0.34 [BC,b]	15.68 ± 1.45 [A,a]
PHE	4	1.30 ± 0.09 [d]	2.04 ± 0.28 [B,c]	2.95 ± 0.12 [D,bc]	3.48 ± 0.77 [C,b]	5.23 ± 0.83 [C,a]
	10	1.30 ± 0.09 [c]	3.78 ± 0.15 [A,b]	4.32 ± 0.49 [C,b]	4.59 ± 1.21 [C,ab]	5.68 ± 0.38 [C,a]
	25	1.30 ± 0.09 [d]	3.05 ± 0.05 [A,c]	5.88 ± 0.28 [B,b]	8.93 ± 1.24 [B,a]	8.98 ± 0.97 [B,a]
	30	1.30 ± 0.09 [e]	3.69 ± 0.89 [A,d]	7.46 ± 0.84 [A,c]	12.73 ± 1.92 [A,b]	18.79 ± 0.91 [A,a]
PUT	4	1.82 ± 0.05 [b]	2.45 ± 0.51 [AB,b]	2.84 ± 0.38 [B,b]	8.24 ± 0.95 [A,a]	8.32 ± 0.59 [B,a]
	10	1.82 ± 0.05 [e]	2.89 ± 0.37 [A,d]	3.77 ± 0.39 [AB,c]	8.48 ± 0.12 [A,b]	9.24 ± 0.74 [B,a]
	25	1.82 ± 0.05 [d]	1.83 ± 0.26 [B,d]	3.50 ± 0.28 [AB,c]	7.72 ± 0.31 [AB,b]	9.34 ± 0.92 [B,a]
	30	1.82 ± 0.05 [d]	3.09 ± 0.87 [A,cd]	4.28 ± 0.86 [A,c]	6.32 ± 1.12 [B,b]	12.27 ± 1.43 [A,a]

Table 2. *Cont.*

BA Contents/(mg kg^{-1})	Temperature (°C)	Time (Month)				
		0	3	6	9	12
HIS	4	ND [c]	6.94 ± 0.74 [C,b]	14.96 ± 1.90 [B,a]	16.82 ± 2.15 [C,a]	17.38 ± 2.01 [D,a]
	10	ND [d]	13.54 ± 1.53 [B,c]	18.58 ± 1.96 [B,b]	20.14 ± 1.69 [C,b]	25.39 ± 2.96 [C,a]
	25	ND [c]	ND [D,c]	29.34 ± 1.53 [A,b]	35.87 ± 3.40 [B,a]	37.86 ± 4.83 [B,a]
	30	ND [c]	28.48 ± 3.19 [A,b]	31.32 ± 3.50 [A,b]	42.02 ± 3.03 [A,a]	46.29 ± 3.01 [A,a]
TYR	4	0.95 ± 0.40 [b]	1.02 ± 0.04 [B,b]	0.93 ± 0.28 [C,b]	1.88 ± 0.13 [C,a]	2.40 ± 0.41 [D,a]
	10	0.95 ± 0.40 [c]	1.33 ± 0.48 [B,c]	1.43 ± 0.53 [BC,c]	4.03 ± 0.21 [B,b]	5.23 ± 0.86 [C,a]
	25	0.95 ± 0.40 [d]	1.97 ± 0.25 [A,c]	1.99 ± 0.20 [B,c]	3.24 ± 0.39 [B,b]	6.38 ± 0.41 [B,a]
	30	0.95 ± 0.40 [d]	1.05 ± 0.09 [B,d]	3.11 ± 0.26 [A,c]	7.62 ± 1.22 [A,b]	9.29 ± 0.46 [A,a]
SPD	4	1.94 ± 0.17 [c]	3.05 ± 0.29 [B,bc]	4.55 ± 0.79 [D,b]	8.99 ± 1.28 [B,a]	10.42 ± 1.06 [B,a]
	10	1.94 ± 0.17 [d]	2.41 ± 0.23 [B,d]	6.30 ± 0.21 [C,c]	8.32 ± 0.92 [B,b]	11.52 ± 2.21 [B,a]
	25	1.94 ± 0.17 [c]	2.52 ± 0.35 [B,c]	9.02 ± 0.88 [B,b]	9.95 ± 1.40 [B,b]	12.75 ± 1.37 [B,a]
	30	1.94 ± 0.17 [d]	6.03 ± 1.92 [A,c]	14.35 ± 1.16 [A,b]	19.01 ± 0.31 [A,a]	21.38 ± 2.22 [A,a]
SPM	4	2.14 ± 0.26 [d]	2.99 ± 0.28 [C,d]	7.24 ± 0.95 [AB,c]	12.49 ± 0.47 [A,b]	13.86 ± 0.77 [A,a]
	10	2.14 ± 0.26 [d]	5.02 ± 0.38 [A,c]	5.94 ± 0.85 [B,bc]	7.92 ± 0.37 [C,b]	10.58 ± 2.57 [B,a]
	25	2.14 ± 0.26 [d]	3.83 ± 0.57 [B,c]	6.26 ± 0.24 [B,b]	8.86 ± 0.62 [BC,a]	9.56 ± 1.35 [B,a]
	30	2.14 ± 0.26 [d]	5.04 ± 0.42 [A,c]	9.06 ± 1.94 [A,b]	10.93 ± 2.01 [AB,b]	14.45 ± 1.47 [A,a]

ND: not detected. Within each column and for each storage time of each amine, different capital letters (A–D) indicate significant differences ($p < 0.05$); within each row and for each storage temperature, different lowercase letters (a–e) indicate significant differences ($p < 0.05$). The absence of a letter indicates that no significant differences were found ($p > 0.05$).

3.3. Changes of BAs in Canned Mantis Shrimp during Storage

The total BAs in the canned mantis shrimp significantly increased with the increase in temperatures for 12 months ($p < 0.05$). A total of seven BAs (TRM, PHE, PUT, CAD, HIS, OCT, and TYR) were detected in canned mantis shrimp during storage (Table 3). No SPD or SPM was detected during storage, which was different from canned mud carp, canned sardine, and canned oyster due to the differences in species. The total BAs content in canned sardine increased from the initial values of 8.59 mg kg^{-1} for 12 months to 50.21, 76.42, 123.57, and 145.14 mg kg^{-1}, respectively, when stored at 4, 10, 25, and 30 °C. The sum of all seven amines was <300 mg kg^{-1} in all samples (maximum value: 145.14 mg kg^{-1}). Among all BAs, HIS showed a significant increase ($p < 0.05$) during storage and reached 12.46, 22.47, 33.74, and 42.18 mg kg^{-1}, respectively. The higher the storage temperature, the higher the HIS content. It was found that PUT, CAD, OCT, and TYR significantly increased ($p < 0.05$) when stored at 4 °C; PHE, PUT, CAD, and TYR significantly increased ($p < 0.05$) when stored at 10 °C; with further increases in temperature, PHE, PUT, CAD, OCT, and TYR significantly increased ($p < 0.05$) at 25 and 30 °C. Therefore, PHE, PUT, CAD, HIS, and OCT could be used as the characteristic BAI of canned mantis shrimp during storage. These seven BAs showed an increasing tendency with increased storage temperature (Table 3). Zhai et al. reported that the total BAs content in different canned fish products ranged from 1.94 to 112.54 mg kg^{-1}, with a mean value of 46.43 mg kg^{-1} [25], which was consistent with the present study result.

3.4. Changes of BAs in Canned Scallop during Storage

As summarized in Table 4, only five BAs (TRM, PHE, PUT, CAD, and HIS) were detected in canned scallops during storage, which was consistent with canned mantis shrimp. Although the total BAs content in canned scallop was much lower than those in canned sardine and canned scallop, they also showed a significantly increasing trend with increasing storage temperature for 12 months ($p < 0.05$). The total BA in canned sardine increased to 33.29, 59.94, 76.68, and 90.79 mg kg^{-1}, respectively, when stored at 4, 10, 25, and 30 °C for 12 months. The sum of all five amines was <300 mg kg^{-1} in all samples (maximum value: 90.79 mg kg^{-1}). HIS showed a significant increase ($p < 0.05$) in all the BAs during storage and reached 23.05, 39.86, 49.57, and 52.39 mg kg^{-1}, respectively, which was more than the limit standard of FDA (50 mg kg^{-1}). TRM, PUT, PHE, and CAD significantly increased ($p < 0.05$) when stored at 10, 25, and 30 °C, while OCT, TYR, SPD, and SPM were not detected. PHE showed no significant difference when stored at 4 °C ($p > 0.05$), maintaining an overall growth trend [15]. These five BAs showed an increasing tendency with increased storage temperature (Table 4). Nevertheless, the values reported for total BAs content in samples are low, well below FDA-recommended levels at 4, 10, and 25 °C [19].

Table 3. Changes of BAs in canned mantis shrimp during storage at different temperatures.

BA Contents/(mg kg^{-1})	Temperature (°C)	Time (Month)				
		0	3	6	9	12
TRM	4	ND [d]	ND [d]	1.03 ± 0.08 [B,c]	1.48 ± 0.39 [C,b]	2.31 ± 0.21 [C,a]
	10	ND [d]	ND [d]	0.56 ± 0.07 [B,c]	1.99 ± 0.53 [BC,b]	3.28 ± 0.17 [B,a]
	25	ND [c]	ND [c]	2.75 ± 0.83 [A,b]	2.92 ± 0.28 [B,b]	3.75 ± 0.36 [B,a]
	30	ND [c]	ND [c]	2.58 ± 0.49 [A,b]	4.39 ± 0.62 [A,a]	4.81 ± 0.75 [A,a]
PHE	4	1.84 ± 0.16 [c]	2.27 ± 0.63 [C,b]	2.26 ± 0.09 [C,b]	2.50 ± 0.24 [C,b]	3.48 ± 0.28 [D,a]
	10	1.84 ± 0.16 [c]	4.77 ± 0.85 [B,b]	4.85 ± 0.37 [B,b]	7.05 ± 0.99 [B,a]	7.54 ± 0.51 [C,a]
	25	1.84 ± 0.16 [e]	6.11 ± 0.84 [AB,c]	3.26 ± 0.49 [C,d]	9.05 ± 0.36 [A,b]	12.48 ± 1.39 [B,a]
	30	1.84 ± 0.16 [d]	6.59 ± 0.82 [A,c]	6.97 ± 1.38 [A,c]	9.33 ± 1.71 [A,b]	14.38 ± 1.13 [A,a]
PUT	4	2.24 ± 0.07 [c]	3.95 ± 0.55 [b]	4.75 ± 0.83 [C,b]	8.39 ± 0.87 [C,a]	8.82 ± 0.53 [C,a]
	10	2.24 ± 0.07 [b]	3.42 ± 0.37 [b]	6.40 ± 0.29 [B,a]	7.82 ± 1.04 [C,a]	8.72 ± 1.38 [C,a]
	25	2.24 ± 0.07 [d]	3.22 ± 0.06 [d]	8.77 ± 0.59 [B,c]	10.92 ± 0.89 [B,b]	20.47 ± 1.21 [B,a]
	30	2.24 ± 0.07 [c]	3.62 ± 1.02 [c]	16.57 ± 3.03 [A,b]	17.10 ± 2.27 [A,b]	25.38 ± 1.79 [A,a]
CAD	4	3.37 ± 0.35 [d]	4.01 ± 0.19 [cd]	4.75 ± 0.27 [D,c]	5.87 ± 0.04 [C,b]	7.29 ± 1.18 [C,a]
	10	3.37 ± 0.35 [d]	4.40 ± 0.59 [d]	8.54 ± 1.43 [C,c]	15.49 ± 3.82 [AB,b]	19.74 ± 2.81 [B,a]
	25	3.37 ± 0.35 [c]	3.83 ± 0.25 [c]	11.82 ± 1.33 [B,b]	12.43 ± 2.01 [B,b]	19.31 ± 1.37 [B,a]
	30	3.37 ± 0.35 [d]	4.02 ± 0.15 [d]	15.12 ± 2.85 [A,c]	18.84 ± 2.57 [A,b]	28.37 ± 2.15 [A,a]
HIS	4	ND [d]	1.31 ± 0.07 [C,d]	8.05 ± 1.01 [C,c]	10.84 ± 1.29 [D,b]	12.46 ± 0.82 [D,a]
	10	ND [d]	10.19 ± 0.93 [A,c]	13.08 ± 3.97 [B,c]	18.23 ± 1.01 [C,b]	22.47 ± 1.82 [C,a]
	25	ND [d]	10.94 ± 1.07 [A,c]	13.10 ± 1.81 [B,c]	28.65 ± 2.98 [B,b]	33.74 ± 2.39 [B,a]
	30	ND [c]	4.58 ± 0.13 [B,c]	22.94 ± 3.22 [A,b]	37.10 ± 4.89 [A,a]	42.18 ± 2.57 [A,a]
OCT	4	ND [c]	ND [B,c]	4.99 ± 0.95 [B,b]	5.03 ± 1.92 [BC,b]	8.47 ± 0.41 [C,a]
	10	ND [d]	ND [B,d]	2.55 ± 0.59 [C,c]	3.90 ± 0.21 [C,b]	9.95 ± 0.18 [C,a]
	25	ND [d]	1.44 ± 0.28 [A,d]	4.56 ± 0.37 [B,c]	8.54 ± 1.93 [B,b]	13.56 ± 0.69 [B,a]
	30	ND [d]	ND [B,d]	8.11 ± 1.05 [A,c]	14.03 ± 2.69 [A,b]	19.47 ± 1.49 [A,a]
TYR	4	1.14 ± 0.17 [d]	1.85 ± 0.84 [B,cd]	2.59 ± 0.65 [C,c]	4.56 ± 0.74 [B,b]	7.38 ± 0.63 [C,a]
	10	1.14 ± 0.17 [d]	1.28 ± 0.21 [B,d]	3.34 ± 0.67 [C,c]	7.95 ± 0.02 [A,b]	10.72 ± 1.58 [B,a]
	25	1.14 ± 0.17 [d]	4.32 ± 0.36 [A,c]	4.88 ± 0.49 [B,c]	7.39 ± 0.96 [A,b]	15.37 ± 2.42 [A,a]
	30	1.14 ± 0.17 [c]	2.10 ± 0.39 [B,c]	6.60 ± 0.58 [A,b]	7.02 ± 1.32 [A,b]	15.46 ± 1.51 [A,a]

ND: not detected. Different capital letters (A–D) indicate significant differences ($p < 0.05$); within each row and for each storage temperature, different lowercase letters (a–e) indicate significant differences ($p < 0.05$). The absence of a letter indicates that no significant differences were found ($p > 0.05$).

3.5. Changes of BAs in Canned Oyster during Storage

The total BAs in canned oyster significantly increased with the increase of storage temperature for 12 months ($p < 0.05$), showing a faster rate at 30 than at 4 °C. As summarized in Table S1, a total of seven BAs (TRM, PHE, PUT, CAD, HIS, SPD, and SPM) were detected in the canned oyster during storage. TRM, PUT, CAD, HIS, OCT, TYR, and SPM were detected in the initial phase. The total BAs content in canned oyster increased from the initial value of 2.54 mg kg^{-1} to 52.11, 77.92, 100.97, and 122.62 mg kg^{-1}, respectively, when stored at 4, 10, 25, and 30 °C. The sum of all seven amines was <300 mg kg^{-1} in all samples (maximum value: 122.62 mg kg^{-1}). HIS significantly increased in all BAs during storage and decreased slowly after 3 months of storage. TRM and SPM significantly increased ($p < 0.05$) when stored at 4 °C; TRM and CAD significantly increased ($p < 0.05$) when stored at 10 °C; TRM, SPD, and CAD significantly increased ($p < 0.05$) when stored at 25 °C; TRM, SPD, PUT, and CAD significantly increased ($p < 0.05$) when stored at 30 °C. However, TYR and OCT were not detected during storage at 4, 10, 25, and 30 °C for 12 months. Notably, CAD was detected, which significantly increased ($p < 0.05$) when stored at 10 °C. This might have happened because CAD in canned oyster could not be decomposed at 10 °C. These seven BAs (TRM, PHE, PUT, CAD, HIS, SPD, and SPM) showed an increasing tendency with increased storage temperature. The total BAs content increased during storage for 12 months, which was consistent with Gómez-Limia, et al. (2020) [7].

Table 4. Changes of BAs in canned scallop during storage at different temperatures.

BA Contents/(mg kg^{-1})	Temperature (°C)	Time (Month)				
		0	3	6	9	12
TRM	4	ND [c]	ND [c]	1.04 ± 0.19 [B,b]	1.65 ± 0.38 [D,a]	1.72 ± 0.32 [C,a]
	10	ND [c]	ND [c]	3.09 ± 0.83 [B,b]	4.23 ± 0.54 [C,b]	9.49 ± 1.23 [B,a]
	25	ND [d]	ND [d]	3.14 ± 1.70 [B,c]	7.02 ± 1.12 [B,b]	10.42 ± 1.95 [B,a]
	30	ND [d]	ND [d]	5.92 ± 1.41 [A,c]	9.32 ± 0.92 [A,b]	13.83 ± 1.21 [A,a]
PHE	4	0.94 ± 0.06 [b]	2.49 ± 0.69 [A,a]	2.52 ± 0.08 [A,a]	3.21 ± 1.02 [B,a]	3.23 ± 0.32 [B,a]
	10	0.94 ± 0.06 [b]	1.06 ± 0.11 [B,b]	1.19 ± 0.74 [C,b]	3.59 ± 0.24 [B,a]	3.77 ± 0.75 [B,a]
	25	0.94 ± 0.06 [d]	2.25 ± 0.20 [A,c]	1.38 ± 0.03 [B,d]	5.46 ± 0.39 [A,b]	6.38 ± 0.92 [A,a]
	30	0.94 ± 0.06 [d]	0.41 ± 0.12 [B,d]	2.04 ± 0.46 [AB,c]	5.85 ± 0.53 [A,b]	7.28 ± 0.48 [A,a]
PUT	4	ND [c]	ND [c]	0.79 ± 0.07 [C,b]	1.32 ± 0.56 [C,ab]	1.98 ± 1.02 [C,a]
	10	ND [d]	ND [d]	0.58 ± 0.24 [C,c]	1.43 ± 0.26 [C,b]	2.32 ± 0.28 [C,a]
	25	ND [d]	ND [d]	1.39 ± 0.03 [B,c]	3.64 ± 0.19 [B,b]	5.38 ± 0.37 [B,a]
	30	ND [d]	ND [d]	3.54 ± 0.39 [A,c]	5.20 ± 0.92 [A,b]	9.37 ± 0.49 [A,a]
CAD	4	ND [e]	0.56 ± 0.13 [d]	1.04 ± 0.05 [C,c]	2.99 ± 0.17 [B,b]	3.31 ± 0.30 [C,a]
	10	ND [e]	0.67 ± 0.09 [d]	1.27 ± 0.16 [C,c]	3.05 ± 0.21 [B,b]	3.57 ± 0.25 [C,a]
	25	ND [d]	0.77 ± 0.20 [d]	2.21 ± 0.53 [B,c]	4.02 ± 0.02 [A,b]	5.86 ± 1.01 [B,a]
	30	ND [c]	0.53 ± 0.18 [c]	4.02 ± 0.72 [A,b]	4.09 ± 0.81 [A,b]	7.92 ± 0.61 [A,a]
HIS	4	ND [b]	ND [C,b]	20.67 ± 2.56 [C,a]	22.43 ± 2.21 [C,a]	23.05 ± 0.83 [C,a]
	10	ND [c]	12.47 ± 1.31 [B,b]	34.75 ± 3.21 [B,a]	37.76 ± 4.18 [B,a]	39.86 ± 4.12 [B,a]
	25	ND [d]	12.68 ± 1.52 [B,c]	43.75 ± 3.97 [A,b]	45.29 ± 4.76 [A,ab]	49.57 ± 2.01 [A,a]
	30	ND [d]	27.13 ± 3.22 [A,c]	38.22 ± 3.05 [AB,b]	42.69 ± 5.29 [AB,b]	52.39 ± 3.59 [A,a]

ND: not detected. Different capital letters (A–D) indicate significant differences ($p < 0.05$); within each row and for each storage temperature, different lowercase letters (a–e) indicate significant differences ($p < 0.05$). The absence of a letter indicates that no significant differences were found ($p > 0.05$).

3.6. Formulas for BAs Content in Canned Seafood under Different Storage Conditions

The Chemical Index and Amine Index of all the samples were determined according to Equations (1) and (2), and the results are summarized in Table S2. It can be seen that the initial Chemical Index and Amine Index of canned fish were less than 2.5 and 60, respectively, indicating good quality. Different varieties have different quality standards due to the differences in raw materials, processing technology, and storage environment. Therefore, the Chemical Index and Amine Index range of less than 55/55 and 75/90 might correspond to good quality of canned shrimp and shellfish, respectively, while the range higher than 30 and 68 might indicate the quality change of the samples. If one of the two indicators meets the limited standards, the sample would be considered poor quality. Therefore, more studies on the indicative correlation of BAs are demanded to gain insights into the quality characteristics of canned seafood.

Moreover, the sensory properties stored at 4 and 10 °C were better than those stored at 25 and 30 °C, indicating that low storage temperatures could better obtain the quality of canned seafood. With prolonged storage, the Chemical Index and Amine Index of canned mud carp, canned sardine, canned scallop, and canned oyster exceed the human consumption limits recommended by FAO/WHO [26], indicating that different degrees of deterioration negatively affect the quality of these samples. Nevertheless, these samples are still safe for human consumption due to BAs' limited content. The canned mantis shrimp and canned scallop had a more obvious response to storage time and temperature, showing significant quality changes. The canned mantis shrimp and canned scallop samples decomposed when stored at 25 and 30 °C for more than 6 months. There was no significant change in the quality of canned oysters when stored at any temperature for one year. Thus, temperature control could effectively inhibit BAs' formation in canned seafood during storage. The changes in BAs' contents in the canned seafood were relatively low and avoided possible harmful effects under low temperature storage. Therefore, further studies on the storage conditions of canned seafood could guarantee the quality and safety of canned seafood. Additionally, different species of aquatic canned products could affect BAs' formation. Canned scallop and canned oyster produced more HIS than canned mud carp, canned sardine, and canned mantis shrimp at 4 °C. The correlation between BAs and other quality characteristics should be studied further to construct the quality and safety prediction model and reveal the changing tendency of BAs content in canned seafood during storage.

4. Discussion

BAs levels of canned seafood depend on the intrinsic food factors, such as pH, a_w values, and nutrients, and the extrinsic factors, including storage time, temperature, etc., [6]. Shalaby (1996) reported that the BAs concentrations differ not only between different food varieties but also within the same variety [27]. With storage time prolonged, BAs will accumulate. Especially CAD and PUT in BAs increased with the increased temperature [28]; however, storage at low temperatures might induce PUT accumulation [29]. Thus, higher BAs contents were detected in the canned seafood with higher storage temperatures. Pinho et al. (2001) reported a higher BAs content at a storage temperature of 21 °C than that stored at 4 °C [30]. Kim et al. (2002) demonstrated that the optimum temperature for HIS production in fish product was 25 °C [31]. These results were consistent with the present study results. This indicates that adjusting the storage environment factors (storage time and temperature, etc.) during storage is the primary means of controlling the quality and safety of canned seafood after manufacturing [32]. Other studies have reported the effect of storage temperature on HIS formation in fish products, such as tuna and anchovies [33,34]. Moreover, low HIS and TYR contents have been reported in crustaceans such as shrimp [35]. These results were consistent with previous reports [36].

FAO/WHO stipulates a level of 200mg kg^{-1} for fish and fishery products, and EFSA stipulates 400 mg kg^{-1} for fishery products, respectively, [19,36,37]. According to a previous study, HIS and TYR exert severe acute effects on human health, while, PUT and CAD have low toxicological properties [6]. According to EFSA, TYR values of 600 mg kg^{-1} or more are good for healthy individuals [36]. The TYR contents in all the samples were less than 600 mg kg^{-1} (Tables 1–4), which was consistent with Prester et al., who reported a dietary value of up to 800 mg kg^{-1} of TYR to be acceptable [35]. Other studies have also reported that the acute toxicity levels of TYR and CAD are significantly greater than 2000 mg kg^{-1} and the oral toxicity level of PUT is 2000 mg kg^{-1} [38]. It is known that TYR has a stronger and more rapid cytotoxic effect than HIS [39], while the bioactivities of PUT and CAD are less potent. Furthermore, PUT and CAD significantly increased with increased storage times, resulting in strong unpleasant decaying odors at very low concentrations [40]. These results were consistent with previous studies that canned seafood should be considered as a low-risk product based on the BAs contents [36]. Poveda (2019) reported that PUT is either converted to SPD or SPM or is catabolized by succinate [41]. Gezginc et al. (2013) found that arginine serves as a precursor of PUT, which can also be converted to SPD [42,43]. Furthermore, for the BAs used as the indicative of food quality, they are mostly produced at the end of shelf-life of food. This index was also used to indicate the evaluation and comparison of BAs contents in food [44]. Therefore, the storage temperature and time of canned seafood for different varieties should be studied more comprehensively to minimize the BAs content and obtain good quality.

5. Conclusions

The present study evaluated the BAs content in canned seafood during storage at different temperatures and times. HIS exhibited the most significant changes under different storage conditions. HIS content in the canned sardine and canned mantis shrimp increased to 50 mg kg^{-1} when stored at 25 and 30 °C for one year, while it increased to more than 50 mg kg^{-1} in the canned mantis shrimp and canned scallop when stored at 30 °C. TRM, PUT, and HIS could be used as the characteristic BAs of canned seafood during storage. These findings provide deeper insights into the influence of storage time and temperature on BAs level, which can be used to improve the quality and safety of canned seafood.

Supplementary Materials: The following supporting information can be downloaded at: https://www.mdpi.com/article/10.3390/foods11182743/s1, Table S1: Changes of BAs in canned oyster during storage at different temperatures title; Table S2 Chemical Index and Amine Index of the canned seafood samples

Author Contributions: Conceptualization, Y.Q. and Z.L.; methodology, J.W.; formal analysis, H.Z.; writing—original draft preparation, Z.L.; writing—review and editing, Z.L.; supervision, X.W.; project administration, Z.L. All authors have read and agreed to the published version of the manuscript.

Funding: This work was financially supported by the National Key Research and Development Program of China (2018YFD0901006).

Institutional Review Board Statement: Not applicable.

Informed Consent Statement: Not applicable.

Data Availability Statement: Data is contained within the article.

Conflicts of Interest: The authors declare no conflict of interest.

References

1. Alva, C.V.; Marsico, E.T.; Ribeiro, R.D.O.R.; Carneiro, C.D.S.; Simoes, J.S.; Ferreira, M.D.S. Concentrations and health risk assessment of total mercury in canned tuna marketed in Southest Brazil. *J. Food. Compos. Anal.* **2020**, *88*, 103357. [CrossRef]
2. Biji, K.B.; Ravishankar, C.N.; Venkateswarlu, R.; Mohan, C.O.; Gopal, T.K. Biogenic amines in seafood: A review. *J. Food Sci. Technol.* **2016**, *53*, 2210–2218. [CrossRef] [PubMed]
3. Wang, J.; Qu, Y.; Liu, Z.; Zhou, H. Formation, analytical methods, change tendency, and control strategies of biogenic amines in canned aquatic products: A systematic review. *J. Food Prot.* **2021**, *84*, 2020–2036. [CrossRef] [PubMed]
4. Jastrzebska, A.; Piasta, A.; Szlyk, E. Application of ion chromatography for the determination of biogenic amines in food samples. *J. Anal. Chem.* **2015**, *70*, 1131–1138. [CrossRef]
5. Visciano, P.; Schirone, M.; Tofalo, R.; Suzzi, G. Biogenic amines in raw and processed seafood. *Front. Microbiol.* **2012**, *3*, 188. [CrossRef]
6. Gardini, F.; Ozogul, Y.; Suzzi, G.; Tabanelli, G.; Ozogul, F. Technological factors affecting biogenic amine content in foods: A review. *Front. Microbiol.* **2016**, *7*, 18. [CrossRef]
7. Gómez-Limia, L.; Cutillas, R.; Carballo, J.; Franco, I.; Martínez, S. Free amino acids and biogenic amines in canned European eels: Influence of processing step, filling medium and storage Time. *Foods* **2020**, *9*, 1377. [CrossRef]
8. Ruiz-Capillas, C.; Herrero, A.M. Impact of biogenic amines on food quality and safety. *Foods* **2019**, *8*, 62. [CrossRef]
9. Wang, J.; Liu, Z.; Qu, Y. Ultrasound-assisted dispersive solid-phase extraction combined with reversed-phase high-performance liquid chromatography-photodiode array detection for the determination of nine biogenic amines in canned seafood. *J. Chromatogr. A* **2021**, *1636*, 461768. [CrossRef]
10. Hadidi, M.; Ibarz, A.; Pagan, J. Optimisation and kinetic study of the ultrasonic-assisted extraction of total saponins from alfalfa (*Medicago sativa*) and its bioaccessibility using the response surface methodology. *Food Chem.* **2020**, *309*, 8. [CrossRef]
11. Ishimaru, M.; Muto, Y.; Nakayama, A.; Hatate, H.; Tanaka, R. Determination of biogenic amines in fish meat and fermented foods using column-switching high-performance liquid chromatography with fluorescence detection. *Food Anal. Method.* **2019**, *12*, 166–175. [CrossRef]
12. Gong, X.; Wang, X.X.; Qi, N.L.; Li, J.H.; Lin, L.J.; Han, Z.P. Determination of biogenic amines in traditional Chinese fermented foods by reversed-phase high-performance liquid chromatography (RP-HPLC). *Food Addit. Contam. Part A* **2014**, *31*, 1431–1437. [CrossRef]

13. Mietz, J.L.; Karmas, E. Polyamine and histamine content of rockfish, salmon, lobster, and shrimp as an indicator of decomposition. *J. Assoc. Off. Anal. Chem.* **1978**, *61*, 139–145. [CrossRef]
14. Bakar, J.; Yassoralipour, A.; Abu Bakar, F.; Rahman, R.A. Biogenic amine changes in barramundi (*Lates calcarifer*) slices stored at 0 degrees C and 4 degrees C. *Food Chem.* **2010**, *119*, 467–470. [CrossRef]
15. Zarei, M.; Najafzadeh, H.; Enayati, A.; Pashmforoush, M. Biogenic amines content of canned tuna fish marketed in Iran. *American-Eurasian J. Toxicol. Sci.* **2011**, *3*, 190–193.
16. Oliveira, R.B.A.; Evangelista, W.P.; Sena, M.J.; Gloria, M.B.A. Tuna fishing, capture and post-capture practices in the northeast of Brazil and their effects on histamine and other bioactive amines. *Food Control* **2012**, *25*, 64–68. [CrossRef]
17. Duflos, G.; Dervin, C.; Malle, P.; Bouquelet, S. Use of biogenic amines to evaluate spoilage in Plaice (*Pleuronectes platessa*) and Whiting (*Merlangus merlang*). *J. AOAC Int.* **1999**, *82*, 1357–1363. [CrossRef]
18. Paulsen, P.; Bauer, S.; Bauer, F.; Dicakova, Z. Contents of polyamines and biogenic amines in canned pet (Dogs and Cats) food on the Austrian market. *Foods* **2021**, *10*, 2365. [CrossRef]
19. FDA (United States Food and Drug Administration). Ch.7: Scombrotoxin (histamine) formation. In *Fish and Fishery Products Hazards and Controls Guidance*, 4th ed.; Department of Health and Human Service, Public Health Service, Food and Drug Administration, Center for Food Safety and Applied Nutrition, Office of Seafood: Washington, DC, USA, 2011; pp. 113–152.
20. Mercogliano, R.; Santonicola, S. Scombroid fish poisoning: Factors influencing the production of histamine in tuna supply chain. A review. *LWT* **2019**, *114*, 7. [CrossRef]
21. Rahmani, J.; Miri, A.; Mohseni-Bandpei, A.; Fakhri, Y.; Bjorklund, G.; Keramati, H.; Moradi, B.; Amanidaz, N.; Shariatifar, N.; Khaneghah, A.M. Contamination and prevalence of histamine in canned tuna from Iran: A systematic review, meta-analysis, and health risk assessment. *J. Food Prot.* **2018**, *81*, 2019–2027. [CrossRef]
22. Kim, M.K.; Mah, J.H.; Hwang, H.J. Biogenic amine formation and bacterial contribution in fish, squid and shellfish. *Food Chem.* **2009**, *116*, 87–95. [CrossRef]
23. Bilgin, B.; Gençcelep, H. Determination of biogenic amines in fish products. *Food Sci. Biotechnol.* **2015**, *24*, 1907–1913.
24. Visciano, P.; Schirone, M.; Paparella, A. An overview of histamine and other biogenic amines in fish and fish products. *Foods* **2020**, *9*, 1795. [CrossRef]
25. Zhai, H.; Yang, X.; Li, L.; Xia, G.; Cen, J.; Huang, H.; Hao, S. Biogenic amines in commercial fish and fish products sold in southern China. *Food Control* **2012**, *25*, 303–308. [CrossRef]
26. Codex Alimentarius Code. Report of the 3rd session of the Codex Committee on Contaminants in Foods. In *FAO/WHO Food Standards Programme, ALINORM*; 09/32/41; FAO: Rome, Italy, 2009; p. 92.
27. Shalaby, A.R. Significance of biogenic amines to food safety and human health. *Food Res. Int.* **1996**, *29*, 675–690.
28. Wunderlichová, L.; Bunková, L.; Koutný, M.; Jăncová, P.; Buňka, F. Formation, degradation, and detoxification of putrescine by foodborne bacteria: A review. *Compr. Rev. Food Sci. Food Saf.* **2014**, *13*, 1012–1033. [CrossRef]
29. Paulsen, P.; Bauer, F. Biogenic amines in fermented sausages: 2. Factors influencing the formation of biogenic amines in fermented sausages. *Fleisch. Int.* **1997**, *77*, 32–34.
30. Pinho, O.; Ferreira, I.M.P.L.V.O.; Mendes, E.; Oliveira, B.M.; Ferreira, M. Effect of temperature on evolution of free amino acid and biogenic amine contents during storage of Azeitao cheese. *Food Chem.* **2001**, *75*, 287–291.
31. Kim, S.H.; Price, B.J.; Morrissey, M.T.; Field, K.G.; Wei, C.I.; An, H. Occurrence of histamine-forming bacteria in albacore and histamine accumulation in muscle at ambient temperature. *J. Food Sci.* **2002**, *67*, 1515–1521.
32. Knope, K.E.; Sloan-Gardner, T.S.; Stafford, R.J. Histamine fish poisoning in Australia, 2001 to 2013. *Commun. Dis. Intell. Q. Rep.* **2014**, *38*, 285–293.
33. Kanki, M.; Yoda, T.; Tsukamoto, T.; Baba, E. Histidine decarboxylases and their role in accumulation of histamine in tuna and dried saury. *Appl. Environ. Microbiol.* **2007**, *73*, 1467–1473. [CrossRef] [PubMed]
34. Visciano, P.; Campana, G.; Annunziata, L.; Vergara, A.; Ianieri, A. Effect of storage temperature on histamine formation in *Sardina pilchardus* and *Engraulis encrasicolus* after catch. *J. Food Biochem.* **2007**, *31*, 577–588. [CrossRef]
35. Prester, L. Biogenic amines in fish, fish products and shellfish: A review. *Food Addit. Contam. Part A Chem. Anal. Control Expo. Risk Assess.* **2011**, *28*, 1547–1560. [PubMed]
36. EFSA. Scientific opinion on risk based control of biogenic amine formation in fermented foods. *EFSA J.* **2011**, *9*, 2393.
37. Food and Agriculture Organization of the United Nations/World Health Organization (FAO/WHO). *Joint FAO/WHO Expert Meeting on the Public Health Risks of Histamine and Other Biogenic Amines from Fish and Fishery Products*; Joint FAO/WHO Expert Meeting Report; FAO/WHO: Rome, Italy, 2012; pp. 1–111.
38. Park, Y.K.; Lee, J.H.; Mah, J.-H. Occurrence and Reduction of Biogenic Amines in Kimchi and Korean Fermented Seafood Products. *Foods* **2019**, *8*, 547. [CrossRef]
39. Linares, D.M.; del Rio, B.; Redruello, B.; Ladero, V.; Martin, M.C.; Fernandez, M.; Ruas-Madiedo, P.; Alvarez, M.A. Comparative analysis of the in vitro cytotoxicity of the dietary biogenic amines tyramine andhistamine. *Food Chem.* **2016**, *197*, 658–663.
40. Stute, R.; Petridis, K.; Steinhart, H.; Biernoth, G. Biogenic amines in fish and soy sauces. *Eur. Food Res. Technol.* **2002**, *215*, 101–107.
41. Poveda, J.M. Biogenic amines and free amino acids in craft beers from the Spanish market: A statistical approach. *Food Control* **2019**, *96*, 227–233.
42. Gezginc, Y.; Akyol, I.; Kuley, E.; Ozogul, F. Biogenic amines formation in Streptococcus thermophiles isolated from home-made natural yogurt. *Food Chem.* **2013**, *138*, 655–662.

43. Yoon, S.H.; Koh, E.; Choi, B.; Moon, B. Effects of soaking and fermentation time on biogenic amines content of Maesil (*Prunus Mume*) extract. *Foods* **2019**, *8*, 592. [CrossRef]

44. Ly, D.; Mayrhofer, S.; Schmidt, J.-M.; Zitz, U.; Domig, K.J. Biogenic Amine Contents and Microbial Characteristics of Cambodian Fermented Foods. *Foods* **2020**, *9*, 198. [CrossRef]

Article

Microbial Decontamination of Red Wine by Pulsed Electric Fields (PEF) after Alcoholic and Malolactic Fermentation: Effect on *Saccharomyces cerevisiae*, *Oenococcus oeni*, and Oenological Parameters during Storage

Carlota Delso, Alejandro Berzosa, Jorge Sanz, Ignacio Álvarez and Javier Raso *

Food Technology, Facultad de Veterinaria, Instituto Agroalimentario de Aragón-IA2, Universidad de Zaragoza-CITA, 50013 Zaragoza, Spain
* Correspondence: jraso@unizar.es; Tel.: +34-976762675; Fax: +34-976761590

Abstract: New techniques are required to replace the use of sulfur dioxide (SO_2) or of sterilizing filtration in wineries, due to those methods' drawbacks. Pulsed electric fields (PEF) is a technology capable of inactivating microorganisms at low temperatures in a continuous flow with no detrimental effect on food properties. In the present study, PEF technology was evaluated for purposes of microbial decontamination of red wines after alcoholic and malolactic fermentation, respectively. PEF combined with SO_2 was evaluated in terms of microbial stability and physicochemical parameters over a period of four months. Furthermore, the effect of PEF on the sensory properties of red wine was compared with the sterilizing filtration method. Results showed that up to 4.0 Log_{10} cycles of *S. cerevisiae* and *O. oeni* could be eradicated by PEF and sublethal damages and a synergetic effect with SO_2 were also observed, respectively. After 4 months, wine treated by PEF after alcoholic fermentation was free of viable yeasts; and less than 100 CFU/mL of *O. oeni* cells were viable in PEF-treated wine added with 20 ppm of SO_2 after malolactic fermentation. No detrimental qualities were found, neither in terms of oenological parameters, nor in the sensory parameters of wines subjected to PEF after storage time.

Keywords: red wine; pulsed electric fields; microorganisms; inactivation; sulfur dioxide; shelf-life

Citation: Delso, C.; Berzosa, A.; Sanz, J.; Álvarez, I.; Raso, J. Microbial Decontamination of Red Wine by Pulsed Electric Fields (PEF) after Alcoholic and Malolactic Fermentation: Effect on *Saccharomyces cerevisiae*, *Oenococcus oeni*, and Oenological Parameters during Storage. *Foods* **2023**, *12*, 278. https://doi.org/10.3390/foods12020278

Academic Editors: Sidonia Martinez and Javier Carballo

Received: 8 December 2022
Revised: 31 December 2022
Accepted: 4 January 2023
Published: 6 January 2023

1. Introduction

Wine is a fermented beverage with particular characteristics, such as a low pH (3.0–3.9) and the presence of ethanol (8–16% *v/v*), which thereby restrain the proliferation of food-borne pathogenic microorganisms and spore-forming bacteria. However, certain groups of microorganisms are able to grow in wines and spoil them in different stages of the winemaking process [1]. Such spoilage microorganisms are usually either the endogenous microbiota of grape skins, or microorganisms stemming from contact surfaces and equipment in wineries [2]. On the other hand, microorganisms that are essential in wine production, such as *Saccharomyces cerevisiae* and lactic acid bacteria (LAB), responsible for alcoholic and malolactic fermentation, respectively, may likewise be involved in wine spoilage. These microorganisms therefore need to be controlled after fermentation to avoid undesired re-fermentation, which, in turn, may lead to off-flavors, increments in volatile acidity, or even the production of biogenic amines [3].

To obtain quality wines and avoid economic losses, it is essential to control all of the involved spoiling microorganisms in grapes, must, and wine, as well as on surfaces in wineries. Apart from rigorous standard cleaning and disinfection plans, the conventional method for microbial control in wineries is the application of sulfur dioxide (SO_2) [4]. Sulfur dioxide has bacteriostatic and antifungal activity but also plays an essential role as an antioxidant: all of these functions make it a thoroughly convenient compound for wine preservation [5]. Sulfur dioxide is usually dosed along the production process: when grapes are received, after alcoholic and malolactic fermentation, and usually prior to bottling in order to ensure wine stability during distribution. When SO_2 is incorporated into must or wine, a fraction of it will react with sugars, aldehydes, or ketones [6]. Consequently, two classes of sulfites are found in wine: free and bound. The free sulfites determine how much SO_2 is available in its most active, molecular form to help protect the wine from oxidation and spoilage. The bound sulfites are those which have reacted with other molecules. Total sulfite concentration is the sum of the free and bound sulfites.

However, the widespread use of sulfur dioxide in winemaking is currently being called into question, due to its potential toxicity for human health, the sensitivity of certain people allergic to SO_2, and an increasing overall consumer rejection of chemical preservatives [7]. Another procedure frequently conducted in wineries to guarantee complete microbial decontamination is sterilizing filtration prior to bottling. However, filters with a nominal pore size of 0.45 μm generally used in sterilizing filtration can be ineffective in retaining certain smaller-sized bacteria. Furthermore, wine microfiltration usually implies fouling, regeneration problems, and high operational costs [8]. Moreover, this procedure is frequently controversial in wineries due to its deleterious effect on the chemical and sensory characteristics of wine, especially red wine [9]. The industry is thus searching for suitable alternative methods for microbial control that do not modify the properties of wine. The capability of PEF for inactivating vegetative forms of microbial cells at lower temperatures than those used in thermal processing [10–12] may prove to be thoroughly useful for wineries as a physical procedure for microbial decontamination.

PEF technology consists in the application of short pulses (microseconds) of high voltages (kV) to a product located between two electrodes. The electric field thereby generated produces the electroporation of cells. Electroporation compromises the permeability of the cytoplasmic membrane of microorganisms, leading to the loss of microbial homeostasis and, ultimately, to cell death. The efficacy of PEF for microbial inactivation, together with the prospect of applying it at high rates of continuous flow thanks to the availability of flexible commercial devices on the market, makes this technology a reasonable potential method for liquid food decontamination [13]. Several studies have demonstrated the effectivity of PEF in inactivating diverse microorganisms in acidic beverages, such as fruit juices, but with a noticeable variability in the amount of reported lethality [14–17]. Some research has been performed regarding the use of PEF for microbial inactivation in wine. Puértolas et al., (2009) [18] characterized the PEF resistance of some of the most common microorganisms involved in wine spoilage in batch conditions; González-Arenzana et al., (2015) [19] extended that approach to a greater number of microorganisms, performing the treatments in continuous flow. These studies demonstrated that PEF treatments at intensities over 20 kV/cm were effective for all of the microorganisms investigated (yeasts, lactic acid bacteria, and acetic acid bacteria) in the range of 3.0 to 4.0 Log_{10} cycles of inactivation. Further studies have demonstrated that PEF treatments do not cause detrimental effects: neither in terms of oenological parameters, nor on the sensory properties of wine, even after long-term storage [20,21]. However, all of these studies were conducted at high electric fields (which would be difficult to apply on an industrial scale with the current commercial PEF units) and/or without reporting the free SO_2 concentration present in wines, which may exert an influence on the efficacy of PEF.

The objective of our study was to characterize the PEF resistance of *Saccharomyces cerevisiae* and *Oenococcus oeni* in red wine after alcoholic and malolactic fermentation, respectively. Once defined the most suitable PEF processing conditions, the impact of these treatments on the microbial population, and the oenological parameters of wines after 4 months of storage were investigated.

2. Material and Methods

2.1. Red Wine Samples

The Cooperative San Juan Bautista (Fuendejalón, Aragón, Spain) provided 50 L of *Grenache* wine immediately after they had undergone alcoholic fermentation (AF), or malolactic fermentation (MLF). Alcoholic fermentation was performed by the *S. cerevisiae var. bayanus* (CHP Levuline, OENOFRANCE, Magenta, France) strain, and malolactic fermentation by *Oenococcus oeni* (Viniferm OE AG-20, Agrovin, Ciudad Real, Spain). The initial oenological parameters of the two wines after AF and MLF are shown in Table 1. An aliquot of 10 L of each wine was used to characterize microbial resistance to PEF, and the effect of combining PEF with SO_2. The rest of the wine samples were kept under refrigeration (4 °C) and no-light conditions in view of conducting an experiment on the impact of PEF on their microbial population, as well as on oenological parameters after AF and MLF during storage. Immediately after PEF treatments, SO_2 was added, and wine samples were distributed in sterilized glass bottles of 500 mL and stored at 18 °C for 4 months. PEF-treated and untreated wines with and without sulfites were monitored in terms of their microbial population and chemical parameters over a period of 4 months.

Table 1. Initial oenological parameters of red wines after alcoholic (AF) and malolactic fermentation (MLF).

	Alcoholic Fermented (AF) Red Wine	Malolactic Fermented (MLF) Red Wine
pH	3.80 ± 0.01	3.54 ± 0.01
Glucose-Fructose (g/L)	0.94 ± 0.02	0.29 ± 0.02
% Ethanol (*v/v*)	14.70 ± 0.04	14.35 ± 0.03
Total Acidity (g/L) [a]	4.34 ± 0.32	5.17 ± 0.14
Volatile Acidity (g/L) [b]	0.59 ± 0.03	0.45 ± 0.02
Free SO_2 (ppm) [c]	19.2 ± 3.2	10.0 ± 3.2
Total SO_2 (ppm) [c]	22.4 ± 3.2	16.0 ± 3.2
CI * (A.U.)	20.48 ± 0.36	13.07 ± 0.41
TPI ** (A.U.) [d]	67.3 ± 0.14	54.62 ± 0.52
TAC *** (mg/L) [e]	982.09 ± 8.05	532.21 ± 7.39

Values represent mean with standard deviation. * Color intensity. ** Total polyphenol index. *** Total anthocyanin content. A.U.: absorbance units. [a]: expressed as tartaric acid. [b]: expressed as acetic acid. [c]: expressed as the mean ± the deviation of the analytical method. [d]: expressed as tartaric acid. [e]: expressed as malvidin-3-glucoside.

2.2. PEF Processing

PEF treatments were applied in a continuous flow by means of a commercial generator (Vitave, Prague, Czech Republic) able to deliver pulses of up to 20 kV. Square waveform monopolar pulses were delivered in a parallel titanium electrode chamber with a 0.4 cm gap (3.0 × 0.5 cm). Wines tempered at 20 °C in a heat exchanger placed prior to the treatment chamber were pumped by a peristaltic pump (BVP, Ismatec, Wertheim, Germany) at different flow rates into the chamber. After the treatments, wines were cooled down in less than 5 s to under 20 °C in a second heat exchanger located after the treatment chamber. The actual voltage during treatments was measured by a high voltage probe (Tektronik, P6015A, Wilsonville, OR, USA) connected to an oscilloscope (Tektronik, TDS 220). Inlet and outlet temperatures were measured by a type K thermocouple inserted in the circuit (Ahlborn, Holzkirchen, Germany).

In preliminary studies, a matrix of different PEF parameters was tested with the aim of identifying optimal conditions for microbial decontamination. Wines after AF (1.9 mS/cm) and MLF (2.0 mS/cm) were pumped at 10 L/h, resulting in a residence time of 0.22 s in the chamber. Pulses of 10 µs were delivered at electric field strengths of 15, 20, and 25 kV/cm, and at a repetition rate ranging from 8.0 to 80 Hz, corresponding to effective treatment times ranging from 20 to 175 µs. These treatments corresponded to total specific energies of 35 to 120 kJ/kg, thereby implying an exit temperature in the range of 30 to 50 °C ± 2 °C. For the storage study, the two PEF treatments selected for each red wine were applied at 15 kV/cm. Wine after AF was treated with total specific energies of 39 and 97.2 kJ/kg (exit temperatures of 30 and 45 °C, respectively), whereas for wine after MLF, energies were 77.8 and 116.7 kJ/kg (exit temperatures of 40 and 50 °C, respectively).

2.3. Sulfite Addition

Immediately after PEF treatments, wine samples were added with the corresponding amount of SO_2. A stock solution of 25 g/L of SO_2 was prepared from potassium bisulfite (Sigma, Burlington, MA, USA). PEF-treated and untreated wines were dosed with 0, 20, and 30 ppm of SO_2. The total free SO_2 content at the starting point of the storage study was the sum of the initial free SO_2 content in wine after AF (19 ppm) and after MLF (10 ppm) plus the corresponding doses of SO_2 added in this step.

2.4. Microbial Analysis

Microbial survivors were measured by the corresponding plating of aliquots of wine samples diluted in peptone water (Oxoid, Basingtok, Hampshire, UK) and plated onto the appropriate agar medium. For yeast enumeration, Potato Dextroxe Agar (Oxoid) was used, and plates were incubated at 25 °C for 48 h. Lactic acid bacteria (LAB) survivors were enumerated in Mann Rogosa Sharpe (MRS) Agar (Oxoid) and the plates were incubated in anaerobic conditions (<1% O_2) at 30 °C for 24 to 72 h. After the plate incubation, the number of counted colonies corresponds with the number of viable microorganisms expressed as a colony form unit per milliliter (CFU/mL) or its decimal logarithm (Log_{10} CFU/mL). The survival fraction was calculated by dividing the number of microorganisms that survived the treatment (N_t) by the initial number of viable cells (N_0).

2.5. Analysis of Oenological Parameters

The initial and final oenological parameters of all of the wine samples were measured. The pH, glucose-fructose, % ethanol, and total and volatile acidity were analyzed by FTIR spectroscopy using MIURA 200 and BACCHUS 3 MultiSpec models (TDI, Barcelona Spain). Absorbance measurements were performed after centrifuging wine samples in an Eppendorf AG centrifuge for 15 min at 3000 rpm (Eppendorf, Hamburg, Germany). All spectrophotometric determinations were measured by spectrophotometer (DS-11, DeNovix, Wilmington, DE, USA). The color index (CI) was determined as the sum of absorbance at 420, 520, and 620 nm. The total polyphenol index (TPI) was determined by measuring the direct absorbance at 280 nm. Total anthocyanins (TAC) expressed in malvidin-3-glucoside (mg/L) were calculated by determining the absorbance at 520 nm of samples diluted 1/10 (v/v) in 1% (v/v) of HCl, adapted from Maza et al., (2019) [22].

The determination of total and free sulfur dioxide (SO_2) was performed by the Ripper method, which is based on an oxidation-reduction titration using iodine as a reagent in an acid medium in the presence of starch. Briefly, 1 mL of starch (1%) and 2 mL of sulfuric acid 1/3 w/v vinikit (PanReac, Barcelona, Spain) were added to 15 mL of wine. This solution was titrated with an iodine solution (0.01 N) until a blue color appeared.

2.6. Sensory Evaluation of PEF Treatments in a Commercial Wine

To evaluate the impact of PEF on sensory properties, independent experiments were performed in a finished red wine prior to being bottled. This red wine was treated by PEF at two different intensities and compared with the same wine after sterilizing filtration in the facilities of the winery. A total amount of 50 L of red wine (1.5 mS/cm) was processed by PEF at a flow rate of 25 L/h. Pulses of 5 μs were applied at 107 and 190 Hz at 15 kV/cm at the total specific energies of 77.8 of 155.6 kJ/kg that corresponded to exit temperatures of 40 and 60 ± 2.0 °C, respectively. The cooling exchanger placed immediately after the PEF chamber reduced the wine's temperature to under 20 °C in less than 5 s before bottling. After 1 month of storage, physicochemical and sensory analyses were performed. Oenological parameters were measured as previously described (Section 2.5), and a sensory evaluation was carried out by a complete sensory triangle discrimination test, performed by 16 panelists from the Campo de Borja Appellation of Origin (nine men/six women ages

26 to 58). Panelists were distributed in individual booths and were given no information regarding the testing samples. Comparisons performed in the triangle test were between the sterilized filtered wine and the PEF-treated wines, using a completely randomized design. Samples were previously tempered to room temperature and 20 mL were served in clear wine glasses (ISO NORM 3591) [23]. Panelists had to distinguish the one different sample among the three samples presented in each batch, by taste and/or aroma.

2.7. Statistical Analysis

Samples were analyzed in three independent replicates and data are expressed as the mean ± the standard deviation. When called for, one-way analysis of variance (ANOVA) and Tukey tests using GraphPad Prism (Graph-Pad Software, San Diego, CA, USA) were performed to evaluate the significance of differences among the mean values. Differences were considered significant at $p \leq 0.05$. The significant difference for the triangular test was determined using statistical tables reported by Roessler et al., (1948) [24].

3. Results and Discussion

3.1. PEF-Resistance of Saccharomyces cerevisiae and Oenococcus oeni

Preliminary experiments were conducted to determine the resistance of *S. cerevisiae* to PEF treatments of *Grenache* red wine after alcoholic fermentation, and of *O. oeni* after malolactic fermentation. The ultimate objective of these experiments was to select PEF processing conditions for subsequent study, aiming to evaluate the evolution of these microbial populations in each wine during storage.

Survival curves corresponding to the inactivation of *S. cerevisiae* and *O. oeni* at different electric fields (15, 20, and 25 kV/cm) are shown in Figure 1A,B, respectively. The numbers next to the symbols indicate the outlet temperature of wine attained during PEF processing. Due to the wine's short residence time in the treatment chamber (0.22 s), no heat is exchanged with the surroundings. Consequently, all of the electrical energy delivered to the treatment chamber to generate the electric field is transformed into heat, thereby increasing the wine's temperature. However, in the experimental approach used in this study, wine was cooled below 20 °C within 5 s after the PEF treatment, independently of the outlet temperature. As can be observed in the two graphs, the temperature increased with the treatment time (number of pulses × pulse width).

Figure 1 shows that the inactivation kinetics of the two microorganisms is different. In the case of *S. cervisiae*, the survival curve's shape is concave upwards, whereas in the case of *O. oeni* a linear behavior can be observed. The shape of the survival curves of *S. cerevisiae* could be explained by the effect of temperature on the membrane's stability as a consequence of the phase transition of the phospholipids from gel to the liquid-crystalline phase. A sudden change in inactivation kinetics can be observed when the output temperature of the wine was over 40 °C. This change would indicate that above 40 °C the cytoplasmic membrane of *S. cerevisiae* is more vulnerable to the pore formation caused by PEF [25]. The positive effect of treatment medium temperature on microbial inactivation has been previously reported by several authors [10,26]. However, in the case of *O. oeni*, no radical changes in the inactivation kinetics were observed in the same range of outlet temperatures. This might indicate differences in the cytoplasmic membrane composition of both microorganisms, which could vary in terms of the manner in which the temperature affects the phase transition of the phospholipids.

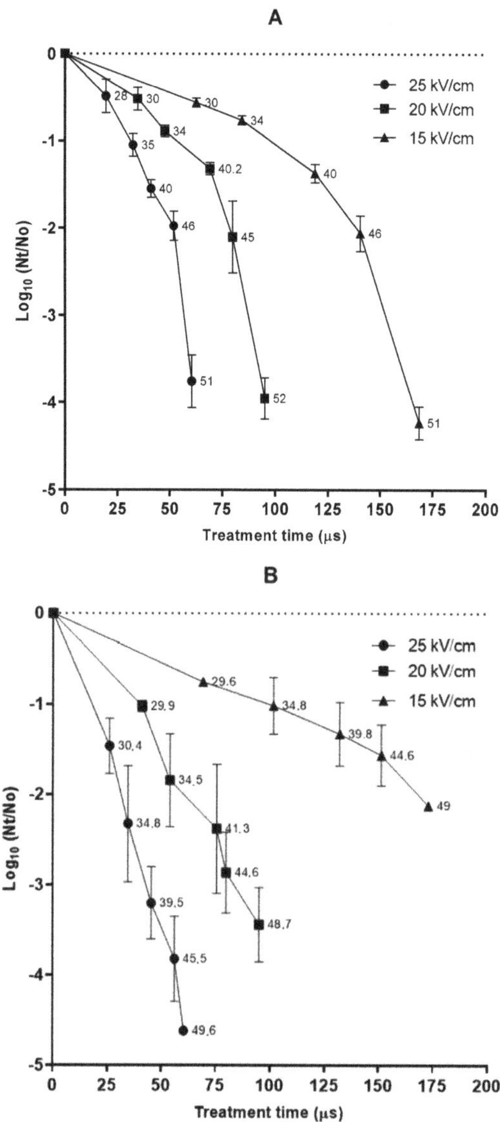

Figure 1. Survival curves in wine of *S. cerevisiae* (**A**) and *O. oeini* (**B**) after alcoholic (AF) and malolactic (MLF) fermentation, respectively, at different electric field strengths. Numbers near the dots indicate the outlet temperature achieved during treatments.

The inactivation of the two microorganisms increased with the electric field, whereby shorter treatments at higher electric fields strengths were required to achieve a given level of inactivation. For example, to inactivate 2.0 Log_{10} cycles of the population of *S. cerevisiae*, the treatment time decreased from 140 to 52 µs when the electric field was increased from 15 to 25 kV/cm. The same electric field increment reduced the treatment time from 180 to 40 µs to achieve a similar inactivation in *O. oeni*. In the case of *S. cerevisiae*, the outlet temperature for both treatments applied at different electric field strengths was the same (45 °C), thereby indicating that the total specific energy of the treatments applied at different electric fields was equivalent (97 kJ/kg). Huang et al. (2014) and Puértolas et al.

(2009) [18,27] also reported that the electric field did not modify PEF lethality on different *Saccharomyces* strains suspended in must or wine when treatments of the same specific energy were applied. However, in the case of *O. oeni*, the total specific energy required for achieving an equivalent lethality was lower when the electric field strength was increased. For example, in order to obtain 2.0 Log_{10} cycles, the total specific energy decreased from 117 to 49 kJ/kg when the electric field was increased from 15 to 25 kV/cm. Consequently, whereas the outlet temperature of the treatment applied at 15 kV/cm was around 50 °C, the outlet temperature of the treatment applied at 25 kV/cm lay in the range of 30.4 to 34.8 °C. The total specific energy of a PEF treatment depends on the applied voltage, total treatment time, and the electrical resistance of the treatment chamber. Since total specific energy is an integrate parameter that involves electric field strength and treatment time, it has been proposed as a single parameter to define the intensity of a PEF treatment [28]. This approach could be considered in the case of the strain of *S. cerevisiae* used in this study, upon which equivalent total specific energy delivered at different electric field strengths had the same lethal effect. However, in most microorganisms observed in the *O. oeni* strain, treatments of the same specific energy are more effective in terms of microbial inactivation when the applied electric field strength is higher [29]. Therefore, in this case, to define the intensity of a given PEF treatment, it is necessary to report both the electric field strength and total delivered specific energy.

It has generally been reported that yeasts are more sensitive to PEF than bacteria [30–32]. The high PEF sensitivity of yeast has been associated with the fact that larger cells require a lower critical electric field to achieve the transmembrane potential threshold for the manifestation of electroporation. However, other intrinsic microbial factors in addition to cell size seem to play a role in microbial resistance to PEF. Whereas at 15 kV/cm the resistance of the two microorganisms to PEF treatments of different duration was similar, at higher electric field strength *O. oeni* was slightly more sensitive than *S. cerevisiae*. For example, a treatment of 20 kV/cm for 70 µs that corresponded with an exit temperature of 40 °C inactivated around 1.5 Log_{10} cycles of *S. cerevisiae*, and around 2.5 Log_{10} of *O. oeni*. The higher resistance of *S. cerevisiae* to PEF might be explained by the fact that, in contrast to other studies in which the wine was contaminated with yeast previously grown in laboratory media, our investigation was conducted with the cells that had performed the alcoholic fermentation. The presence of ethanol during the growth of those cells could provoke changes in the composition or structure of their cytoplasmic membrane [33,34], which, in turn, could lead to a cross-protection against the electroporation brought about by PEF.

It is difficult to compare the PEF resistance of the strains used in this investigation with reported data in view of the widely varying protocols employed for PEF application, the differences among the investigated strains, their physiological state, and the variabilities in the composition of wines. The PEF resistance of yeast and lactic acid bacteria in wine has been reported in previous studies. Treatments at an electric field above 17 kV/cm and 90 kJ/kg were required to achieve between 2.0 and 3.0 Log_{10} cycles of the inactivation of the *Saccharomyces* population [18,19]. After alcoholic fermentation, intense PEF treatments (33 kV/cm; 158 kJ/kg; 105 µs) inactivated between 3.7 and 7.2 Log_{10} cycles of the population of yeast cells involved in the fermentation process [35]. Meanwhile, lethality reported for *O. oeni* ranged from 1.5 to 3.2 Log_{10} when PEF treatments of 17–23 kV/cm and 60–100 kJ/kg were applied [19,20].

3.2. Inactivation of Saccharomyces cerevisiae and Oenococcus oeni by Combining Moderate PEF Treatments with SO_2

Several authors have reported that when PEF treatments are applied to a microbial population, a proportion of cells is sublethally injured, and this is more evident when moderate conditions are applied [36]. The final recovery or death of that injured population is directly dependent on either optimal or adverse recovery conditions [37,38]. Preventing the reparation of sublethal injuries caused by PEF by adding preservatives is a strategy that can increase the lethality of moderate PEF treatments [39–42]. Figure 2 illustrates the inactivation of *S. cerevisiae* (Figure 2A) and *O. oeni* (Figure 2B) in the respective wines after fermentation by combining PEF treatments of different durations at 15 kV/cm with the addition of 20 ppm of SO_2 evaluated 24 h after processing. The lethal effects of the individual treatments are also shown in order to identify if the effect of the combined treatments was additive or synergic. Figure 2A,B show that the added SO_2 scarcely affected the viability of the untreated cells of *S. cerevisiae* and *O. oeni*. On the other hand, the inactivation of PEF-treated cells of *S. cerevisiae* maintained for 24 h in wine without added SO_2 increased significantly. For example, the inactivation detected just after the 140 μs PEF treatment increased from 2.0 to 4.0 Log_{10} cycles after 24 h. This effect could be explained by the supposition that the presence of ethanol and the content of free sulfites in the wine (14.7%; 19.2 ppm, see Table 1) prevented the recovery of a proportion of the yeast cells that had been sublethally injured as a consequence of the PEF treatment. The addition of 20 ppm of SO_2 barely increased lethality compared with wines treated solely with PEF; the difference is not statistically significant ($p > 0.05$). The yeast cells that survived in PEF-treated wine during 24 h were thus unaffected by the addition of 20 ppm of extra sulfites.

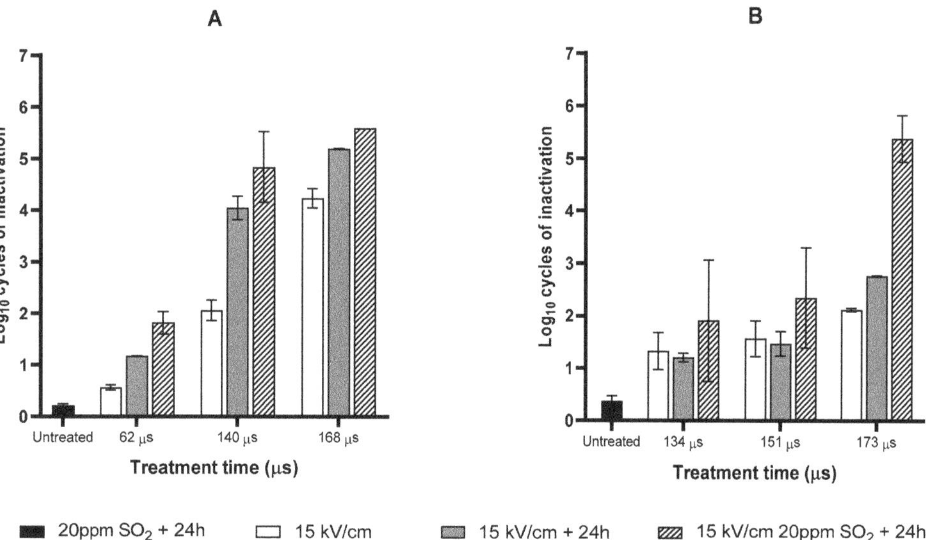

Figure 2. Comparison of the Log_{10} cycles of inactivation of *S. cerevisiae* (**A**) and *O. oeini* (**B**) after alcoholic (AF) and malolactic (MLF) fermentation, respectively, treated at 15 kV/cm and plated immediately after PEF treatment, or plated after 24 h of incubation with 0 or 20 ppm of SO_2. PEF treatments 62 μs, 39 kJ/kg, exit temperature: 30 °C; 134 μs, 78 kJ/kg, exit temperature: 40 °C; 140 μs; 97 kJ/kg, exit temperature: 45 °C; 168 μs, 116,7 kJ/kg; exit temperature: 50 °C; 173 μs, 116.7 kJ/kg; exit temperature: 50 °C.

The incubation of PEF-treated *O. oeni* cells for 24 h in wine without added SO_2 did not significantly increase the lethality of the treatments applied over different time intervals. This observation confirms results obtained by other authors who reported that sublethal injury did not occur when Gram-positive bacteria, such as *O. oeni*, were treated by PEF in media of low pH [43,44]. Regarding the combination of PEF with SO_2, the most effective combination was observed 24 h after mixing 20 ppm of SO_2 with wine treated for 173 μs. The PEF treatment thus sensitized a proportion of the surviving population to SO_2, whereby the lethality of the combined treatment was over 2.0 Log_{10} cycles more than the sum of the single treatments. The main mechanism involved in the antimicrobial effect of SO_2 on yeast is related to the diffusion of SO_2 into the cytoplasm and the subsequent disturbance of metabolic processes by the binding of SO_2 to essential molecules (proteins, nucleic acids, coenzymes, cofactors, vitamins, etc.). Although the activity of SO_2 against bacteria is still unclear, the electroporation phenomenon triggered by PEF could enhance the diffusion of SO_2 into the cytoplasm of *O. oeni*, thereby exerting an effect similar to the one described for yeast [4].

3.3. Evolution of the Microbial Population in Wine Treated by PEF after Alcoholic and Malolactic Fermentation during 4 Months of Storage

Based on preliminary results obtained on the resistance of *S. cerevisiae* and *O. oeni* to PEF treatments of different intensities and their combination with SO_2, treatment conditions with the aim of evaluating the evolution of the microbial population in wines during 4 months of storage were chosen.

Treatments at 15 kV/cm of a shorter (62 µs) and longer (140 µs) duration that corresponded to a total specific energy of 39 and 97.2 kJ/kg and an outlet temperature of 30 and 45 °C, respectively, were selected for wine obtained after alcoholic fermentation. Table 2 shows the evolution of yeast populations in wine treated by the two selected PEF treatments after alcoholic fermentation without adding SO_2 or with the addition of 20 ppm of SO_2. For comparative purposes, the evolution of the yeast populations in the untreated wines with and without 20 ppm of added SO_2 is also included. The initial number of viable yeast cells in the untreated wines was 3×10^6 CFU/mL. This initial number was maintained during the first month of storage, even in the untreated wine dosed with 20 ppm of sulfites. After 4 months of storage, the yeast cell population decreased by about 3.0 logarithmic units. These results indicate that the addition of 20 ppm of SO_2 did not compromise the viability of yeast cells that fermented a wine which already had 19 ppm of free SO_2 by the end of fermentation. The reduction of the yeast cell population observed between 1 and 4 months might be the loss of viability that occurs in a microbial population when it is maintained under conditions in which multiplication does not occur. In the case of PEF-treated wines, a similar reduction in yeast viability as in the control wine after 4 months was achieved just after the application of the most intense PEF_2 treatment (97.2 kJ/kg; 171 µs), or after 1 day of incubation in the wine treated at lower intensity PEF_1 (39 kJ/kg; 83 µs). After 4 months of incubation, no viable yeasts were detected in the two wines, even when SO_2 was not added to them.

Table 2. Evolution of Log_{10} CFU/mL of *S. cerevisiae* cells during the storage time of red wine after alcoholic fermentation (AF) treated by PEF and added with different SO_2 concentrations (0 or 20 ppm). PEF1 (15 kV/cm, 39 kJ/kg, exit temperature: 30 °C) and PEF2 (15 kV/cm, 97.2 kJ/kg, exit temperature 45 °C).

	0 Days	1 Day	7 Days	15 Days	1 Month	4 Months
Control	6.46 ± 0.04 *a*	6.47 ± 0.03 *a*	6.21 ± 0.05 *a*	6.28 ± 0.21 *a*	5.93 ± 0.06 *a*	3.69 ± 0.04 *a*
Control 20 ppm SO_2	6.40 ± 0.02 *a*	6.46 ± 0.04 *a*	6.20 ± 0.06 *a*	6.18 ± 0.12 *a*	5.97 ± 0.10 *a*	3.93 ± 0.15 *b*
PEF_1	5.94 ± 0.04 *b*	3.39 ± 0.01 *b*	3.24 ± 0.10 *b*	3.24 ± 0.04 *b*	3.10 ± 0.01 *b*	n.d. *c*
PEF_1 20 ppm SO_2	5.89 ± 0.01 *b*	3.12 ± 0.03 *c*	3.08 ± 0.02 *c*	3.25 ± 0.11 *b*	2.32 ± 0.76 *bc*	n.d. *c*
PEF_2	2.36 ± 0.04 *c*	2.35 ± 0.04 *d*	2.31 ± 0.01 *d*	2.34 ± 0.08 *c*	2.38 ± 0.04 *bc*	n.d. *c*
PEF_2 20 ppm SO_2	2.79 ± 0.08 *d*	3.03 ± 0.11 *c*	2.96 ± 0.03 *c*	2.39 ± 0.03 *c*	2.19 ± 0.01 *c*	n.d. *c*

Values represent mean with standard deviation. Different letters within the same column indicate significant differences ($p \leq 0.05$). n.d. = not detected. <1.5 Log_{10} CFU/mL = below the quantification limit (30 CFU/mL).

In the case of the wine that had undergone malolactic fermentation, as sublethal injury was not detected in *O. oeni* cells after PEF treatment, the two longer treatments (132 and 173 µs) that correspond to a total specific energy of 77.8 and 116.7 kJ/kg, respectively (outlet temperatures of 40 and 50 °C), were selected. Moreover, in order to obtain synergetic effects, the low-intensity treatment (PEF_3) was combined with the addition of 30 ppm of sulfites, and the high-intensity treatment (PEF_4) was combined with the addition of 20 ppm of sulfites.

As the presence of yeast in addition to *O. oeni* was detected in the wine obtained after malolactic fermentation, the evolution of the two microorganisms under the different assayed conditions is shown in Table 3. It shows that the initial concentration of *O. oeni* after malolactic fermentation ($\approx 10^5$ CFU/mL) slowly decreased during storage in untreated wines with no added SO_2, as well as in those with 20 or 30 ppm of SO_2 added; the decrease was slightly higher in the latter ones. After 4 months of storage, the differences in concentration of viable cells between the untreated wine without added SO_2 and with 30 ppm added was lower than 1.0 Log_{10} cycle. On the other hand, as had been observed in the case of the yeast after alcoholic fermentation, the population of *O. oeni* decreased more rapidly in wines treated by PEF. After 4 months of storage, the concentration of viable cells was around 2×10^3 and 6×10^2 CFU/mL in wines without added SO_2 but treated with low and high intensity PEF, respectively. However, in wines with 20 or 30 ppm of SO_2, the number of viable cells of *O. oeni* was lower than 1×10^2 CFU/mL, independently of the intensity of the PEF treatment applied. It is remarkable that in both PEF-treated wines combined with added SO_2, a pronounced synergetic effect could be observed, obtaining a reduction of 2.0–3.0 Log_{10} cycles of the initial *O. oeni* population within just 24 h.

Table 3. Evolution of the Log_{10} CFU/mL of O. oeni cells and yeast cells during the storage time of wine after malolactic fermentation (MLF) treated by PEF and added with different SO_2 concentrations (0, 20 or 30 ppm). PEF3 (15 kV/cm, 77.8 kJ/kg, exit temperature: 40 °C) and PEF4 (15 kV/cm, 116.7 kJ/kg, exit temperature: 50 °C).

	0 Days		1 Day		15 Days		1 Month		4 Month	
	O. oeni	Yeast	*O. oeni*	Yeast	*O. oeni*	Yeast	*O. oeni*	Yeast	*O. oeni*	Yeast
Control	5.14 ± 0.08 *a*	5.97 ± 0.08 *a*	4.98 ± 0.01 *a*	5.94 ± 0.02 *a*	4.64 ± 0.08 *a*	3.43 ± 0.01 *a*	4.42 ± 0.04 *a*	2.79 ± 0.21 *a*	4.31 ± 0.08 *a*	1.51 ± 0.03 *ab*
Control 20 ppm SO_2	5.13 ± 0.04 *a*	5.98 ±0.09 *a*	4.52 ± 0.06 *b*	4.55 ± 0.04 *b*	4.04 ± 0.05 *a*	3.10 ± 0.01 *b*	4.43 ± 0.02 *a*	1.99 ± 0.11 *b*	3.95 ± 0.19 *ab*	1.43 ± 0.07 *a*
Control 30 ppm SO_2	5.10 ± 0.06 *a*	5.96 ± 0.08 *a*	4.08 ± 0.01 *c*	3.43 ± 0.04 *c*	2.98 ± 0.14 *b*	3.04 ± 0.27 *b*	2.80 ± 0.20 *b*	1.80 ± 0.01 *b*	3.62 ± 0.06 *bc*	1.75 ± 0.23 *b*
PEF_3	4.50 ± 0.01 *b*	2.02 ± 0.03 *b*	4.43 ± 0.01 *b*	1.78 ± 0.11 *d*	4.39 ± 0.54 *a*	0.80 ± 0.14 *c*	4.48 ± 0.06 *a*	n.d. *c*	3.37 ± 0.06 *c*	n.d. *c*
PEF_3 30 ppm SO_2	4.49 ± 0.01 *b*	1.94 ± 0.13 *b*	2.90 ± 0.07 *d*	<1.5 *e*	2.35 ± 0.49 *bc*	n.d. *d*	1.54 ± 0.22 *c*	n.d. *c*	1.73 ± 0.31 *d*	n.d. *c*
PEF_4	3.57 ± 0.04 *c*	1.89 ± 0.16 *b*	2.15 ± 0.18 *e*	<1.5 *e*	2.55 ± 0.10 *b*	n.d. *d*	3.61 ± 0.17 *d*	n.d. *c*	2.82 ± 0.25 *e*	n.d. *c*
PEF_4 20 ppm SO_2	3.55 ± 0.05 *c*	1.83 ± 0.06 *b*	<1.5 *f*	<1.5 *e*	1.73 ± 0.02 *c*	n.d. *d*	1.59 ± 0.16 *c*	n.d. *c*	1.90 ±0.23 *d*	n.d. *c*

Values represent mean with standard deviation. Different letters within the same column indicate significant differences ($p \leq 0.05$). n.d. = not detected. <1.5 Log_{10} CFU/mL = below the quantification limit (30 CFU/mL).

Regarding the evolution of yeast, Table 3 shows that the concentration of yeast populations in wine after malolactic fermentation was around 0.5 log cycles lower than in wine after alcoholic fermentation (Table 2). This confirms that yeast viability after alcoholic fermentation decreases along time. Untreated yeast populations in wine after malolactic fermentation decreased along storage time: the number of viable yeasts after 4 month of storage was less than 1×10^2 CFU/mL. As occurred in the case of wine after alcoholic fermentation, the addition of 20 ppm of SO_2 did not significantly affect the yeasts' viability along storage as compared to the wine without added SO_2 that contained 10 ppm of free SO_2 at the moment of the trial. After 15 days of incubation, viable yeast cells were not detected in the wines treated by PEF at the two assayed intensities, independently of whether SO_2 had been added or not.

These results thus evidence the capacity of PEF, even applied at quite moderate intensities, as a potential alternative alone or in combination with sulfites, for microbial control at different steps of the red wine production process. The inactivation of yeast by means of PEF after alcoholic fermentation would allow for the achievement of a lower free SO_2 content, thereby facilitating the implementation of LAB for further malolactic fermentation. González-Arenzana et al., (2018) [35] reported a shortening of the malolactic fermentation time for wines pre-treated by PEF, attributing this reduction of fermentation time to the decrease in competitive pressure for lactic acid bacteria. Furthermore, the complete decontamination of yeasts after alcoholic fermentation would be of special interest for sweet or semi-sweet wines with residual sugars, in order to prevent re-fermentation [45]. On the other hand, in addition to preventing re-fermentation by yeast, PEF alone, or in combination with SO_2 after malolactic fermentation, could contribute to the prevention spoilage caused by lactic acid bacteria during storage in bottles, which generally leads to phenomena such as a slimy appearance, undesirable off-flavors, and/or the production of biogenic amines [3].

3.4. Effect of PEF Treatments on the Oenological Parameters of Wine after 4 Months of Storage

Any new technique that might be potentially introduced in wineries should guarantee zero drawbacks in terms of physicochemical parameters of wine. Furthermore, as a potential alternative to SO_2, which acts as an antioxidant, PEF technology should not trigger oxidative reactions that compromise wine quality. Thus, once the effectivity of PEF treatments for controlling different microbial populations along the winemaking process had been observed, the effect of PEF on the oenological parameters of wines was evaluated. After 4 months of storage, the oenological parameters of the untreated and PEF-treated wines, with and without the addition of SO_2 after alcoholic and malolactic fermentation, are shown in Tables 4 and 5, respectively.

Table 4. Oenological parameters after 4 months of storage of red wine after alcoholic fermentation (AF) treated by PEF and added with different SO_2 concentrations (0 or 20 ppm). PEF1 (15 kV/cm, 39 kJ/kg, exit temperature: 30 °C) and PEF2 (15 kV/cm, 97.2 kJ/kg, exit temperature 45 °C).

	Control		PEF$_1$		PEF$_2$	
	SO_2	SO_2	SO_2	SO_2	SO_2	SO_2
	0 ppm	20 ppm	0 ppm	20 ppm	0 ppm	20 ppm
pH	3.81 ± 0.02	3.80 ± 0.01	3.80 ± 0.00	3.82 ± 0.02	3.77 ± 0.02	3.80 ± 0.01
Glucose-Fructose (g/L)	0.92 ± 0.01	0.96 ± 0.02	1.33 ± 0.03	1.41 ± 0.03	1.31 ± 0.02	1.35 ± 0.02
% Ethanol (*v/v*)	14.74 ± 0.15	14.71 ± 0.04	14.69 ± 0.07	14.70 ± 0.03	14.72 ± 0.03	14.70 ± 0.02
Total Acidity (g/L) [a]	4.00 ± 0.22	4.40 ± 0.19	4.55 ± 0.10	4.40 ± 0.17	4.48 ± 0.22	4.48 ± 0.08
Volatile Acidity (g/L) [b]	0.61 ± 0.03	0.56 ± 0.03	0.63 ± 0.01	0.57 ± 0.03	0.58 ± 0.06	0.56 ± 0.02
Free SO_2 (ppm) [c]	9.6 ± 3.2	12.8 ± 3.2	9.6 ± 3.2	12.8 ± 3.2	9.6 ± 3.2	12.8 ± 3.2
Total SO_2 (ppm) [c]	22.4 ± 3.2 a	32 ± 3.2 b	22.4 ± 3.2 a	32 ± 3.2 b	22.4 ± 3.2 a	32 ± 3.2 b
CI * (A.U.)	16.17 ± 0.23	16.45 ± 1.18	17.77 ± 0.62	16.59 ± 0.49	17.33 ± 0.00	18.22 ± 1.00
TPI ** (A.U.) [d]	60.75 ± 1.48	61.10 ± 0.42	61.10 ± 0.85	60.65 ± 0.49	60.05 ± 0.07	60.60 ± 0.14
TAC *** (mg/L) [e]	645.25 ± 11.27 a	691.81 ± 9.18 b	633.87 ± 1.61 a	683.28 ± 11.36 b	636.14 ± 1.61 a	691.80 ± 10.68 b

Values represent mean with standard deviation. Different letters within the same row indicate significant differences ($p \leq 0.05$). * Color intensity. ** Total polyphenol index. *** Total anthocyanin content. A.U.: absorbance units. [a]: expressed as tartaric acid. [b]: expressed as acetic acid. [c]: expressed as the mean ± the deviation of the analytical method. [d]: expressed as tartaric acid. [e]: expressed as malvidin-3-glucoside.

Table 5. Oenological parameters after 4 months of storage of red wine after malolactic fermentation (MLF) treated by PEF and added with different SO_2 concentrations (0, 20 or 30 ppm). PEF3 (15 kV/cm, 77.8 kJ/kg, exit temperature: 40 °C) and PEF4 (15 kV/cm, 116.7 kJ/kg, exit temperature: 50 °C).

	Control			PEF$_3$		PEF$_4$	
	SO_2	SO_2	SO_2	SO_2	SO_2	SO_2	SO_2
	0 ppm	20 ppm	30 ppm	0 ppm	30 ppm	0 ppm	20 ppm
pH	3.54 ± 0.01	3.54 ± 0.01	3.54 ± 0.01	3.53 ± 0.00	3.52 ± 0.01	3.51 ± 0.01	3.52 ± 0.01
Glucose-Fructose (g/L)	0.27 ± 0.02	0.29 ± 0.02	0.30 ± 0.02	0.27 ± 0.03	0.30 ± 0.02	0.27 ± 0.02	0.29 ± 0.02
% Ethanol (*v/v*)	14.39 ± 0.05	14.35 ± 0.03	14.33 ± 0.04	14.37 ± 0.01	14.30 ± 0.04	14.32 ± 0.02	14.29 ± 0.04
Total Acidity (g/L) [a]	5.37 ± 0.14	5.15 ± 0.12	5.07 ± 0.14	5.22 ± 0.09	5.07 ± 0.07	5.15 ± 0.05	5.15 ± 0.00
Volatile Acidity (g/L) [b]	0.53 ± 0.02	0.44 ± 0.02	0.42 ± 0.04	0.48 ± 0.04	0.42 ± 0.02	0.48 ± 0.05	0.42 ± 0.02
Free SO_2 (ppm) [c]	6.4 ± 3.2	12.8 ± 3.2	12.8 ± 3.2	9.6 ± 3.2	12.8 ± 3.2	6.4 ± 3.2	12.8 ± 3.2
Total SO_2 (ppm) [c]	16 ± 3.2 a	22.4 ± 3.2 ab	25.6 ± 3.2 b	19.2 ± 3.2 ab	25.6 ± 3.2 b	16 ± 3.2 a	22.4 ± 3.2 ab
CI * (A.U.)	12.80 ± 0.12	12.16 ± 0.03	12.35 ± 0.18	13.51 ± 1.19	13.24 ± 0.28	13.40 ± 0.47	13.87 ± 0.22
TPI ** (A.U.) [d]	50.07 ± 2.11	50.25 ± 1.07	49.92 ± 0.80	50.89 ± 3.84	50.00 ± 0.40	49.63 ± 1.43	49.08 ± 0.03
TAC *** (mg/L) [e]	327.69 ± 10.41	345.12 ± 19.97	345.46 ± 11.41	309.72 ± 11.88	306.33 ± 22.89	314.95 ± 7.76	323.52 ± 8.16

Values represent mean with standard deviation. Different letters within the same row indicate significant differences ($p \leq 0.05$). * Color intensity. ** Total polyphenol index. *** Total anthocyanin Content. A.U.: absorbance units. [a]: expressed as tartaric acid. [b]: expressed as acetic acid. [c]: expressed as the mean ± the deviation of the analytical method. [d]: expressed as tartaric acid. [e]: expressed as malvidin-3-glucoside.

The results in Tables 4 and 5 show no significant differences between the PEF-treated wines and the untreated wines from a practical point of view in terms of pH, glucose-fructose, % ethanol, and total and volatile acidity. Neither did PEF treatments significantly affect indexes related to polyphenol content, such as color index, total polyphenol index, and total anthocyanin content. These results support previous studies which reported that PEF treatments applied at moderate intensities did not impair wine properties [20,46].

3.5. Evaluation of the Effect of PEF Versus the Sterilizing Filtration Method on the Oenological and Sensory Properties of a Commercial Red Wine

Red wine is a valuable product in which, in addition to physicochemical properties, it is especially vital to maintain sensory characteristics. Color, flavor, and aroma characteristics are essential in wine quality. In order to prevent undesirable effects on the sensory properties of wine, thermal treatments have traditionally been avoided in the wine industry for microbial decontamination in comparison with other food industries. Consequently, before bottling, wine frequently undergoes a sterilizing filtration treatment to avoid microbial spoilage during its subsequent shelf life.

As pointed out above, when a liquid food is processed by PEF in a continuous flow, a temperature increment in the food occurs because all of the electrical energy required to generate the electric field is transformed into heat. In order to ascertain whether that temperature increment affected the oenological and sensory properties of wine, we compared a commercial wine ready for bottling after sterilizing filtration with the same wine treated by PEF at two different intensities that corresponded to outlet temperatures of 40 and 60 °C. The oenological parameters of wines either untreated, PEF-treated, or sterilized by filtration after 1 month of storage are presented in Table 6. In comparison to the untreated wine, PEF treatments did not affect oenological parameters, even in the case of the more intense treatment applied (155.6 kJ/kg; 60 °C). In contrast, the only difference found in Table 6 corresponds with the color index (CI) of the wine treated by means of sterilizing filtration, which was significantly lower ($p \leq 0.05$). This finding confirms that microbial sterilization by means of a filtration process is a harsh procedure that affects wine color [47,48].

Table 6. Oenological parameters after 1 month of storage of untreated, PEF-treated, or sterilizing filtrated red wine. PEF treatments: PEF_A (15 kV/cm, 195 µs, 84.5 kJ/kg exit temperature: 40 °C) or PEF_B (15 kV/cm, 310 µs, 155.6 kJ/kg, exit temperature: 60 °C).

	Untreated	PEF_A 84.5 kJ/kg; 40 °C	PEF_B 155.6 kJ/kg; 60 °C	Sterilizing Filtration
pH	3.54 ± 0.05	3.53 ± 0.02	3.53 ± 0.04	3.53 ± 0.02
% Ethanol (*v/v*)	13.73 ± 0.12	13.75 ± 0.20	13.76 ± 0.14	13.70 ± 0.12
Total Acidity (g/L) [a]	4.85 ± 0.19	4.70 ± 0.22	4.70 ± 0.18	4.80 ± 0.22
Volatile Acidity (g/L) [b]	0.49 ± 0.02	0.49 ± 0.01	0.49 ± 0.01	0.49 ± 0.02
Malic Acid (g/L)	0.07 ± 0.01	0.08 ± 0.01	0.08 ± 0.02	0.07 ± 0.03
Free SO2 (ppm) [c]	32.0 ± 3.2	32.0 ± 4.2	35.2 ± 5.3	36.0 ± 3.2
Total SO2 (ppm) [c]	80.0 ± 7.2	80.0 ± 3.2	80.0 ± 5.24	80.0 ± 3.1
CI * (A.U.)	11.6 ± 1.4 a	11.3 ± 0.2 a	12.9 ± 0.8 a	8.0 ± 1.2 b
TPI ** (A.U.) [d]	59.3 ± 1.4	59.8 ± 1.2	60.2 ± 1.0	57.9 ± 1.2
TAC *** (mg/L) [e]	291.4 ± 5.1	282.3 ± 10.1	282.9 ± 12.4	284.8 ± 14.1

Values represent mean with standard deviation. Different letters within the same row indicate significant differences ($p \leq 0.05$). * Color Intensity. ** Total Polyphenol Index. *** Total Anthocyanin Content. A.U.: absorbance units. [a]: expressed as tartaric acid. [b]: expressed as acetic acid. [c]: expressed as the mean ± the deviation of the analytical method. [d]: expressed as tartaric acid. [e]: expressed as malvidin-3-glucoside.

Regarding the effect of PEF on sensory properties of wine, Table 7 shows the results of the triangle test comparing the aroma and taste of wines sterilized by filtration and treated by PEF at two intensities. Results shown in Table 7 indicate that panelists were not able to detect differences in aroma and taste between wine after sterilizing filtration and the PEF-treated wines, even those treated at the highest intensity. More than 56%

of correct responses are required for differences to be considered statistically significant ($p < 0.05$) [24].

Table 7. Percentage of correct responses in a triangle test comparing untreated, PEF-treated, and sterilizing filtrated red wine after 1 month of storage. PEF treatments: PEF_A (15 kV/cm, 195 μs, 84.5 kJ/kg exit temperature: 40 °C) or PEF_B (15 kV/cm, 310 μs, 155.6 kJ/kg, exit temperature: 60 °C).

Triangle Test (Percentage of Correct Responses)	
PEF_A/PEF_B	50.0%
PEF_A/Sterilizing Filtration	22.2%
PEF_B/PEF_B	33.3%
PEF_B/Sterilizing Filtration	38.9%
Sterilizing Filtration/PEF_A	50.0%
Sterilizing Filtration/PEF_B	44.4%

Therefore, results obtained in the sensory study demonstrated that the rapid, brief temperature increment that occurred in a red wine as a consequence of the application of a PEF treatment did not negatively impair its oenological and sensory properties, even when the wine's outlet temperature was 60 °C. This result is particularly relevant and underscores PEF's potential for implementation as a microbial control method in the wine industry.

4. Conclusions

This study reinforces the growing evidence that PEF is a technology that can be potentially successful in helping to reduce the number of sulfites used in wineries, not only due to its effectivity in inactivating different microorganisms, but also due to its zero detrimental effect on wine quality.

Author Contributions: Conception and design of the study: C.D., I.Á. and J.R. Acquisition of data: C.D., A.B. and J.S. Analysis and interpretation of the data: C.D., I.Á. and J.R. Drafting of the article C.D., A.B. and J.S. Critical revision of the article I.Á. and J.R. All authors have read and agreed to the published version of the manuscript.

Funding: This research received no external funding.

Institutional Review Board Statement: Not applicable.

Informed Consent Statement: Informed consent was obtained from all subjects involved in the study.

Data Availability Statement: Data is contained within the article.

Acknowledgments: C. Delso gratefully acknowledges the financial support for her doctoral studies provided by the Department of Innovation, Investigation and the University of the Government of Aragón.

Conflicts of Interest: The authors declare no conflict of interest.

References

1. Vaudano, E.; Costantini, A.; Garcia-Moruno, E. Microbial Ecology of Wine. In *Quantitative Microbiology in Food Processing*; John Wiley & Sons, Ltd.: Chichester, UK, 2016; pp. 547–559, ISBN 9781118823071.
2. Jeon, S.H.; Kim, N.H.; Shim, M.B.; Jeon, Y.W.; Ahn, J.H.; Lee, S.H.; Hwang, I.G.; Rhee, M.S. Microbiological Diversity and Prevalence of Spoilage and Pathogenic Bacteria in Commercial Fermented Alcoholic Beverages (Beer, Fruit Wine, Refined Rice Wine, and Yakju). *J. Food Prot.* **2015**, *78*, 812–818. [CrossRef] [PubMed]
3. Campaniello, D.; Sinigaglia, M. Wine Spoiling Phenomena. In *The Microbiological Quality of Food*; Elsevier: Amsterdam, The Netherlands, 2017; pp. 237–255, ISBN 9780081005033.
4. Tedesco, F.; Siesto, G.; Pietrafesa, R.; Romano, P.; Salvia, R.; Scieuzo, C.; Falabella, P.; Capece, A. Chemical Methods for Microbiological Control of Winemaking: An Overview of Current and Future Applications. *Beverages* **2022**, *8*, 58. [CrossRef]
5. Ribéreau-Gayon, P.; Dubourdie, D.; Donèche, B.; Lovaud, A.A.; Darriet, P.; Towey, J. The Use of Sulfur Dioxide in Must and Wine Treatment. In *Handbook of Enolgy*; John Wiley Sons, Ltd.: Hoboken, NJ, USA, 2006; ISBN 0-470-01034-7.

6. Giacosa, S.; Río Segade, S.; Cagnasso, E.; Caudana, A.; Rolle, L.; Gerbi, V. SO2 in Wines. In *Red Wine Technology*; Elsevier: Amsterdam, The Netherlands, 2019; pp. 309–321, ISBN 9780128144008.

7. Lisanti, M.T.; Blaiotta, G.; Nioi, C.; Moio, L. Alternative Methods to SO_2 for Microbiological Stabilization of Wine. *Compr. Rev. Food Sci. Food Saf.* **2019**, *18*, 455–479. [CrossRef]

8. Girard, B.; Fukumoto, L.R.; Koseoglu, S.S. Membrane Processing of Fruit Juices and Beverages: A Review. *Crit. Rev. Biotechnol.* **2000**, *20*, 109–175. [CrossRef] [PubMed]

9. Ruiz-Rico, M.; García-Ríos, E.; Barat, J.M.; Guillamón, J.M. Microbial stabilisation of white wine by filtration through silica microparticles functionalised with natural antimicrobials. *LWT* **2021**, *149*, 111783. [CrossRef]

10. Timmermans, R.; Mastwijk, H.; Berendsen, L.; Nederhoff, A.; Matser, A.; Van Boekel, M.; Groot, M.N. Moderate intensity Pulsed Electric Fields (PEF) as alternative mild preservation technology for fruit juice. *Int. J. Food Microbiol.* **2019**, *298*, 63–73. [CrossRef]

11. Raso, J.; Álvarez, I. Pulsed Electric Field Processing: Cold Pasteurization. In *Reference Module in Food Science*; Elsevier: Amsterdam, The Netherlands, 2016; pp. 1–8.

12. Toepfl, S.; Siemer, C.; Saldaña-Navarro, G.; Heinz, V. Overview of Pulsed Electric Fields Processing for Food. In *Emerging Technologies for Food Processing*; Elsevier: Amsterdam, The Netherlands, 2014; pp. 93–114.

13. Timmermans, R.; Nierop Groot, M.; Matser, A. Liquid Food Pasteurization by Pulsed Electric Fields. In *Pulsed Electric Fields Technology for the Food Industry*; Raso, J., Heinz, V., Alvarez, I., Toepfl, S., Eds.; Springer International Publishing: Cham, Switzerland, 2022; pp. 299–323, ISBN 978-3-030-70586-2.

14. Agcam, E.; Akyildiz, A.; Evrendilek, G.A. A comparative assessment of long-term storage stability and quality attributes of orange juice in response to pulsed electric fields and heat treatments. *Food Bioprod. Process.* **2016**, *99*, 90–98. [CrossRef]

15. Katiyo, W.; Yang, R.; Zhao, W. Effects of combined pulsed electric fields and mild temperature pasteurization on microbial inactivation and physicochemical properties of cloudy red apple juice (*Malus pumila* Niedzwetzkyana (Dieck)). *J. Food Saf.* **2017**, *37*, e12369. [CrossRef]

16. Marsellés-Fontanet, R.; Puig, A.; Olmos, P.; Mínguez-Sanz, S.; Martín-Belloso, O. Optimising the inactivation of grape juice spoilage organisms by pulse electric fields. *Int. J. Food Microbiol.* **2009**, *130*, 159–165. [CrossRef]

17. Mosqueda-Melgar, J.; Raybaudi-Massilia, R.M.; Martín-Belloso, O. Non-thermal pasteurization of fruit juices by combining high-intensity pulsed electric fields with natural antimicrobials. *Innov. Food Sci. Emerg. Technol.* **2008**, *9*, 328–340. [CrossRef]

18. Puértolas, E.; López, N.; Condón, S.; Raso, J.; Álvarez, I. Pulsed electric fields inactivation of wine spoilage yeast and bacteria. *Int. J. Food Microbiol.* **2009**, *130*, 49–55. [CrossRef] [PubMed]

19. González-Arenzana, L.; Portu, J.; López, R.; López, N.; Santamaría, P.; Garde-Cerdán, T.; López-Alfaro, I. Inactivation of wine-associated microbiota by continuous pulsed electric field treatments. *Innov. Food Sci. Emerg. Technol.* **2015**, *29*, 187–192. [CrossRef]

20. Delsart, C.; Grimi, N.; Boussetta, N.; Sertier, C.M.; Ghidossi, R.; Vorobiev, E.; Peuchot, M.M. Impact of pulsed-electric field and high-voltage electrical discharges on red wine microbial stabilization and quality characteristics. *J. Appl. Microbiol.* **2015**, *120*, 152–164. [CrossRef] [PubMed]

21. González-Arenzana, L.; Portu, J.; López, N.; Santamaría, P.; Gutiérrez, A.R.; López, R.; López-Alfaro, I. Pulsed Electric Field treatment after malolactic fermentation of Tempranillo Rioja wines: Influence on microbial, physicochemical and sensorial quality. *Innov. Food Sci. Emerg. Technol.* **2019**, *51*, 57–63. [CrossRef]

22. Maza, M.A.; Martínez, J.M.; Hernández-Orte, P.; Cebrián, G.; Sánchez-Gimeno, A.C.; Álvarez, I.; Raso, J. Influence of pulsed electric fields on aroma and polyphenolic compounds of Garnacha wine. *Food Bioprod. Process.* **2019**, *116*, 249–257. [CrossRef]

23. *ISO NORM 3591*; Sensory Analysis—Apparatus—Wine-Tasting Glass. International Organization for Standarization (ISO): Geneva, Switzerland, 1977.

24. Roessler, E.B.; Warren, J.; Guymon, J.F. Significance in Triangular Taste Tests. *J. Food Sci.* **1948**, *13*, 503–505. [CrossRef]

25. Coster, H.G.L.; Zimmermann, U. The mechanism of electrical breakdown in the membranes of Valonia utricularis. *J. Membr. Biol.* **1975**, *22*, 73–90. [CrossRef]

26. Saldaña, G.; Monfort, S.; Condón, S.; Raso, J.; Álvarez, I. Effect of temperature, pH and presence of nisin on inactivation of *Salmonella* Typhimurium and *Escherichia coli* O157:H7 by pulsed electric fields. *Food Res. Int.* **2012**, *45*, 1080–1086. [CrossRef]

27. Huang, K.; Yu, L.; Wang, W.; Gai, L.; Wang, J. Comparing the pulsed electric field resistance of the microorganisms in grape juice: Application of the Weibull model. *Food Control* **2014**, *35*, 241–251. [CrossRef]

28. Heinz, V.; Alvarez, I.; Angersbach, A.; Knorr, D. Preservation of liquid foods by high intensity pulsed electric fields—basic concepts for process design. *Trends Food Sci. Technol.* **2001**, *12*, 103–111. [CrossRef]

29. Álvarez, I. The influence of process parameters for the inactivation of *Listeria monocytogenes* by pulsed electric fields. *Int. J. Food Microbiol.* **2003**, *87*, 87–95. [CrossRef] [PubMed]

30. Hülsheger, H.; Potel, J.; Niemann, E.G. Killing of bacteria with electric pulses of high electric field strength. *Radiat. Environ. Biophys.* **1981**, *20*, 53–65. [CrossRef] [PubMed]

31. Sale, A.; Hamilton, W. Effects of high electric fields on microorganisms: I. Killing of bacteria and yeasts. *Biochim. Biophys. Acta BBA Gen. Subj.* **1967**, *148*, 781–788. [CrossRef]

32. Wouters, P.C.; Alvarez, I.; Raso, J. Critical factors determining inactivation kinetics by pulsed electric field food processing. *Trends Food Sci. Technol.* **2001**, *12*, 112–121. [CrossRef]

33. Mannazzu, I.; Angelozzi, D.; Belviso, S.; Budroni, M.; Farris, G.A.; Goffrini, P.; Lodi, T.; Marzona, M.; Bardi, L. Behaviour of *Saccharomyces cerevisiae* wine strains during adaptation to unfavourable conditions of fermentation on synthetic medium: Cell lipid composition, membrane integrity, viability and fermentative activity. *Int. J. Food Microbiol.* **2008**, *121*, 84–91. [CrossRef]
34. Aguilera, F.; Peinado, R.; Millán, C.; Ortega, J.; Mauricio, J. Relationship between ethanol tolerance, H+-ATPase activity and the lipid composition of the plasma membrane in different wine yeast strains. *Int. J. Food Microbiol.* **2006**, *110*, 34–42. [CrossRef]
35. González-Arenzana, L.; López-Alfaro, I.; Garde-Cerdán, T.; Portu, J.; López, R.; Santamaría, P. Microbial inactivation and MLF performances of Tempranillo Rioja wines treated with PEF after alcoholic fermentation. *Int. J. Food Microbiol.* **2018**, *269*, 19–26. [CrossRef]
36. García, D.; Gómez, N.; Raso, J.; Pagán, R. Bacterial resistance after pulsed electric fields depending on the treatment medium pH. *Innov. Food Sci. Emerg. Technol.* **2005**, *6*, 388–395. [CrossRef]
37. Saulis, G. Electroporation of Cell Membranes: The Fundamental Effects of Pulsed Electric Fields in Food Processing. *Food Eng. Rev.* **2010**, *2*, 52–73. [CrossRef]
38. Delso, C.; Martínez, J.M.; Cebrián, G.; Álvarez, I.; Raso, J. Understanding the occurrence of tailing in survival curves of *Salmonella* Typhimurium treated by pulsed electric fields. *Bioelectrochemistry* **2020**, *135*, 107580. [CrossRef]
39. Novickij, V.; Švediené, J.; Paškevičius, A.; Markovskaja, S.; Girkontaitė, I.; Zinkevičienė, A.; Lastauskienė, E.; Novickij, J. Pulsed electric field-assisted sensitization of multidrug-resistant *Candida albicans* to antifungal drugs. *Futur. Microbiol.* **2018**, *13*, 535–546. [CrossRef]
40. Somolinos, M.; García, D.; Condón, S.; Mañas, P.; Pagán, R. Relationship between Sublethal Injury and Inactivation of Yeast Cells by the Combination of Sorbic Acid and Pulsed Electric Fields. *Appl. Environ. Microbiol.* **2007**, *73*, 3814–3821. [CrossRef]
41. Smith, K.; Mittal, G.; Griffiths, M. Pasteurization of Milk Using Pulsed Electrical Field and Antimicrobials. *J. Food Sci.* **2002**, *67*, 2304–2308. [CrossRef]
42. Wu, Y.; Mittal, G.; Griffiths, M. Effect of Pulsed Electric Field on the Inactivation of Microorganisms in Grape Juices with and without Antimicrobials. *Biosyst. Eng.* **2005**, *90*, 1–7. [CrossRef]
43. García, D.; Gómez, N.; Mañas, P.; Condón, S.; Raso, J.; Pagán, R. Occurrence of sublethal injury after pulsed electric fields depending on the microorganism, the treatment medium pH and the intensity of the treatment investigated. *J. Appl. Microbiol.* **2005**, *99*, 94–104. [CrossRef]
44. García, D.; Gómez, N.; Mañas, P.; Raso, J.; Pagán, R. Pulsed electric fields cause bacterial envelopes permeabilization depending on the treatment intensity, the treatment medium pH and the microorganism investigated. *Int. J. Food Microbiol.* **2007**, *113*, 219–227. [CrossRef]
45. Loureiro, V.; Malfeito-Ferreira, M. Spoilage yeasts in the wine industry. *Int. J. Food Microbiol.* **2003**, *86*, 23–50. [CrossRef]
46. Abca, E.E.; Evrendilek, G.A. Processing of Red Wine by Pulsed Electric Fields with Respect to Quality Parameters. *J. Food Process. Preserv.* **2014**, *39*, 758–767. [CrossRef]
47. Arriagada-Carrazana, J.; Sáez-Navarrete, C.; Bordeu, E. Membrane filtration effects on aromatic and phenolic quality of Cabernet Sauvignon wines. *J. Food Eng.* **2005**, *68*, 363–368. [CrossRef]
48. Buffon, P.; Heymann, H.; Block, D.E. Sensory and Chemical Effects of Cross-Flow Filtration on White and Red Wines. *Am. J. Enol. Vitic.* **2014**, *65*, 305–314. [CrossRef]

 *foods*

Article

Freeze-Dried Cooked Chickpeas: Considering a Suitable Alternative to Prepare Tasty Reconstituted Dishes

M. Isabel Cambero, Gonzalo Doroteo García de Fernando, M. Dolores Romero de Ávila, Víctor Remiro, Luis Capelo and José Segura *

Department of Food Technology, Faculty of Veterinary Medicine, Complutense University of Madrid, Av. Puerta de Hierro s/n., 28040 Madrid, Spain; icambero@ucm.es (M.I.C.); mingui@ucm.es (G.D.G.d.F.); mdavilah@ucm.es (M.D.R.d.Á.); vremiro@ucm.es (V.R.)
* Correspondence: josesegu@ucm.es; Tel.: +34-913-943-746

Abstract: The current trend in food consumption is toward convenience, i.e., fast food. The present work aims to study the potential of incorporating freeze-dried cooked chickpeas into a complex and traditional dish in Spanish gastronomy, such as *Cocido*, which has this legume as the main ingredient. *Cocido* is a two-course meal: a thin-noodle soup and a mix of chickpeas, several vegetables, and meat portions. The textural properties, sensory qualities, and rehydration kinetics of chickpeas of three Spanish varieties were investigated to select the most suitable cooking conditions to obtain freeze-dried chickpeas of easy rehydration whilst maintaining an adequate sensory quality for the preparation of the traditional dish. The sensory quality of various vegetables and meat portions, cooked under different conditions, was evaluated after freeze-drying and rehydration. It was possible to reproduce the sensory quality of the traditional dish after rehydration with water, heating to boiling in a microwave oven for 5 min, and resting for 10 min. Therefore, it is possible to commercialize complex dishes based on pulses and other cooked and freeze-dried ingredients as reconstituted meals with a wide nutrient profile. Nevertheless, additional research is required on the shelf life, together with other economic and marketing issues such as design of a proper packaging, that would allow consumption as a two-course meal.

Keywords: chickpea; freeze-drying; rehydration; instant meal; *Cocido*; *easy-to-prepare*; reconstituted meal

Citation: Cambero, M.I.; García de Fernando, G.D.; Romero de Ávila, M.D.; Remiro, V.; Capelo, L.; Segura, J. Freeze-Dried Cooked Chickpeas: Considering a Suitable Alternative to Prepare Tasty Reconstituted Dishes. *Foods* **2023**, *12*, 2339. https://doi.org/10.3390/foods12122339

Academic Editors: Javier Carballo and Sidonia Martinez

Received: 20 May 2023
Revised: 7 June 2023
Accepted: 8 June 2023
Published: 10 June 2023

1. Introduction

Today, consumer lifestyles require foods that can be prepared quickly. More than ever, healthy, nutritional, and appetizing "instant/reconstituted meals", meaning convenience meals, are sought out by consumers. Instant or reconstituted meals ("*easy-to-prepare*") are characterized by minimal processing before eating. Mostly, these products are constituted by a dehydrated base (soup, vegetables, pasta, etc.) and consumed in the container itself after rehydration with boiling water [1].

In the current market, reconstituted meals based on instant noodles are widespread. For decades, dried noodles and instant fried noodles have been the most consumed products due to their ease of preservation and cooking [2]. Nevertheless, the offer of a reconstituted meal based on a complex dish is almost non-existent. Research is being carried out to increase both the quality and the variety of such products [2–4]. Chen et al. [5] observed that the consumption of dry solid ingredients in instant soups in the modern era is increasing rapidly. Therefore, the authors reviewed the recent changes in the quality of the solid ingredients of instant soups. Pieniazek and Messina [6] analyzed the microstructure and texture of legumes and other vegetables for instant meals using scanning electron microscopy combined with an image processing technique. Anecdotally, even America's space agency, NASA (National Aeronautics and Space Administration) is interested in the preservation of vitamins, a low weight/volume ratio in packaging, and the retention of the sensory quality of the instant/reconstituted meals [7].

In this line of work, freeze-drying is one of the ideal drying methods for the preservation of foods that are susceptible to alterations due to heat and/or oxidation. Freeze-dried products have low moisture content and usually a porous structure and can be reconstituted with a high rehydration speed [8,9]. The freeze-drying method allows food to preserve its original color, flavor, aroma, and appearance to the maximum degree possible while protecting its composition [10]. Nutritional value is not altered since protein denaturation and vitamin loss do not occur [11].

Chickpeas are the most consumed leguminous crop in many countries around the world, including Spain. They are characterized by having a low amount of lipids and a significant amount of all the essential amino acids, except sulfur-containing ones, followed by tryptophan and threonine [12,13].

Nevertheless, from a culinary point of view, they are usually combined with other ingredients due to their mild flavor [14].

The chickpea is a legume that can be used together with other vegetables and different types of meat to fulfill the nutrients of a balanced diet [15]. An example of such a combination is a traditional Spanish dish known as *Cocido*, of which there are different varieties, such as *Cocido Madrileño*, *Cocido Extremeño*, *Cocido Andaluz*, *Cocido Maragato*, etc. All these dishes have chickpeas in common as the main ingredient. This pulse is cooked together with different types of vegetables and meat (beef, chicken, and pork). *Cocido Madrileño* is usually eaten in two or three courses: once the chickpeas, meat portions, and vegetables are cooked, the broth is separated and used to make a soup with thin noodles, which is served as the first course. The remaining ingredients are then made into the second course. Often, the chickpeas and vegetables are eaten first, followed by meat pieces [16,17].

This work studied the effect of cooking conditions on the rehydration kinetics of different varieties of cooked and freeze-dried chickpeas. In addition, the effect of cooking on the textural attributes and sensory characteristics of rehydrated chickpeas was studied. Conversely, the sensory attributes of different vegetables and portions of meat of different species, freeze-dried after cooking and subsequent rehydration, were analyzed. The final objective of this work was to determine the potential use of chickpeas, vegetables and meat portions, freeze-dried after cooking, for the assembly of a complex dish such as *Cocido Madrileño* for its consumption as an *easy-to-prepare* dish, which only requires rehydration and heating for its consumption. The aim is to offer an alternative to bring traditional gastronomy closer to the preparation of *easy-to-prepare* food.

2. Materials and Methods

2.1. Ingredients (Raw Products)

Regarding chickpea (*Cicer arietinum* L.), the three most known varieties in Spain were considered: *Pedrosillano* (PD, ecotype guarantee mark from *Pedrosillo el Ralo*, in the region of *La Armuña*, Salamanca, Spain), *Castellano* (CA, *La Moraña*, Ávila, Spain), and *Blanco Lechoso* (BL, *Marchena*, Sevilla, Spain). In agreement with the literature, PD showed a caliber range of 320–360 beans/100 g, whereas CA and BL showed 190–210 beans/100 g [18–21].

The procedure followed in this study is summarized in Figure 1. According to traditional *Cocido* recipes [16,17], the following ingredients, together with chickpeas, were used (Figure 2): beef shank and flank, chicken leg-quarters, cow leg-bone, pork belly, pig dry-cured ham bone, *chorizo* (spicy pork sausage), *morcilla* (pork blood sausage), turnip (*Brassica napus*), potato (*Solanum tuberosum*), carrot (*Daucus carota sativus*), leek (*Allium ampeloprasum var. porrum*), and cabbage (*Brassica oleracea*). Meat products were obtained from a local butcher shop and vegetables from a local greengrocer (Madrid, Spain).

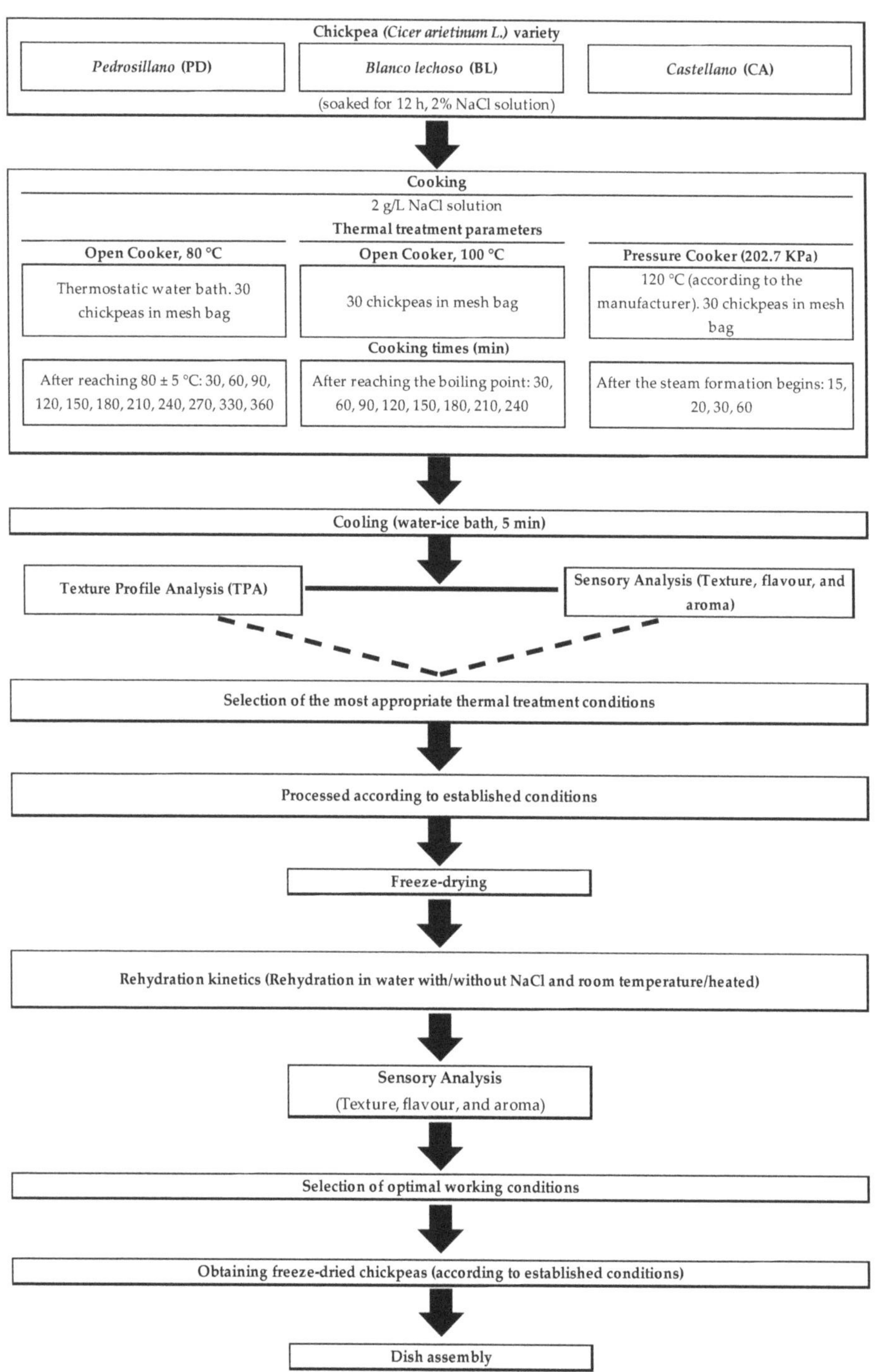

Figure 1. Process followed for the selection of the preparation conditions of different Spanish varieties of chickpeas (*Cicer arietinum*).

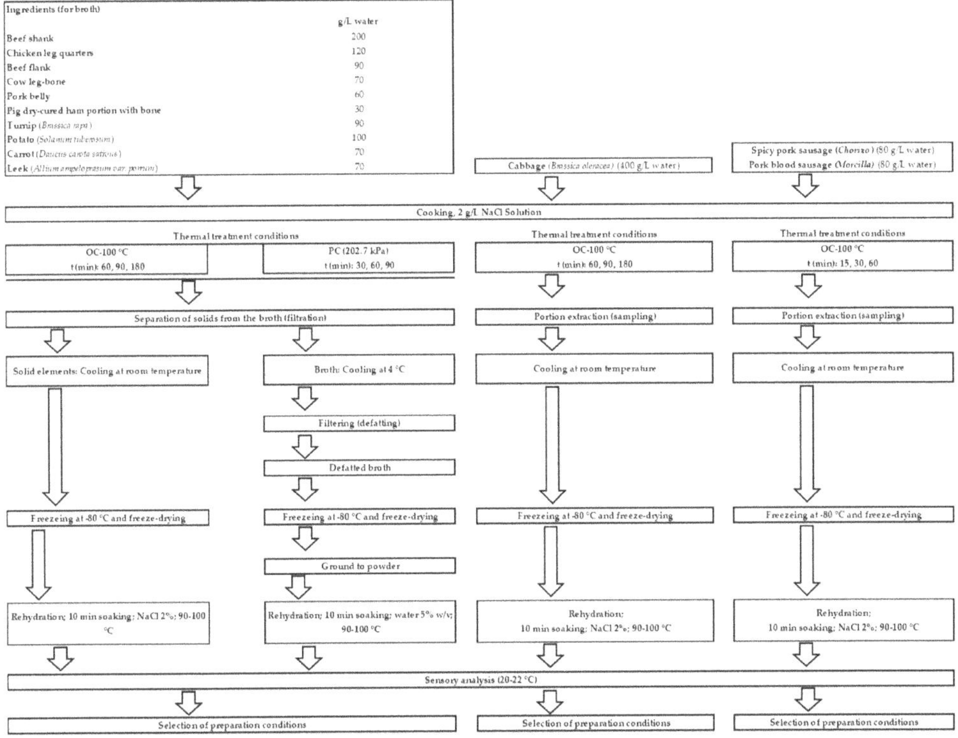

Figure 2. Process followed for the selection of the preparation conditions of different ingredients (meat and vegetables) of the dish. OC = open cooker; PC = pressure cooker.

2.2. Cooking Conditions

2.2.1. Chickpea Processing

The chickpeas were soaked in a 2% NaCl aqueous solution for 12 h and rinsed with fresh water before the thermal treatment. Afterward, mesh bags with 30 chickpeas were prepared to be extracted from the cooker at each evaluated temperature and time: Open cooker (OC, atmospheric pressure): temperature of 80 °C with times of 30 (OC-30), 60 (OC-60), 90 (OC-90), 120 (OC-120), 150 (OC-150), 180 (OC-180), 210 (OC-210), and 240 (OC-240) min; and temperature of 100 °C with the same times. Pressure cooker (PC, 202.7 KPa, 120 °C, according to manufacturer) with times of 15 (PC-15), 20 (PC-20), 30 (PC-30) and 60 (PC-60) min. In all cases, after cooking, the mesh bags were submerged for 5 min in an ice-water solution to stop the cooking process. Each group was split into three portions: two were prepared for textural and sensorial analyses, and the third was frozen and freeze-dried (see below).

2.2.2. Processing of Meat and Vegetables as well as Broth Preparation

Both OC and PC processed/cooked, with 1 L of distilled water with 2 g NaCl, about 200 g beef shank, 120 g chicken leg-quarters, 90 g beef flank, 70 g cow leg-bone, 60 g pork belly, 30 g pig dry-cured ham portion with bone, 90 g turnip (*Brassica napus*), 100 g potato (*Solanum tuberosum*), 70 g carrot (*Daucus carota sativus*) and 70 g leek (*Allium ampeloprasum var. porrum*). After the designated cooking times, the ingredients were quickly extracted, separated, and cooled to room temperature (20–22 °C) and then packed in plastic bags to be frozen for freeze-drying. The broth was cooled to 4 °C and filtered (1 mm mesh) to separate the solidified fat. The defatted broth was also packed in plastic bags and frozen for subsequent freeze-drying (Figure 2).

In addition, cabbage (*Brassica oleracea*) was cooked alone (400 g/L of distilled water with 2 g of NaCl, for 60, 90, and 180 min). Spicy pork sausage (*Chorizo*) and pork blood sausage (*Morcilla*) were cooked together but separately from the other ingredients (80 g of each meat product/L distilled water, no NaCl, for 15, 30, and 60 min) (Figure 2). After the cooking times, these products were extracted, and they were cooled and frozen using the same method as indicated above. In these cases, the cooking water was discarded. Ten repetitions were carried out for each cooking condition, processing the amounts corresponding to 4 L of distilled water in each of them.

2.3. Texture Profile Analysis of Chickpeas

Texture profile analysis (TPA) was carried out using a TA.XT2i SMS Stable Micro Systems Texture Analyzer (Stable Microsystems Ltd., Surrey, England) with the Texture Expert program. Textural tests were carried out at 20–22 °C. A double compression cycle test was performed up to 50% compression of the original portion height with an aluminum cylinder probe of 2 cm diameter (5 s were allowed to elapse between two compression cycles). Force–time deformation curves were obtained with a 50-kg load cell applied at a crosshead speed of 2 mm/s. Hardness (the maximum force required to compress the sample, N), springiness (the ability of the sample to recover its original form after deforming force was removed, m), adhesiveness (the area under the abscissa after the first compression, $N\times s$) and cohesiveness (the extent to which the sample could be deformed before rupture) were quantified according to Segura et al. [22].

2.4. Freeze-Drying and Rehydration Studies

Frozen samples (chickpeas, broth, and the other vegetable and meat ingredients) at −80 °C were freeze-dried under vacuum in a Lyoquest Lyophilizer (Telstar, Terrasa, Spain) with a chamber pressure of 0.03 bar, −70 °C condenser temperature, and 20 °C shelf temperature. During sublimation drying, the temperature was kept below eutectic. In the desorption drying, the temperature was kept below 24 °C.

Freeze-dried samples were placed inside a glass container and vacuum-packed (20 kPa) in plastic bags (10 cm^2; laminated film: polyamide and polyethylene; thickness of 90 μm; low gas permeability: O_2 and CO_2 transmission rates of 35 cm^3/24 h m^2 bar) to avoid any damage to the structure of the solid ingredients, and stored at −20 °C.

Chickpeas were rehydrated by soaking in two different media: distilled water and 2% (*w/v*) NaCl solution. In both cases, the studies were carried out at room (20–22 °C) and close to boiling (90–100 °C) temperatures. The chickpea rehydration kinetics were developed by weighing them every 2 min for 25 min (the time at which no statistically significant differences ($p > 0.05$) were detected between consecutive measurements). At specific times, 10 chickpeas were extracted and blotted free of excess surface moisture by being placed separately on a stainless-steel mesh (1 mm hole diameter) with a Reshma paper filter at room temperature for 3 min and then weighed. For rehydration of the broth, the substrate obtained after freeze-drying was reduced to a powder using a mortar. This operation was performed immediately before rehydration using distilled hot water (90–100 °C). The ratio of powder to water was 2.5% (*w/v*). Rehydration of the other cooked ingredients was performed with 2% (*w/v*) NaCl water solution (90–100 °C) or in the hot rehydrated broth.

2.5. Sensory Evaluation

Eight semi-trained panelists (four females and four males) with knowledge of the product were asked to carry out a rank order test [23,24] in a taste panel area equipped with individual booths according to the methodology guidelines for sensory analysis from the International Standards Organization 6658:2017 [25]. Panelists analyzed the chickpeas, the vegetables, and the meat ingredients separately, as well as the broth, after different cooking and rehydration procedures. The samples were maintained in a water bath at 35 ± 5 °C for no more than 1 h. For the sensory evaluation of these products, the panelists were instructed to rank samples by their liking/satisfaction level considering the texture, the taste/flavor, and the aroma, by taking as reference the expected characteristics in a whole *Cocido* dish or each of its ingredients individually. Each sample was given a different score; the sample with the least appropriate sensory characteristics was allocated a score of 1, and the sample with the most satisfactory sensory characteristics and close to the *Cocido* dish was assigned the highest score, which coincided with the number of samples tasted. To obtain the final score of each sample, the sum of ranks was calculated corresponding to the sum of scores of the sample sensory characteristics. This was estimated by calculating the sum of the products of the values given to each sample by the number of times each sample was allocated to a specific score: $(1 \times n_1) + (2 \times n_2) + (3 \times n_3) + \ldots + (g \times n_g)$, where n_1 = the frequency that the sample receives score 1 (worst sensory characteristics), n_2 = frequency of score 2, etc., and n_g = frequency of the highest score (g, best sensory characteristics).

Together with the ranking, panelists were asked to indicate the causes of each score and were offered the possibility to point out possible weaknesses or defects in a "descriptive" section.

The complete culinary preparation was tasted after rehydration using a descriptive test of each sensory attribute (flavor, aroma, texture) of the *Cocido Madrileño* dishes (soup, chickpeas, vegetables, and meats).

2.6. Statistical Analysis

The statistical analysis was carried out using Statgraphics Plus version 5.1. Data were brought forward as the mean $\pm$ the standard deviation (SD) of each treatment. After determining the goodness of fit of the data to a normal distribution (95% confidence) using the Shapiro–Wilks test, Duncan's test for multiple mean comparisons was applied to ascertain differences among means. Multifactor ANOVA was conducted to determine the significant effects produced by chickpea variety, cooking time and temperature associated with the culinary treatment (OC and PC), and their combinations, on the textural parameters of chickpeas. Analyses were carried out in triplicate. The significance level ($p < 0.05$) of data obtained in the rank order test was determined by Friedman's rank addition using the tables for multiple comparison procedures for the analysis of ranked data [26].

3. Results and Discussion

3.1. Texture Profile and Sensory Analyses of Cooked Chickpeas

The statistical analysis of hardness, adhesiveness, springiness, and cohesiveness of chickpeas, depending on the cooking temperature, time, and chickpea variety, together with the cooking procedure (OC and PC), is shown in Table 1. The multifactor ANOVA results indicated significant interactions ($p < 0.05$) between the effect of cooking conditions (temperature and time) and chickpea variety on hardness, adhesiveness, and springiness. It should be noted that, as customary, before cooking, a 12 h soaking in a NaCl solution was carried out. According to Fabbri et al. [27] and Wood [28], soaking with salt, discarding the water, rinsing, and boiling was the best procedure to remove anti-nutrients and increase iron absorption, reduce cooking time and gastric issues, and improve the protein quality, texture, and appearance of pulses.

For OC-30 at 80 °C, CA chickpeas showed the highest hardness value, followed by BL and PD. Still considering cooking at 80 °C, the highest decrease with time in harness values was detected for BL (89.5%), followed by CA (81.5%) and PD (65.9%). Consequently, at OC-240 at 80 °C, BL showed a lower value of hardness than CA and PD, but no difference was observed between CA and PD (3.64 N and 11.2 N, respectively). When cooking at 100 °C, no statistical differences in hardness values, which were lower than 1 N, were observed after 120 min (0.22 N; $p > 0.05$). When using PC, no statistical differences were detected in hardness (0.18 N; $p > 0.05$) among the chickpea varieties. Nevertheless, cooking time showed a statistical tendency to lower hardness values when the cooking time was longer ($p = 0.095$; 0.19 N vs. 0.17 N; 10.7% difference).

Table 1. Texture profile analysis of chickpeas cooked under different systems and times.

Cooking System	Temperature (°C)	Time (min)	Hardness (N) BL	Hardness (N) CA	Hardness (N) PD	Adhesiveness (N×s, ×10¹) BL	Adhesiveness (N×s, ×10¹) CA	Adhesiveness (N×s, ×10¹) PD	Springiness (m, ×10²) BL	Springiness (m, ×10²) CA	Springiness (m, ×10²) PD	Cohesiveness BL	Cohesiveness CA	Cohesiveness PD
OC	80	30	34.8 ± 6.76 [a,β]	65.6 ± 9.01 [a,α]	29.8 ± 2.92 [a,γ]	−0.04 ± 0.027 [a]	−0.03 ± 0.007 [a]	−0.02 ± 0.021 [a]	0.20 ± 0.020 [b,α]	0.25 ± 0.050 [c,α]	0.11 ± 0.039 [c,β]	0.15 ± 0.041 [ab]	0.11 ± 0.008 [ab]	0.14 ± 0.021 [ab]
		60	21.6 ± 2.50 [b,β]	42.1 ± 5.57 [b,α]	21.9 ± 3.15 [b,β]	−0.05 ± 0.019 [a]	−0.03 ± 0.015 [a]	−0.03 ± 0.023 [a]	0.21 ± 0.031 [b,β]	0.28 ± 0.011 [b,α]	0.13 ± 0.058 [c,γ]	0.23 ± 0.039 [a]	0.11 ± 0.009 [ab]	0.13 ± 0.020 [ab]
		90	19.8 ± 3.52 [b,α]	21.0 ± 9.00 [c,α]	13.8 ± 3.48 [c,β]	−0.09 ± 0.044 [a]	−0.06 ± 0.018 [a]	−0.04 ± 0.013 [a]	0.22 ± 0.039 [b,β]	0.30 ± 0.043 [b,α]	0.13 ± 0.039 [c,γ]	0.17 ± 0.049 [a]	0.17 ± 0.008 [a]	0.18 ± 0.011 [a]
		120	12.8 ± 4.31 [b,β]	17.2 ± 6.30 [c,α]	13.7 ± 6.43 [c,β]	−0.04 ± 0.031 [a]	−0.06 ± 0.017 [a]	−0.06 ± 0.030 [a]	0.27 ± 0.051 [ab,α]	0.33 ± 0.066 [b,α]	0.21 ± 0.038 [b,β]	0.18 ± 0.038 [a]	0.19 ± 0.014 [a]	0.19 ± 0.031 [a]
		150	10.5 ± 1.88 [b,β]	19.3 ± 7.14 [c,α]	8.90 ± 1.79 [c,β]	−0.03 ± 0.020 [a]	−0.02 ± 0.011 [a]	−0.03 ± 0.013 [a]	0.30 ± 0.033 [a,β]	0.41 ± 0.019 [a,α]	0.23 ± 0.078 [ab,γ]	0.15 ± 0.048 [ab]	0.15 ± 0.041 [a]	0.14 ± 0.021 [ab]
		180	6.72 ± 1.52 [c,β]	18.3 ± 4.01 [c,α]	11.8 ± 5.33 [c,β]	−0.09 ± 0.041 [a]	−0.05 ± 0.019 [a]	−0.03 ± 0.023 [a]	0.33 ± 0.074 [a,α]	0.37 ± 0.033 [a,α]	0.23 ± 0.030 [ab,β]	0.19 ± 0.090 [a]	0.16 ± 0.071 [a]	0.18 ± 0.009 [a]
		210	5.24 ± 1.22 [c,β]	8.83 ± 4.01 [d,β]	15.8 ± 2.98 [c,α]	−0.04 ± 0.026 [a]	−0.03 ± 0.016 [a]	−0.02 ± 0.023 [a]	0.30 ± 0.064 [a,α]	0.37 ± 0.071 [a,α]	0.23 ± 0.041 [ab,β]	0.24 ± 0.048 [a]	0.17 ± 0.043 [a]	0.13 ± 0.041 [ab]
		240	3.64 ± 1.12 [c,β]	12.2 ± 3.00 [d,α]	10.2 ± 3.74 [c,α]	−0.05 ± 0.022 [a]	−0.03 ± 0.007 [a]	−0.04 ± 0.011 [a]	0.25 ± 0.014 [b,α]	0.35 ± 0.099 [ab,α]	0.18 ± 0.058 [b,β]	0.15 ± 0.014 [ab]	0.20 ± 0.032 [a]	0.17 ± 0.011 [a]
	100	30	14.1 ± 4.70 [b,α]	3.25 ± 0.51 [e,β]	9.98 ± 2.92 [c,α]	−0.16 ± 0.061 [b,β]	−0.11 ± 0.001 [b,γ]	−0.60 ± 0.101 [b,α]	0.23 ± 0.059 [b,α]	0.25 ± 0.010 [c,α]	0.17 ± 0.060 [ab,γ]	0.16 ± 0.021 [ab]	0.12 ± 0.041 [ab]	0.17 ± 0.020 [a]
		60	7.45 ± 1.50 [c,α]	2.08 ± 0.03 [f,β]	6.49 ± 0.15 [d,α]	−0.19 ± 0.091 [b,β]	−0.12 ± 0.003 [b,γ]	−1.07 ± 0.121 [c,α]	0.27 ± 0.058 [ab,β]	0.32 ± 0.004 [b,β]	0.23 ± 0.059 [ab,β]	0.15 ± 0.011 [ab]	0.14 ± 0.033 [ab]	0.17 ± 0.019 [a]
		90	1.42 ± 0.700 [d,α]	1.02 ± 0.011 [g,β]	1.14 ± 0.137 [e,α]	−0.20 ± 0.081 [bc,β]	−0.18 ± 0.081 [c,β]	−1.90 ± 0.301 [d,α]	0.33 ± 0.060 [a,α]	0.35 ± 0.061 [ab,α]	0.23 ± 0.049 [ab,β]	0.17 ± 0.023 [ab]	0.15 ± 0.021 [ab]	0.17 ± 0.031 [a]
		120	0.39 ± 0.018 [e]	0.28 ± 0.013 [h]	0.27 ± 0.010 [f]	−0.20 ± 0.071 [bc,β]	−0.20 ± 0.023 [c,β]	−1.60 ± 0.173 [d,α]	0.35 ± 0.060 [a,α]	0.37 ± 0.059 [a,α]	0.27 ± 0.058 [a,β]	0.15 ± 0.011 [ab]	0.17 ± 0.041 [ab]	0.16 ± 0.021 [a]
		150	0.35 ± 0.019 [e]	0.26 ± 0.140 [h]	0.20 ± 0.014 [f]	−0.21 ± 0.055 [bc,β]	−0.21 ± 0.004 [c,β]	−1.57 ± 0.151 [d,α]	0.37 ± 0.059 [a,α]	0.40 ± 0.080 [a,α]	0.27 ± 0.059 [a,β]	0.17 ± 0.022 [ab]	0.14 ± 0.023 [ab]	0.14 ± 0.019 [ab]
		180	0.19 ± 0.013 [e]	0.20 ± 0.010 [h]	0.18 ± 0.010 [f]	−0.21 ± 0.039 [c,β]	−0.19 ± 0.090 [c,β]	−1.43 ± 0.423 [d,α]	0.37 ± 0.039 [a,αβ]	0.45 ± 0.030 [a,α]	0.33 ± 0.021 [a,β]	0.15 ± 0.014 [ab]	0.16 ± 0.018 [ab]	0.16 ± 0.009 [ab]
		210	0.19 ± 0.020 [e]	0.18 ± 0.011 [h]	0.16 ± 0.015 [f]	−0.29 ± 0.061 [d,α]	−0.22 ± 0.013 [bc,β]	−1.60 ± 0.464 [d,α]	0.37 ± 0.044 [a,β]	0.47 ± 0.029 [a,α]	0.30 ± 0.018 [a,γ]	0.14 ± 0.013 [b]	0.15 ± 0.019 [ab]	0.14 ± 0.019 [ab]
		240	0.15 ± 0.017 [e]	0.16 ± 0.015 [h]	0.11 ± 0.031 [f]	−0.25 ± 0.071 [c,β]	−0.26 ± 0.043 [c,β]	−1.24 ± 0.321 [bc,α]	0.23 ± 0.014 [b,β]	0.35 ± 0.027 [ab,α]	0.23 ± 0.013 [ab,β]	0.11 ± 0.011 [b]	0.12 ± 0.009 [b]	0.12 ± 0.011 [b]
PC	120	15	0.19 ± 0.011	0.18 ± 0.020	0.19 ± 0.019	−0.17 ± 0.018 [a]	−0.16 ± 0.013 [a]	−0.14 ± 0.018 [a]	0.33 ± 0.039 [b,α]	0.30 ± 0.031 [b,α]	0.23 ± 0.022 [b,β]	0.16 ± 0.012 [a,β]	0.16 ± 0.010 [a,β]	0.19 ± 0.020 [a,α]
		20	0.18 ± 0.013	0.18 ± 0.010	0.17 ± 0.010	−0.22 ± 0.010 [b,β]	−0.17 ± 0.010 [a,α]	−0.26 ± 0.013 [b,β]	0.37 ± 0.045 [ab,α]	0.37 ± 0.060 [ab,α]	0.30 ± 0.013 [a,β]	0.14 ± 0.020 [a,β]	0.14 ± 0.011 [a,β]	0.17 ± 0.010 [a,α]
		30	0.17 ± 0.010	0.17 ± 0.011	0.16 ± 0.070	−0.30 ± 0.020 [c,β]	−0.27 ± 0.017 [b,α]	−0.44 ± 0.011 [c,γ]	0.40 ± 0.011 [a,α]	0.39 ± 0.020 [a,α]	0.31 ± 0.010 [a,β]	0.05 ± 0.010 [b]	0.04 ± 0.010 [b]	0.05 ± 0.009 [b]

OC = open cooker; PC = pressure cooker. Chickpea varieties from Spain: PD = *Pedrosillano*; BL = *Blanco Lechoso*; CA = *Castellano*. a,b … Different letters within the same column for the same cooking system differ significantly ($p < 0.05$). α,β … Different letters within the same row for the same texture parameter differ significantly ($p < 0.05$).

Chickpea seeds are usually cooked above the gelatinization temperature to soften the grain and improve the nutritional quality and aroma development, resulting in the improvement of overall palatability [29,30]. The hydrothermal cooking process of chickpeas involves water absorption through the seed coat, until reaching an equilibrium condition, followed by structural changes in the seed by heat, such as separation of cells, gelatinization of cell starch of the cotyledons, protein denaturation, and deformation of the spherical granules [31]. Starch gelatinization occurs between 60 and 95 °C [32]; however, it will only occur when the cotyledon moisture content is sufficiently high.

The main conventional cooking method for legume seeds is boiling them in water for an extended period in either OC or PC conditions. During PC, heat is evenly, deeply, and quickly distributed, and faster cooking than OC has been described [33]. The higher intensity of the heat treatment in PC may be the reason for not finding significant differences in hardness among the chickpea varieties. However, in OC, the heating at lower temperatures and the slower heating would allow observing the effect of the thermal treatment on the chickpea structures and, consequently, a dependence of hardness on the chickpea variety.

Güzel and Sayar [33] found a higher percentage of seed coat splits in OC than PC, although the amounts of solid lost were higher when the legumes were PC processed. In addition, higher levels of resistant starch and lower levels of slowly digestible starch were detected when PC was used.

No differences in adhesiveness were detected in OC at 80 °C for the different cooking times and chickpea varieties (-0.004 N×s). Nevertheless, higher absolute values were detected when cooking at 100 °C than at 80 °C (77.0% difference in BL and CA but 97.5% in PD; $p < 0.0001$). When cooking at 100 °C, adhesiveness absolute values tended to increase with cooking time. Furthermore, PD showed six times higher absolute values than BL and CA (-0.138 vs. -0.023 N×s, respectively; $p < 0.0001$). In the case of PC, higher (in absolute value) values were detected in adhesiveness directly related to cooking time. In addition, such an increase was higher for PD than for BL and CA (68.2% vs. 42.0% difference, respectively; $p < 0.0001$). Accordingly, Wani et al. [34] described that swelling capacity had a positive correlation with cooking time and water uptake ratio but a negative correlation with cooked seed hardness and adhesiveness.

Related to springiness, a dependence on the cooking temperature was observed; OC at 100 °C showed higher values than OC at 80 °C (18.7% difference; $p = 0.0401$). In addition, CA springiness of OC at 80 °C showed 21.8% higher values than BL and 45.5% higher values than PD, and BL OC at 80 °C values were 30.3% higher than PD ($p < 0.0001$). Whereas CA and BL values of OC at 100 °C were 25.9% higher than PD (0.34 vs. 0.25 cm; $p < 0.0001$). Therefore, the highest value of springiness was obtained for CA OC-180 and OC-210 when cooking at 100 °C (0.46 cm), and the lowest for PD-30 at 80 °C (0.11 cm).

In terms of cohesiveness, lower values at 100 °C than at 80 °C ($p = 0.0016$) and a value decrease with cooking time ($p = 0.0157$) were observed. Specifically, OC-240 at 100 °C cohesiveness showed the lowest value (0.12; $p = 0.0163$).

Considering PC, no differences were detected in springiness related to the cooking time between 20 and 30 min, but values at 15 min were lower than others (0.36 vs. 0.29 cm; $p < 0.0001$). In addition, PD springiness was lower than BL and CA (0.28 vs. 0.36 cm, respectively; $p < 0.0001$). With regard to cohesiveness, a higher value of PD was detected, but only when cooking for 15 and 20 min (15.8% and 17.6% difference, respectively; $p < 0.05$). In addition, the cohesiveness of chickpeas cooked for 30 min showed a 70.7% lower value than those cooked for 15 or 20 min (0.16 vs. 0.05, respectively; $p < 0.05$; Table 1).

The differences found in the effect of heat treatment on the textural characteristics of the different chickpea varieties used in this study can be attributed to their different structural, morphological, and compositional characteristics. The PD chickpeas are categorized as *desi* (small, wrinkled, and dark-colored) [35,36], whereas BL and CA are *kabuli* varieties (large, smooth-coated, and white to cream-colored). PD chickpeas, due to their smaller diameter, reached the temperature of the cooking medium more quickly than larger chickpeas, thus affecting their structure and textural attributes, such as adhesiveness (greater in PD).

However, PD chickpeas presented greater hardness in OC and required longer cooking times for softening, which is attributed to the characteristics of their seed coat and cotyledon structure, where the starch aggregates are embedded in the protein structure [37].

As is well known, the seed coat can prevent seed swelling during cooking; it is a physical obstacle to hydration capacity [38,39]. Wood et al. [28] reported that *kabuli* seeds had a thinner seed coat associated with the thinner palisade and parenchyma layers, which contained fewer pectic polysaccharides and less protein. The outer palisade layer varied in thickness from one to two cells, leading to a textured and sometimes wrinkled appearance of the seed surface. In contrast, the *desi* palisade layers were rigid and extensively thickened.

It is worth mentioning that Kaur et al. [40] and Singh et al. [41] studied the dependence on genotype and cultivar conditions of the physicochemical and textural properties of soaked and cooked chickpeas. They concluded that the lower the seed weight and/or volume, the more compact the structure (seed coat and cotyledon) and the lower the hydration ability. Several authors [40,41] have reported that, together, the swelling capacity and the swelling index are directly correlated to hardness and cohesiveness and indirectly to springiness. An inverse relationship was observed between hydration and cohesiveness. This agrees with our results and could explain the lower elasticity and higher hardness of PD vs. CA and BL.

In addition to the differences attributable to morphology, especially of the seed coat and the cotyledon, a strong relationship between changes in TPA variables and the chickpea biochemical composition has been described. A dependence of rheological parameters on hydration amount, protein and starch contents, and amylose/amylopectin ratio has been previously described by Gómez-Favela et al. [42] and Costa et al. [43]. However, disparity among studies has been also found. Kalefetoğlu et al. [44] described that *kabuli* varieties are characterized by a higher protein amount than *desi* varieties. Whereas Ipekesen et al. [45] and Sahu et al. [46] found no difference in protein content, only in selected amino acids, when comparing *desi* and *kabuli* varieties. Remarkably, Xiao et al. [36] described a higher starch amount in *kabuli* than in *desi*, but, again, no difference in protein concentration.

According to the existing literature, the analysis of the rheological behavior of different chickpea varieties seems to be a complex problem involving multiple factors. Koskosidis et al. [47] and Ozaktan et al. [48] described a strong dependence of chickpea composition with the genotype and sowing, and with climate and soil composition, respectively. Therefore, food processors should be aware of the different components of various chickpea species in order to select specific species for better preparation of pulse-based food products [36].

The cooked chickpeas were evaluated in a sensory analysis to establish the most suitable cooking conditions for the development of the most satisfactory sensory attributes without affecting the integrity of the legume. Wood [28] carried out a literature review on the evaluation of cooking time in pulses, observing a lack of a standard method. The major direct measures used to evaluate cooking time include sensory analysis, tactile methods, spread area ratio, use of the Mattson bean cooker, and white core and glass slide methods.

In the sensory analysis, by a rank order test and descriptive observation, the panelists evaluated the level of satisfaction, taking into account the saltiness and aroma of the chickpeas and considering the integrity of chickpeas in terms of preserved skin. Concerning the texture characteristics, hardness values were used as limiting factors, together with low cohesiveness (sample disintegration while cooking) but not adhesiveness, which has been described as a less limiting factor for the acceptance of the products by consumers [27]. Overall, the softness in the mouth and the soft inner mass of the chickpeas were mainly considered.

In the first sensory analyses, cooked chickpeas that presented a hardness higher than 1.5 N were considered unsuitable for consumption, due to their high chewing resistance. OC chickpeas cooked at 80 °C showed hardness values higher than 3 N, even when cooking was prolonged up to 360 min (data not shown), resulting in a product that was far from "buttery" and flavorful; this may be associated with an incomplete starch gelatinization in cotyledons. Klamczynska et al. [49] reported that the hardness of seeds decreased continuously with increased cooking time. These authors found that all legume starch was fully gelatinized after heating at the boiling point for 70 min. Retrogradation of starch in the legume seeds occurs relatively quickly during cooking and is promoted by extended cooking.

Table 2 shows the results of the sensory analysis corresponding to cooking conditions (discarding OC chickpeas cooked at 80 °C).

Table 2. Sensory analysis of chickpeas cooked under different systems and times.

Cooking Parameters [1]		Sum of Ranks [2]					
		Chickpea Variety [3]					
System	t (Min)	PD		BL		CA	
OC	90	10	c	11	c	10	c
	120	17	bc	17	bc	18	bc
	150	35	ab	27	ab	26	ab
	180	49	a	53	a	51	a
	210	45	a	39	ab	43	ab
	240	33	ab	42	ab	41	ab
PC	15	27	ab	28	ab	28	ab
	20	36	a	35	a	34	a
	30	18	bc	17	bc	19	bc
	60	9	c	10	c	9	c

[1] OC = open cooker; PC = pressure cooker. [2] Sum of ranks; values without a common letter within the same column for the same cooking system differ significantly ($p < 0.05$). [3] Chickpea varieties from Spain: PD = *Pedrosillano*; BL = *Blanco Lechoso*; CA = *Castellano*.

The highest score was assigned, independently of the chickpea variety, to OC-180 at 100 °C ($p < 0.05$). With regard to PC, the highest score was given to 20 min, also independently of the chickpea variety ($p < 0.05$, Table 2). Interestingly, no panelist issued a negative judgment related to the texture or taste. In addition, in the case of PC chickpeas cooked for longer than 30 min, skin and/or integrity loss was observed in a high percentage; therefore, they were also eliminated from further study. In addition, PC-30 and PC-60 chickpeas showed the lowest cohesiveness values.

Besides cooking, other factors, such as growing conditions and variety/cultivar, can affect sensory parameters [27]. Cobos et al. [50], together with Chigwedere et al. [51], observed that the genotype effect was the greatest source of variability in all grain quality studies; hence, a high genetic gain of grain quality is expected for nutritional, physical, and sensory traits in chickpea. In addition, no clear relationship was found between the sensory and physicochemical properties. The authors concluded that buttery texture, graininess, and hardness were the specific variables adequate to evaluate sensory quality in chickpeas to minimize the required time for panel testing.

The objective of this part of the study is to select cooked chickpeas that are intact, have sensory characteristics suitable for consumption, and could be subjected to freeze-drying for their preservation and subsequent use in various preparations after rehydration. Therefore, considering the sensory evaluation results, it was decided to continue the study with the three varieties of chickpeas, but only with the treatments: OC-180 at 100 °C and PC-20, whose TPA characteristics were as follows: hardness: $\leq 0.18 \pm 0.010$ N, springiness: $\geq 0.30 \pm 0.010$ cm, and cohesiveness: $\leq 0.15 \pm 0.012$. Nevertheless, chickpeas cooked under OC-210 conditions also presented satisfactory sensory characteristics, although the bean cohesiveness and integrity tended to decrease with increasing cooking time. This same tendency was also observed in PC chickpeas cooked longer than 20 min.

3.2. Rehydration Kinetics of Cooked Freeze-Dried Chickpeas

The freeze-dried chickpeas were kept vacuum packed at $-20\ ^\circ$C to avoid changes in the remaining moisture content and in the oxidative stability. The rehydration process, including weight gain and its time dependence, was non-linear. Independently of the chickpea variety, selected cooking conditions (OC-180 at 100 $^\circ$C and PC-20), and the rehydration conditions (distilled water or 2% NaCl solution at 20–22 $^\circ$C or 90–100 $^\circ$C), three stages of rehydration were seen (Figure 3). Firstly, a sudden weight gain can be observed (R_1; $t_1 = 0$–2 min), related to a quick inclusion of water by the external structure of the chickpea (the highest sorption velocity corresponding to the highest slope value). Secondly, a medium velocity of rehydration was seen, which could be related to a subsequent inclusion of water within the internal structure of the chickpea, with the degree of incorporation of external water lower than R_1 but still considerable (R_2; $t_2 = 2$–10 min). Thirdly, a low weight gain period was observed, which could be related to a structural reorganization to a higher extent and to water inclusion as a secondary process (R_3; $t_3 = 10$–20 min). The third period was followed by an equilibrium ($t_4 > 20$ min), in which no significant changes in weight gain were detected. Consequently, the maximum weight gain at the end of rehydration can be estimated at 20 min. Similar curves of rehydration have been previously observed by several authors in different vegetables and pulses [42,52]. It has been reported that the rehydration is associated with hysteresis due to cellular and structural disruption [52].

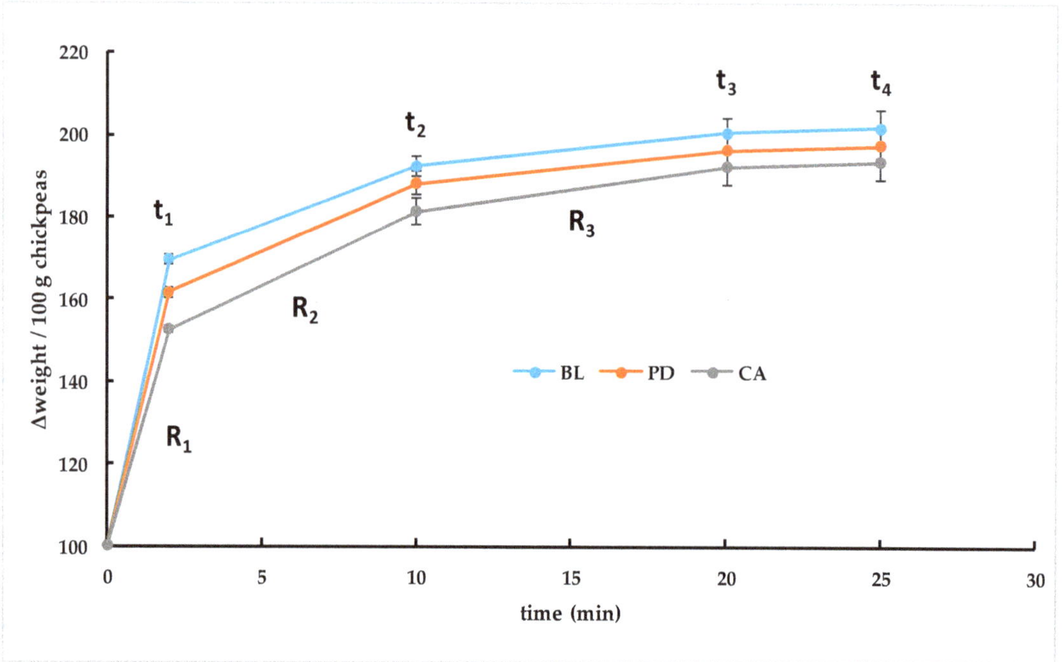

Figure 3. Rehydration curve model of different varieties of cooked and freeze-dried chickpeas. Chickpea varieties from Spain: PD = *Pedrosillano*; BL = *Blanco Lechoso*; CA = *Castellano*.

When only considering the chickpea variety, it was observed that the percentage weight gain of the BL variety was 7.77% and 6.05% higher than CA and PD at R_1 ($p < 0.05$) and 4.89% and 2.86% at R_2 ($p = 0.0713$), respectively. At the end of the rehydration process (R3), CA chickpeas showed slightly lower rehydration values (% weight gain) than the other chickpea varieties (Figure 3).

Water absorption kinetics have been widely studied to evaluate the changes in the microstructure and physicochemical properties of food matrices. In fact, specifically for chickpeas, the rehydration process has been previously described by several authors who established Peleg's [42,53], Weibull's [54,55], or Fick's [56] models as the most accurate for rehydration process prediction. Nevertheless, to compare results, it is necessary to consider the differences between experimental models. For legumes, most published studies analyze the rehydration of seeds during soaking or the cooking process [42], whereas the present work deals with the rehydration of previously cooked and then freeze-dried chickpeas.

Figure 4 compares, for each chickpea variety, the mean value of weight gain at the end of stage R1 (2 min), R2 (10 min), and R3 (20 min), considering the selected cooking conditions (OC-180 at 100 °C vs. PC-20, Figure 4a), the temperature (20–22 °C vs. 90–100 °C, Figure 4b), and the rehydration medium composition (distilled water vs. 2% (*w/v*) NaCl solution, Figure 4c). In all chickpea varieties, PC rehydrated at a faster rate at the R1 stage than OC, although the differences in weight gain of each variety at the end of R3 were not statistically significant (Figure 4a).

Independently of the chickpea variety and the cooking procedure, considering R_1, higher values of weight gain were observed when the rehydration was carried out with 2% NaCl (16.6% difference, Figure 4c) solution and/or with a heated medium (11.0% difference; $p < 0.05$, Figure 4b). With regard to R_2, higher values of weight gain were obtained when using a heated (90–100 °C) rehydration media as compared to a room-temperature (20–22 °C) one (12.4% difference; $p < 0.05$), and a statistical tendency was detected regarding rehydrating solution (14.4% difference; $p = 0.088$) (Figure 4c).

The differences among chickpea varieties could be related to differences in seed morphology and physico-chemical characteristics such as the size and shape of the seed, rate of starch gelatinization, the ratio of the soluble/insoluble pectates, lignification of middle lamella, seed coat characteristics, and lipid structure, among others [42]. In addition, the nature and amounts of non-starch constituents, which act as a physical barrier to the swelling of starch granules, may influence the rate of water uptake during the cooking and, perhaps, also during rehydration after a freeze-drying process. Additionally, genotype and cultivar conditions have been described to affect rehydration parameters [42], which is in line with our results from different chickpea varieties from different locations.

The differences between the rehydration rate of chickpeas treated in OC and PC can be attributed to the higher intensity and homogeneity of the cooking treatments performed at high pressure [33]. Thus, the rate of hydration, hydrogen bond disruption, and other phenomena associated with the cooking process of the seeds will be enhanced in PC and, consequently, in subsequent rehydration after freeze-drying of the cooked product. In addition, the thickness of the palisade layer and the lignin and cellulose content of the seed coat and the cotyledon cell walls, relevant structures involved in the cooking quality, will be affected in a higher proportion in PC. A more intense heat treatment should imply a higher degree of hydrolysis (or other breaking-down processes of macromolecules). The more the macromolecules are degraded, the easier the water inclusion (diffusion) should be. An equilibrium exists between water inclusion (depending on membrane/skin permeability, starch structure, and seed size) and loss of components [43,56].

Regarding the impact of the temperature of the rehydration medium, our results agree with Krokida and Marinos-Kouris [57]. These authors established that, in general, the water temperature positively influenced the rehydration rates and the equilibrium moisture content of dehydrated products. In this line, Monteiro et al. [58] described a linear relationship between water temperature and weight gain of chickpeas during rehydration.

Concerning the increased rehydration rate of chickpeas rehydrated with 2% NaCl, it has been reported that sodium salts increase water absorption, leach solids and result in softness in the cooked beans, and increase pectin solubilization, decreasing Ca and Mg (ions that bind pectin), when compared to distilled water or the other salts [59].

a) Considering cooking parameters [Open cooker (OC) *vs.* Pressure cooker (OC), without considering the type of rehydration solution or the temperature].

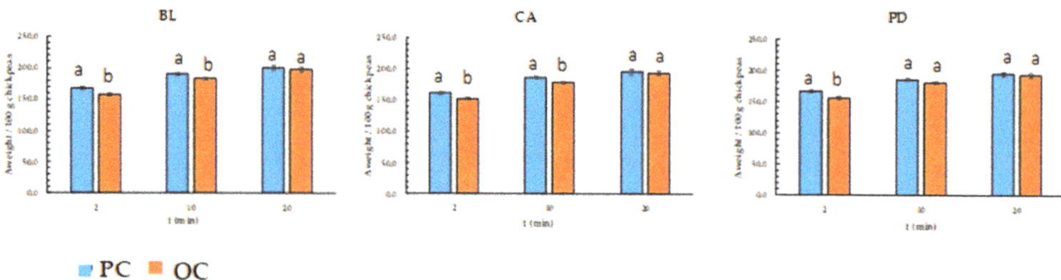

b) Considering rehydration temperature [90-100 °C *vs.* 20-22 °C, without considering the type of rehydration solution or the cooking parameters].

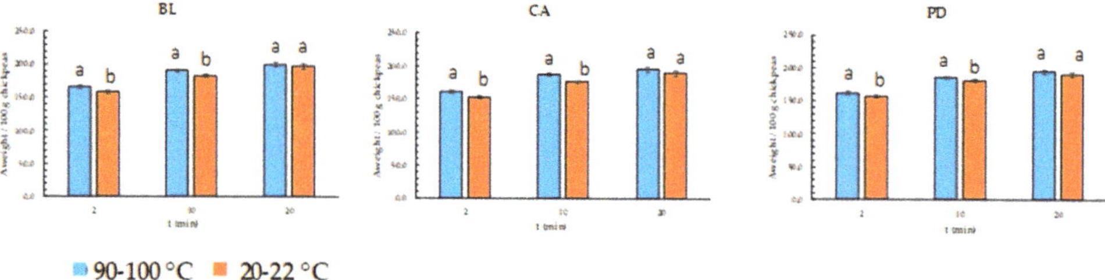

c) Considering rehydration solution (water *vs.* 2% NaCl, without considering the rehydration temperature or the cooking parameters).

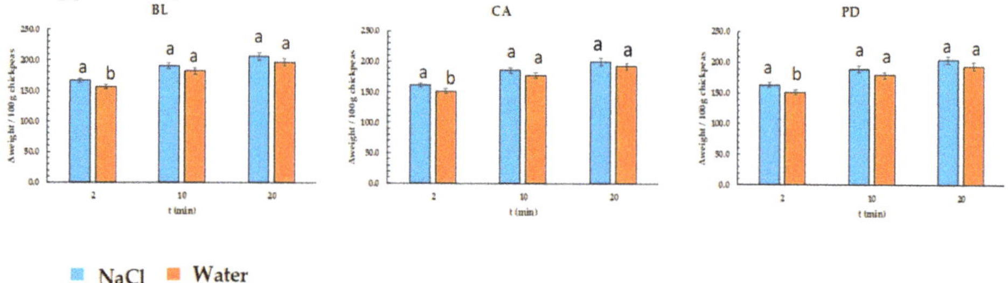

Figure 4. Weight gain of freeze-dried chickpeas considering cooking conditions, rehydration temperature, and solution. Chickpea varieties from Spain: PD = *Pedrosillano*; BL = *Blanco Lechoso*; CA = *Castellano*. Values without a common letter within the same rehydration time differ significantly ($p < 0.05$).

From the data obtained in the present work, it can be deduced that, although differences were observed in the water sorption rate at the beginning of rehydration, in general, all seeds showed a high degree of water uptake at the end of the rehydration process. This fact could be due to the high porosity, low density, and spongy structure induced by the freeze-dying process, facilitating rehydration. On the other hand, the freeze-drying process induces limited changes in the cell network, which would also facilitate rehydration [60]. Therefore, it is considered that, although there are differences in seed structure derived from factors specific to chickpea variety and the applied thermal process, the freeze-drying process would result in suitable matrices for water uptake during rehydration. Aravindakshan et al. [60] reported that the type of drying process together with the temperature of rehydration, the nature of pores, and the state of water-imbibing constituents influenced the rehydration characteristics of dried beans. Liu et al. [8] reported that freeze-drying was the best drying procedure as compared to others.

The results of the sensory analysis are shown in Table 3. In terms of rehydration weight gain, rehydration at 0–100 °C and with NaCl solution, independently of the cooking conditions (OC and PC), were given higher scores than room temperature and only-water rehydration protocols ($p < 0.01$), as samples had a softer texture and a more intense flavor. However, all rehydrated chickpeas were described as having the characteristics of cooked legumes. These results can be related to the findings of Ulloa et al. [61], who reported that the rate of absorbed water determines the sensory properties.

Table 3. Sensory analysis of chickpeas cooked under selected conditions, freeze-dried, and rehydrated under different conditions.

Cooking Parameters [1]		Rehydration Parameters		Sum of Ranks [2]					
				Chickpea Variety [3]					
System	t (Min)	% NaCl	T (°C)	PD		BL		CA	
OC	180	0	20–22	9	c	10	c	10	c
		2	20–22	19	bc	19	bc	17	bc
		0	90–100	26	ab	26	ab	28	ab
		2	90–100	36	a	35	a	35	a
PC	20	0	20–22	11	c	10	c	12	c
		2	20–22	16	bc	18	bc	18	bc
		0	90–100	29	ab	26	ab	25	ab
		2	90–100	34	a	36	a	35	a

[1] OC = open cooker; PC = pressure cooker. [2] Sum of ranks; values without a common letter within the same column for the same cooking system differ significantly ($p < 0.05$). [3] Chickpea varieties from Spain: PD = *Pedrosillano*; BL = *Blanco Lechoso*; CA = *Castellano*.

As is well known, the basic objective of drying food products is the elimination of water from solids to a level at which microbial spoilage is minimized. In this line, freeze-drying has considerable advantages. Chen et al. [62] observed that freeze-drying-based processing technologies are especially suitable for producing products containing abundant nutrients and functional ingredients due to a drying environment with lower temperature and oxygen content. In fact, several studies regarding the development of dehydrated chickpea snacks obtained by sequential hydration, cooking, and drying processes have been found in the literature [58,63]. Ulloa et al. [61] established that instant whole beans obtained by drying could represent a good-quality bean-based product for new market opportunities in the functional food and nutraceutical industry. The authors described that the rehydration characteristics of a dehydrated product reflect the physical and chemical changes that occurred during dehydration, thus being suitable to be used as a quality index. These changes are influenced by the composition, the conditions during drying, and any pre-treatment to which the products have been subjected. Remarkably, the authors also established that the amount and rate of absorbed water determine the sensorial properties and the preparation time required by the consumer.

Therefore, more research is required to determine alternative ways to incorporate chickpea-based ingredients into products and understand their behavior when used in different food categories (beverages, snacks, dairy-like products, etc.). In the present study, after testing the acceptance of cooked freeze-dried, and rehydrated chickpeas, their use is proposed in the preparation of a quick cooking dish based on a traditional culinary preparation in Spain, *Cocido Madrileño*. However, this use is only intended as an example of how complex traditional dishes with several ingredients could be prepared from freeze-dried cooked chickpeas.

3.3. Analysis of the Cooking and Rehydration Parameters and Sensory Characteristics of the Other Components of the Cocido Dish

Table 4 shows the results of the sensory analysis of other ingredients of a traditional *Cocido Madrileño* dish [16,17] (different pieces of beef and chicken meat, dry-cured ham and pork belly, and vegetables) after cooking (OC 100 °C and PC, at different times), freeze-drying, and rehydration (2% NaCl solution) (Figure 2).

Table 4. Sensory analysis of the several ingredients cooked under different conditions, freeze-dried, and rehydrated.

Ingredients Cooked Together for Broth Obtention	Sensory Analysis *	Open Cooker, 100 °C t (min)			Pressure Cooker t (min)		
		60	90	180	30	60	90
Chicken (quarters)	Sum of ranks	9b	19ab	26a	12b	19ab	23a
	Descriptive after cooking	Tough texture. Poor development of taste and aroma	Cohesive and maintains integrity. Not juicy. Mild taste and aroma but pleasant	Texture: soft and juicy. Chicken broth flavor. Pleasant aroma and appearance	Texture: tough. Mild flavor and aroma to chicken broth. Pleasant appearance	Texture: soft but compact and mildly dry. Pleasant taste and aroma to chicken broth	Texture: soft and disintegrated, lack of cohesion. Pleasant taste and aroma
	Descriptive after freeze-drying and rehydration	Texture, flavor and appearance similar to the cooked product	Texture: compact and dry. Taste and aroma similar to the cooked product	Texture: soft and juicy. Flavor, aroma and appearance similar to the cooked product	Texture, flavor and appearance similar to the cooked product	Texture: compact and dry. Taste and aroma: similar to the cooked product	Texture: soft and disintegrated, lack of cohesion. Meat crumbled during mastication. Flavor and appearance similar to the cooked product
Beef: flank and shank cuts	Sum of ranks	9b	18ab	27a	10b	17ab	27a
	Descriptive after cooking	Excessively tough meat, hard to chew. Poor flavor development, very slightly umami. Pleasant appearance	Tough meat. Pleasant taste, aroma and appearance, slightly meaty, umami	Doneness desired. Soft and juicy meat. Intense sapid and aromatic development, very pleasant, umami and meaty touches	Hard, compact meat. Little but pleasant sapid and aromatic development	Tender meat. Pleasant taste, smell and appearance. Soft sapid and aromatic development	Doneness desired. Very soft meat. Soft sapid and aromatic development
	Descriptive after freeze-drying and rehydration	Tough and leathery meat, hard to chew. Nice appearance. Taste and aroma similar to the cooked product	Hard, compact meat. Nice look. Little sapid and aromatic development. Pleasant	Maintains structure. Smooth and juicy meat. Very pleasant tasty and aromatic development, similar to the cooked product	Texture, flavor and appearance similar to the cooked product	Texture, flavor, aroma and appearance similar to the cooked product	Doneness desired. Soft and juicy meat. Disintegrates easily, frayed. Good flavorful and aromatic development
Turnip (*Brassica napus*), Carrot (*Daucus carota sativus*), Leek (*Allium ampeloprasum var. porrum*) and Potato (*Solanum tuberosum*)	Sum of ranks	9b	19ab	26a	9b	18ab	27a
	Descriptive after cooking	Texture: tough. Pleasant appearance, flavor and aroma	Smooth, somewhat rough texture. Pleasant taste, smell and appearance	Texture: soft and juicy. Pleasant flavor, aroma and appearance	Hard texture. Little sapid and aromatic development. Pleasing appearance	Slightly rough and fibrous texture. Mild flavor and aroma. Pleasing appearance	Smooth texture. Good flavor and aromatic development. Pleasing appearance
	Descriptive after freeze-drying and rehydration	Dry texture and appearance similar to the cooked product	Rubbery texture, but juicy. Flavor and aroma characteristics very similar to those of the cooked product	Maintains structure. Soft and juicy. Maintains flavor and aroma	Texture, flavor and appearance similar to the cooked product	Fibrous, juicy texture. Similar flavor and aroma to the freshly cooked product	Maintains tissue integrity. Soft and juicy texture. Taste and aroma similar to the freshly cooked product
Broth	Sum of ranks	9b	18ab	27a	9b	20a	25a
	Descriptive after cooking	Little sapid and aromatic development. Light umami and meaty flavor	Pleasant sapid and aromatic development, but mild umami and meaty flavor. Aroma and taste associated with plain beef broth	Very nice sapid development. Meaty, tasty, similar to the traditional product. Dense	Little sapid and aromatic development. Light umami and meaty flavor	Pleasant sapid and aromatic development, but mild umami and meaty flavor. Aroma and taste associated with plain beef broth	Meaty, tasty, similar to the traditional product. Dense. Certain "overheated" aftertaste
	Descriptive after freeze-drying and rehydration	N/A	Maintains the sapid and aromatic characteristics of the cooked product	Maintains the sapid and aromatic characteristics of the cooked product	N/A	Maintains the sapid and aromatic flavor and aroma characteristics of the cooked product	Maintains the sapid and aromatic characteristics of the cooked product

Table 4. *Cont.*

Ingredients cooked independently		Open Cooker, 100 °C			Pressure Cooker		
		t (min)			t (min)		
		15	30	60	15	30	60
Cabbage (*Brassica Oleracea*)	Sum of ranks	10b	26a	18ab	25a	18ab	11b
	Descriptive after cooking	Texture: tough. Flavor and aroma: poorly developed. Pleasant appearance	Smooth texture. Nice taste and appearance	Pleasant flavor and appearance but soft texture	Smooth texture. Maintains integrity. Nice taste and appearance	Texture: disintegrated, extremely soft	Texture: disintegrated, extremely soft
	Descriptive after freeze-drying and rehydration	Similar texture and flavor to the cooked product	Maintains integrity. Similar flavor and aroma to the cooked product	Maintains integrity. Similar flavor and aroma to the cooked product	Similar flavor and aroma to the cooked product	N/A	N/A
Spicy pork sausage (Spanish *chorizo*)	Sum of ranks	27a	17ab	10b			
	Descriptive after cooking	Texture: compact, soft and juicy. Appearance, flavor and aroma typic of the product	Texture: compact and juicy. Typic (characteristic) flavor and aroma of the product	Compact but dry texture. Characteristic flavor of the product but with aftertastes with slight touches of rancid. Nice appearance			
	Descriptive after freeze-drying and rehydration	Soft texture but fatty. Characteristic flavor and aroma of the product, with rancid nuances. Nice appearance	Texture: compact and dry. flavor and aroma: similar to the cooked product	Texture: compact and dry. flavor and aroma: similar to the cooked product			
Pork blood sausage (Spanish *morcilla*)	Sum of ranks	27a	15ab	12b			
	Descriptive after cooking	Texture: compact, soft and juicy. Appearance, flavor and aroma typic of the product	Texture: disintegrated, lack of cohesion. Appearance, flavor and aroma typic of the product	Texture: dry and disintegrated, lack of cohesion. Appearance, flavor and aroma typic of the product but with slight hints of rancidity			
	Descriptive after freeze-drying and rehydration	Texture: soft. Pleasant appearance, flavor and aroma; typic of the product	Texture: compact and dry. flavor and aroma: similar to the cooked product	Texture: compact and dry. flavor and aroma: similar to the cooked product			

* The sum of ranks corresponds to the analysis of the freshly cooked samples. The same ranking was obtained after freeze-drying and subsequent rehydration. Values without a common letter within the same row for the same cooking system differ significantly ($p < 0.05$).

Juicy chicken meat with a good appearance and flavor was obtained with OC-180; however, OC-60 led to poor development of aroma, taste, and appearance, and the meat texture was defined as tough. Similar to chickpeas, PC cooking shortened the time required for the development of adequate organoleptic characteristics. Samples cooked for 30 min presented a tough texture, poor development of flavor and aroma, and significantly ($p < 0.05$) lower sensory evaluation values than samples cooked for 90 min, which presented a soft but dry and disintegrated texture, with a lack of cohesion and less pleasant flavor and aroma. In all cases, the cooked freeze-dried samples presented sensory characteristics after hydration very similar to those shown when freshly cooked. In all cases, a greater development of flavor and aroma was shown with longer cooking times. Additionally, after 60 min in PC and 90 min in OC, rehydrated chicken meat showed satisfactory sensory qualities.

Overall, in agreement with our results, related to cooking conditions, it has been described that the higher the temperature and the lower the moisture, the more heterocyclic flavor compounds that will be formed. In addition, PC chicken meat has been described to be closer to roasted or grilled chicken meat than boiled [64]. It should be noted that our results agree with the higher temperature and time, but that, in the case of PC, although the literature has reported better flavor compounds, the panelists described the chicken meat as less succulent both before and after freeze-drying.

Regarding beef flank and shank sensory analyses, shorter cooking times (OC-60 and PC-30) led to tough and barely chewable meat. Whereas meat cooked for longer times (OC-180 and PC-90) was soft and juicy and had greater flavor development. Unlike PC-90 meat, the panelists described an intense sapid and aromatic development of meat flavor for OC-180 products. However, in the descriptive analysis of PC-90, a tender, juicy, and smooth texture was described. OC-60 and OC-90 led to poor development of aroma, taste, and appearance, and the meat texture was defined as tough and barely chewable. The differences in flavor intensity and cooking times could be due to the progress of the primary lipid degradation and Maillard reaction products, associated with beef flavor [65]. The results obtained are in line with Juárez et al. [66] regarding the importance of endpoint temperature in beef cookery, which greatly impacts consumer satisfaction. Again, the cooked freeze-dried samples presented sensory characteristics after hydration that were very similar to those shown when freshly cooked. In agreement with the literature, the obtained results strengthen the consideration of freeze-drying as a high-quality preservation method that maintains the organoleptic characteristics of meat and meat products [60,67,68].

Regarding the meat texture, in both OC and PC, deficient collagen hydrolysis could explain the described behavior, according to Etheritong and Sims [69] and Sims and Bailey [70].

When meat is cooked at a slow rate of heating, tenderness and juiciness are improved, although cooking losses increase. Juiciness is affected by exudation and diffusion in moist heat cookery (e.g., stewing or braising). In general, the muscle structural composition determines the cooking effect on tenderness. In the case of muscles with high connective tissue content, such as the meat pieces used in this study, this meat profits from moist cooking due to collagen gelatinization [66].

Although the purpose of incorporating a portion of bone-in cured ham in the traditional dish [16,17] is less about consumption than about contributing to the sapid and aromatic development of the cooking broth, the cured ham cooked at longer times ($\geq$90 min in OC and 60 min in PC) was more palatable and softer than that cooked at shorter times, and both the appearance and the integrity of this product were maintained when rehydrated after freeze-drying. Similar results were obtained for pork belly. A piece (about 60 g/L) of this meat portion is also added in the traditional *Cocido Madrileño* [16,17], both to add flavor and to be consumed if desired. It is worth noting the good rehydration of this portion, despite its fat content when rehydrated with 2% NaCl solution (90–100 °C) after freeze-drying.

To increase the broth flavor [16,17], cow leg-bone (Figure 2) was included. Such a portion was added for all methods during the described cooking times but was not considered in sensory analysis nor freeze-dried.

Potato, carrot, leek, and turnip showed very similar behavior. Although the size of the pieces should be considered (in our case, potato and turnip less than 70 mm and carrot 2–3 cm in diameter), OC-60 and PC-30 vegetables showed harder texture and lower sapid and aromatic development but a more compact appearance and a lower sensory score than OC-180 and PC-90 ($p < 0.05$). In additional tests carried out with longer cooking times (200 min), a disintegration of the matrix structure was detected, especially in potatoes and carrots. Again, the cooked freeze-dried samples presented sensory characteristics after hydration very similar to those shown when freshly cooked (Table 4). These results are in line with the findings of Liu et al. [8] on the suitability of freeze-drying in maintaining the optimal sensory and nutritional properties of rehydrated vegetables. In addition, the results are in synchrony with the findings of several authors who studied the effect of heat treatment on vegetables such as potatoes. Changes in the potato tuber microstructure and texture during cooking have been mainly associated with the gelatinization behavior of starch through the cell wall, but the middle lamella structural components also play a role [71,72]. A study by Paulus and Saguy [73] on the effect of heat treatment on the quality of cooked carrots suggested that boiling for 70 min led to the best sensory evaluation. In addition, other authors have attributed the softening to cellular dehydration and separation in the tissue [74] as well as the membrane disruption associated with the loss of turgor [75].

Parallel to the aromatic and sapid development of the meat and vegetable portions, the cooking water (broth) presented a flavor with higher intensity, closer to that of the broth of the traditional product, at longer cooking times (Table 4), both in OC and PC cooking.

It should be noted that, in the PC-90 broth, although qualified with the highest scores in the ordination analysis, the detection of a slight "overheated" aftertaste was mentioned in the descriptive analysis. Such an aftertaste was not described for OC-180. In all cases, the freeze-dried broths after rehydration presented sensory characteristics very similar to those described in the freshly cooked broth and, in the case of the broths obtained at longer times (OC-180 and PC-90), sapid matrices very close to those of traditional broth were described.

Broth-related results were in synchrony with those obtained in previous works. Cambero et al. [23,24] and Wang et al. [76] observed that the sensory quality of beef broths increased with the increase in the concentration of nitrogen in peptides (molecular weight > 600 Da), small non-amino acid nitrogen compounds (<600 Da), creatine, and inosine, adenosine and guanosine 5′-monophosphate, and substances from the earlier steps of Maillard reactions, thus being involved in the obtaining of suitable beef broth flavor. In fact, the development of off-flavors, especially warmed-over flavors, was related to an intensification of the cooking treatment. Cambero et al. [23,24] and Pereira-Lima et al. [77] for beef, and Pérez-Palacios et al. [78] and Zhang et al. [79] for chicken, described an implication of over-cooking, with bitter amino acids and peptides, among others.

In line with tradition, cabbage, *chorizo*, and *morcilla* were cooked separately (Figure 2) so as to not interfere with the broth taste [16,17]. OC-30 and PC-15 cabbage showed the best sapid characteristics (Table 4). Shorter cooking times resulted in texture toughening, and longer cooking times, especially PC, led to good flavors but, to a higher or lower extent, the disintegration of the product. As with other ingredients, freeze-drying did not produce a significant effect on the sensory characteristics of cabbage cooked at the established times, and after rehydration, sensory attributes similar to those of the freshly cooked product were described (Table 4).

Regarding cabbage, it has been described that mild cooking minimizes tissue damage and maximizes beneficial isothiocyanate formation and the retention of the active myrosinase enzyme, whose activity on the cabbage glucosinolates produces several compounds with chemo-protective effects [80,81]. On the other hand, it has also been reported that freeze-drying is the method in which the contents of chlorophyll and saponins are kept at a high level in vegetables [8].

In the case of *chorizo* and *morcilla*, to maintain their integrity, only OC cooking was considered. The highest rating was achieved when both products were cooked for 15 min. When longer cooking times were considered, dry texture and rancidity aftertaste, which became more evident after the product was freeze-dried and rehydrated, were detected. It must be considered that they are processed ground meat products with a high content of fat and, consequently, very susceptible to oxidation. Accordingly, Toldrá and Reig [82] and Lorenzo et al. [83] described a dependence on the heating extent, moisture content, proteolysis degree, and the content of fat and connective tissue on the sensory attributes of pork sausages. A higher surface of the components in contact with the atmospheric oxygen and light must be expected; therefore, oxidative processes could develop more rapidly. Thus, the freeze-drying and the rehydration optimization of *chorizo* and *morcilla* require further study.

From the results obtained from the sensory analyses to determine the most convenient conditions for the development of satisfactory attributes (Table 4), it is deduced that the cooking process of the different ingredients of *Cocido Madrileño* could be unified so that the vegetable ingredients, bone-in cured ham, chicken, and beef would be OC-180 cooked, obtaining a broth with sensory attributes similar to the traditional dish. An equally valid alternative would be cooking in PC-90. These ingredients maintain their sensory quality by rehydrating after freeze-drying. To these ingredients, chickpeas would be added (preferably of the BL variety, although other studied varieties could also be used) and freeze-dried after OC-180 or PC-20 cooking. Additional ingredients, cooked separately, require shorter times, such as cabbage (OC-30 or PC-15) and pork sausage (OC-30).

3.4. Preparation of a Freeze-Dried Cooked Cocido as an Easy-to-Prepare Dish: Dish Assembly and Packaging Proposal

Once the best cooking conditions for each ingredient, considering the development of appropriate sensory attributes, were selected (Tables 2–4), experiments were carried out to prepare a complete *Cocido Madrileño* dish from cooked and freeze-dried ingredients with the objective of obtaining an easy-to-prepare dish.

As mentioned, *Cocido Madrileño* is a traditional meal, currently restricted to dishes in which chickpeas play the leading role and different vegetables and meats, although essential, enhance the dish and confer the specific flavor [16]. This proposal aims to analyze the potential of formulating complex dishes from cooked and then freeze-dried ingredients of a different nature to offer healthy dishes that require minimal preparation.

In Spain, when the word "*Cocido*" is heard, memories and emotions are evoked. Historically, the root of the *Cocido* dish was assessed to be a plate cooked by Jewish families each Friday, to be eaten the next day, on the *Sabbath*, the sacred day of the Jewish week. It was named *Adafina* and included lamb meat, chickpeas, and other vegetables. In 711, the Arabs revolutionized the *Cocido* with their more sophisticated agriculture: softer and pulpier chickpeas and several vegetables were used in their *Cocido*, named "*Tajine*" [17]. Currently, the name has been restricted to dishes in which chickpeas play the leading role and different vegetables and meats, although essential, enhance the dish and confer the specific flavor [16]. Strictly, the consumption protocol of the *Cocido* establishes that the first course must be a soup of thin noodles, followed by the degustation of the chickpeas, as the second course. Meat and other vegetables constitute the third course [17]. However, nowadays, chickpeas, vegetables, and meat are served together as a second course, after the soup, which was considered in the assembly of the freeze-dried dish under study.

In these experiments, an attempt was made to establish the processing conditions that could be replicated in the industrial processing of this dish. In this case, in contrast to traditional culinary processing, and taking into account the more appropriate processing conditions established, the chickpeas were cooked separately for greater control of the heating time and temperature. Meat portions and vegetables (except cabbage, *morcilla*, and *chorizo*, which were cooked separately) were cooked (with 2% salt in an open cooker) together to obtain a suitable broth (Table 4). Each ingredient was freeze-dried individually. Trying to closely imitate the customary protocol at a laboratory level, for rehydration of the ingredients, the freeze-dried broth and the quick-to-cook thin noodles were placed at the bottom of a beaker and, over them, a perforated basket (hole diameter lower than chickpea diameter, 3 mm^2) containing chickpeas, vegetables, and meat. A total of 500 mL of water was added and heated in a microwave oven (800 W) for 5 min. As shown in Figure 5, the added water was calculated by considering both the rehydration water of each component and the obtainment of a bowl of soup (about 300 mL).

After 5 min of heating, the beaker was removed from the microwave and left to stand for 10 min. The basket with the ingredients was removed, and the noodles and broth (up to the consumer's preference) were placed in a dish. Then, the basket was placed back in the beaker with the remaining broth. Thus, the rest of the ingredients (chickpeas, vegetables, and meat) were kept juicy until they were eaten as a second course.

Sensory analysis was immediately carried out through a descriptive analysis with panelists who were familiar with the traditional dish. The panelists were asked for an evaluation of the whole reconstituted dish compared to an original *Cocido Madrileño* (firstly the taste and aroma of the soup, and secondly the taste, aroma, and texture of the other ingredients) and punctuation of each ingredient individually. In all cases, the rehydrated products were evaluated with sensory characteristics similar to those of the traditional dish, although for *chorizo* and *morcilla*, the panelists described a slightly rancid taste and a lightly dry and disintegrated texture.

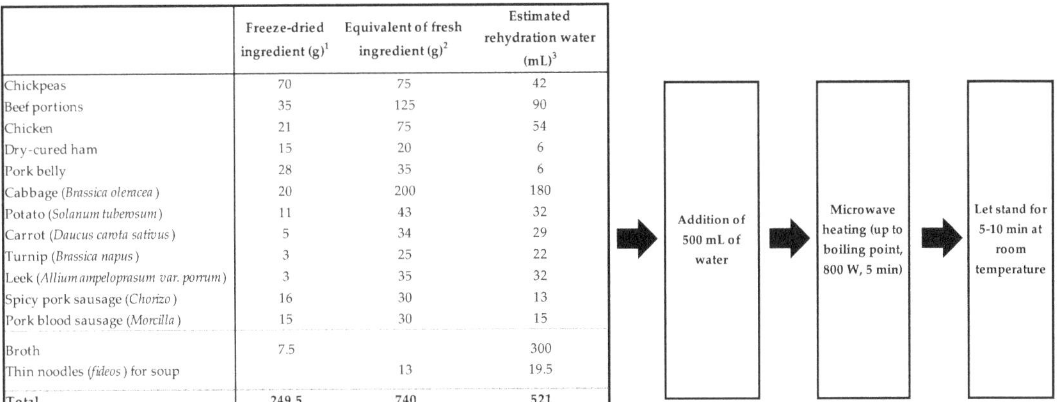

	Freeze-dried ingredient (g)[1]	Equivalent of fresh ingredient (g)[2]	Estimated rehydration water (mL)[3]
Chickpeas	70	75	42
Beef portions	35	125	90
Chicken	21	75	54
Dry-cured ham	15	20	6
Pork belly	28	35	6
Cabbage (*Brassica oleracea*)	20	200	180
Potato (*Solanum tuberosum*)	11	43	32
Carrot (*Daucus carota sativus*)	5	34	29
Turnip (*Brassica napus*)	3	25	22
Leek (*Allium ampeloprasum var. porrum*)	3	35	32
Spicy pork sausage (*Chorizo*)	16	30	13
Pork blood sausage (*Morcilla*)	15	30	15
Broth	7.5		300
Thin noodles (*fideos*) for soup		13	19.5
Total	249.5	740	521

Figure 5. Proposal for the assembly of a *Cocido* dish from cooked and freeze-dried components and the rehydration procedure to prepare a reconstituted tasty meal. [1] Freeze-dried products after cooking in an open cooker for selected times. [2] The equivalent weight of the fresh product before culinary treatment (in the case of chickpeas, this refers to the weight of the dry legume). [3] Estimation of the rehydration water required for each product. In the case of chickpeas, the rehydration water for cooking was considered. For the replacement of the broth, the dry extract of the cooking water of the ingredients was considered (2.5%).

The fast food that can be found on the market is not always as healthy as desirable. If a freeze-dried complex dish (noodle soup, vegetables, and meat) with easy rehydration could be achieved, it would offer the consumer a novel product (there is no similar product on the market) that is easy to preserve (it can be stored at room temperature), requires minimal preparation (limited to the addition of water and short heating in a microwave), has reputed sensory quality and richness of ingredients (legumes, vegetables, and meat products), and is nutritionally well valued, in coherence with the Mediterranean diet [84].

The authors are aware that further studies related to the shelf-life and stability of the product when stored in a container are mandatory. Nevertheless, it must be remembered that one of the objectives of this work was to offer plausible alternatives of an easy-to-prepare reconstituted *Cocido* dish that could be widely produced. We are working on the design of a container that facilitates rehydration, heating, and, above all, the consumption of the ingredients in the traditional way: in two or three separate courses.

4. Conclusions

It is possible to elaborate *easy-to-prepare* dishes containing chickpeas and various ingredients (vegetables and different meat products) prepared with products cooked at their optimum point and then freeze-dried. These dishes can be a commercial alternative to "fast food", since they are easily rehydrated and ready for consumption in a few minutes, providing a varied and complete supply of nutrients. As proof, in the case of the preparation of the traditional Spanish dish "*Cocido Madrileño*", a product with a sensory quality similar to the traditional dish ready requiring about 15 min of preparation was obtained.

However, it is necessary to carry out more research to study the stability and shelf-life of these products, as well as to determine the economic viability.

Author Contributions: Conceptualization, M.I.C. and J.S.; methodology, M.I.C., L.C., M.D.R.d.Á., G.D.G.d.F. and V.R.; formal analysis, M.I.C. and J.S.; investigation, all authors; data curation, L.C., M.I.C., V.R. and J.S.; writing—original draft preparation, M.I.C. and J.S.; writing—review and editing, all authors; visualization, M.I.C., G.D.G.d.F. and J.S. All authors have read and agreed to the published version of the manuscript.

Funding: This research received no external funding.

Data Availability Statement: All related data and methods are presented in this paper. Additional inquiries should be addressed to the corresponding author.

Acknowledgments: The authors express their gratitude to the Spanish Ministry of Science and Innovation (Project PID2010-107542RB-C22).

Conflicts of Interest: The authors declare no conflict of interest.

References

1. Okrent, A.M.; Kumcu, A. U.S. Households' Demand for Convenience Foods. Economic Research Report Number 211. 2016. Available online: https://www.ers.usda.gov/webdocs/publications/80654/err-211.pdf?v=9803.9 (accessed on 4 June 2023).
2. Meenu, M.; Padhan, B.; Zhou, J.; Ramaswamy, H.; Pandey, J.; Patel, R.; Yu, Y.A. Detailed review on quality parameters of functional noodles. *Food Rev. Int.* **2022**, 1–37. [CrossRef]
3. Rani, S.; Singh, R.; Kaur, B.P.; Upadhyay, A.; Kamble, D.B. Optimization and evaluation of multigrain gluten-enriched instant noodles. *Appl. Biol. Chem.* **2018**, *61*, 531–541. [CrossRef]
4. World Instant Noodle Association (WINA). Available online: https://instantnoodles.org/ (accessed on 4 June 2023).
5. Chen, C.; Zhang, M.; Xu, B.; Chen, J. Improvement of the quality of solid ingredients of instant soups: A review. *Food Rev. Int.* **2023**, *39*, 1333–1358. [CrossRef]
6. Pieniazek, F.; Messina, V. Scanning electron microscopy combined with image processing technique: Microstructure and texture analysis of legumes and vegetables for instant meal. *Microsc. Res. Technol.* **2016**, *79*, 267–275. [CrossRef] [PubMed]
7. Barrett, A.H.; Richardson, M.J.; Froio, D.F.; O'Connor, L.F.; Anderson, D.J.; Ndou, T.V. Long-term vitamin stabilization in low moisture products for NASA: Techniques and three-year vitamin retention, sensory, and texture results. *J. Food Sci.* **2018**, *83*, 2183–2190. [CrossRef]
8. Liu, Y.; Zhang, Z.; Hu, L. High efficient freeze-drying technology in food industry. *Crit. Rev. Food Sci. Nutr.* **2022**, *62*, 3370–3388. [CrossRef]
9. Sagar, V.R.; Suresh Kumar, P. Recent advances in drying and dehydration of fruits and vegetables: A review. *J. Food Sci. Technol.* **2010**, *47*, 15–26. [CrossRef] [PubMed]
10. Berk, Z. Freeze drying (lyophilization) and freeze concentration. In *Food Process Engineering and Technology*, 3rd ed.; Academic Press: San Diego, CA, USA, 2018; pp. 567–581. [CrossRef]
11. Ma, Y.; Yi, J.; Jin, X.; Li, X.; Feng, S.; Bi, J. Freeze-drying of fruits and vegetables in food industry: Effects on phytochemicals and bioactive properties attributes—A comprehensive review. *Food Rev. Int.* **2022**, 1–19. [CrossRef]
12. Elma Mathew, S.; Shakappa, D. A review of the nutritional and antinutritional constituents of chickpea (*Cicer arietinum*) and its health benefits. *Crop. Pasture Sci.* **2022**, *73*, 401–414. [CrossRef]
13. Jukanti, A.K.; Gaur, P.M.; Gowda, C.L.L.; Chibbar, R.N. Nutritional quality and health benefits of chickpea (*Cicer arietinum* L.): A review. *Br. J. Nutr.* **2012**, *108* (Suppl. S1), S11–S26. [CrossRef]
14. Polak, R.; Phillips, E.M.; Campbell, A. Legumes: Health benefits and culinary approaches to increase intake. *Clin. Diabetes* **2015**, *33*, 198–205. [CrossRef]
15. Figueira, N.; Curtain, F.; Beck, E.; Grafenauer, S. Consumer understanding and culinary use of legumes in Australia. *Nutrients* **2019**, *11*, 1575. [CrossRef] [PubMed]
16. Iglesias, P.I. Cocidos de España. 2015. Available online: http://www.enciclopediadegastronomia.es/articulos/costumbres-populares/cocidos-de-espana.html (accessed on 4 June 2023).
17. Iglesias, P.I. Historia del Cocido. 2015. Available online: http://www.enciclopediadegastronomia.es/articulos/historias-de-los-alimentos/manufacturados/historia-del-cocido.html (accessed on 4 June 2023).
18. Oficina Española de Patentes y Marcas (OEPM). Listado de Marcas Registradas, Boletin Oficial de la Propiedad Industrial. 2015. Available online: https://www.imarcas.com/registrodemarcas/patentes-y-marcas-oepm-09-12-2015/ (accessed on 4 June 2023).
19. Lescure Beruete, L.F. *Diccionario Gastronómico, Términos, Refranes, Citas y Poemas*, 1st ed.; Vision Libros: Madrid, Spain, 2007; p. 105.
20. Barreras, F.L. *Preelaboracion y Conservacion de Alimentos*; LibrosEnRed: Madrid, Spain, 2007.
21. Argumosa, G.; Franco, F. *Una Receta de Garbanzos*; Academia Castellano y Leonesa de Gastronomía y Alimentación: Valladolid, Spain, 2014.
22. Segura, J.; Escudero, R.; Romero de Ávila, M.D.; Cambero, M.I.; López-Bote, C.J. Effect of fatty acid composition and positional distribution within the triglyceride on selected physical properties of dry-cured ham subcutaneous fat. *Meat Sci.* **2015**, *103*, 90–95. [CrossRef]
23. Cambero, M.I.; Pereira-Lima, C.I.; Ordóñez, J.A.; García de Fernando, G.D. Beef broth flavor: Relation of components with the flavor developed at different cooking temperatures. *J. Sci. Food Agric.* **2000**, *80*, 1519–1528. [CrossRef]
24. Cambero, M.I.; Pereira-Lima, C.I.; Ordóñez, J.A.; García de Fernando, G.D. Beef broth flavor: Study of flavor development. *J. Sci. Food Agric.* **2000**, *80*, 1510–1518. [CrossRef]
25. *ISO 6658:2017*; Sensory Analysis—Methodology—General Guidance. ISO: Geneva, Switzerland, 2017.

26. Christensen, Z.T.; Ogden, L.V.; Dunn, M.L.; Eggett, D.L. Multiple comparison procedures for analysis of ranked data. *J. Food Sci.* **2006**, *71*, S132–S143. [CrossRef]
27. Fabbri, A.D.T.; Crosby, G.A. A review of the impact of preparation and cooking on the nutritional quality of vegetables and legumes. *Int. J. Gastron. Food Sci.* **2016**, *3*, 2–11. [CrossRef]
28. Wood, J.A. Evaluation of cooking time in pulses: A review. *Cereal Chem.* **2017**, *94*, 32–48. [CrossRef]
29. Clemente, A.; Sánchez-Vioque, R.; Vioque, J.; Bautista, J.; Millán, F. Effect of processing on water absorption and softening kinetics in chickpea (*Cicer arietinum* L.) seeds. *J. Sci. Food Agric.* **1998**, *78*, 169–174. [CrossRef]
30. Tharanathan, R.N.; Mahadevamma, S. Grain legumes—A boon to human nutrition. *Trends Food Sci. Technol.* **2003**, *14*, 507–518. [CrossRef]
31. Sefa-dedeh, S.; Stanley, D.W.; Voisey, P.W. Effects of soaking time and cooking conditions on texture and microstructure of cowpeas (*Vigna unguiculata*). *J. Food Sci.* **1978**, *43*, 1832–1838. [CrossRef]
32. Hood-Niefer, S.D.; Warkentin, T.D.; Chibbar, R.N.; Vandenberg, A.; Tyler, R.T. Effect of genotype and environment on the concentrations of starch and protein in, and the physicochemical properties of starch from, field pea and fababean. *J. Sci. Food Agric.* **2012**, *92*, 141–150. [CrossRef]
33. Güzel, D.; Sayar, S. Effect of cooking methods on selected physicochemical and nutritional properties of barlotto bean, chickpea, faba bean, and white kidney bean. *J. Food Sci. Technol.* **2012**, *49*, 89–95. [CrossRef]
34. Wani, I.A.; Sogi, D.S.; Wani, A.A.; Gill, B.S. Physical and cooking characteristics of some Indian kidney bean (*Phaseolus vulgaris* L.) cultivars. *J. Saudi Soc. Agric. Sci.* **2017**, *16*, 7–15. [CrossRef]
35. Acevedo Martínez, K.A.; Yang, M.M.; González de Mejia, E. Technological properties of chickpea (*Cicer arietinum*): Production of snacks and health benefits related to type-2 diabetes. *Compr. Rev. Food Sci. Food Saf.* **2021**, *20*, 3762–3787. [CrossRef] [PubMed]
36. Xiao, S.; Li, Z.; Zhou, K.; Fu, Y. Chemical composition of Kabuli and Desi chickpea (*Cicer arietinum* L.) cultivars grown in Xinjiang, China. *Food Sci. Nutr.* **2023**, *11*, 236–248. [CrossRef]
37. Sayar, S.; Turhan, M.; Gunasekaran, S. Analysis of chickpea soaking by simultaneous water transfer and water–starch reaction. *J. Food Eng.* **2001**, *50*, 91–98. [CrossRef]
38. Wang, T.L.; Domoney, C.; Hedley, C.L.; Casey, R.; Grusak, M.A. Can we improve the nutritional quality of legume seeds? *Plant Physiol.* **2003**, *131*, 886–891. [CrossRef]
39. Avola, G.; Patanè, C. Variation among physical, chemical and technological properties in three Sicilian cultivars of Chickpea (*Cicer arietinum* L.). *Int. J. Food Sci. Technol.* **2010**, *45*, 2565–2572. [CrossRef]
40. Kaur, M.; Singh, N.; Sodhi, N.S. Physicochemical, cooking, textural and roasting characteristics of chickpea (*Cicer arietinum* L.) cultivars. *J. Food Eng.* **2005**, *69*, 511–517. [CrossRef]
41. Singh, N.; Kaur, S.; Isono, N.; Noda, T. Genotypic diversity in physico-chemical, pasting and gel textural properties of chickpea (*Cicer arietinum* L.). *Food Chem.* **2010**, *122*, 65–73. [CrossRef]
42. Gómez-Favela, M.A.; García-Armenta, E.; Reyes-Moreno, C.; Garzón-Tiznado, J.A.; Perales-Sánchez, J.X.K.; Caro-Corrales, J.J.; Gutiérrez-Dorado, R. Modelling of water absorption in chickpea (*Cicer arietinum* L.) seeds grown in Mexico's northwest. *Rev. Mex. Ing. Quim.* **2017**, *16*, 179–191. [CrossRef]
43. Costa, R.; Fusco, F.; Gândara, J.F.M. Mass transfer dynamics in soaking of chickpea. *J. Food Eng.* **2018**, *227*, 42–50. [CrossRef]
44. Kalefetoğlu Macar, T.; Macar, O.; İnci Mart, D. Variability in some biochemical and nutritional characteristics in Desi and Turkish Kabuli chickpea (*Cicer arietinum* L.) types. *Celal Bayar Üniv. Fen Bilim. Derg.* **2017**, *13*, 677–680. [CrossRef]
45. Ipekesen, S.; Basdemir, F.; Tunc, M.; Bicer, B.T. Minerals, vitamins, protein and amino acids in wild Cicer species and pure line chickpea genotypes selected from a local population. *J. Elem.* **2022**, *27*, 127–140. [CrossRef]
46. Sahu, V.K.; Tiwari, S.; Gupta, N.; Tripathi, M.K.; Yasin, M. Evaluation of physiological and biochemical contents in Desi and Kabuli chickpea. *Legume Res.* **2022**, *45*, 1197–1208. [CrossRef]
47. Koskosidis, A.; Khah, E.; Mavromatis, A.; Irakli, M.; Vlachostergios, D.N. Effect of genotype and sowing period on chickpea quality, bioactive and antioxidant traits. *Legume Res.* **2021**, *44*, 1026–1031. [CrossRef]
48. Ozaktan, H.; Uzun, S.; Uzun, O.; Yasar Ciftci, C. Change in chemical composition and morphological traits of chickpea (*Cicer arietinum* L.) genotypes grown under natural conditions. *Gesunde Pflanz.* **2022**, 1–16. [CrossRef]
49. Klamczynska, B.; Czuchajowska, Z.; Baik, B.-K. Composition, soaking, cooking properties and thermal characteristics of starch of chickpeas, wrinkled peas and smooth peas. *Int. J. Food Sci. Technol.* **2001**, *36*, 563–572. [CrossRef]
50. Cobos, M.J.; Izquierdo, I.; Sanz, M.A.; Tomás, A.; Gil, J.; Flores, F.; Rubio, J. Genotype and environment effects on sensory, nutritional, and physical traits in chickpea (*Cicer arietinum* L.). *Span. J. Agric. Res.* **2016**, *14*, e0709. [CrossRef]
51. Chigwedere, C.M.; Wanasundara, J.P.D.; Shand, P.J. Sensory descriptors for pulses and pulse-derived ingredients: Toward a standardized lexicon and sensory wheel. *Compr. Rev. Food Sci. Food Saf.* **2022**, *21*, 999–1023. [CrossRef] [PubMed]
52. Mohammadi, X.; Deng, Y.; Matinfar, G.; Singh, A.; Mandal, R.; Pratap-Singh, A. Impact of Three Different Dehydration Methods on Nutritional Values and Sensory Quality of Dried Broccoli, Oranges, and Carrots. *Foods* **2020**, *9*, 1464. [CrossRef]
53. Johnny, S.; Razavi, S.M.A.; Khodaei, D. Hydration kinetics and physical properties of split chickpea as affected by soaking temperature and time. *J. Food Sci. Technol.* **2015**, *52*, 8377–8382. [CrossRef]
54. Prasad, K.; Vairagar, P.R.; Bera, M.B. Temperature dependent hydration kinetics of *Cicer arietinum* splits. *Food Res. Int.* **2010**, *43*, 483–488. [CrossRef]

55. Mannam, V. Role of hydration in grain processing: A review. In *Developing Technologies in Food Science: Status, Applications, and Challenges*, 1st ed.; Meghwal, M., Goyal, M.R., Eds.; Apple Academic Press: New York, NY, USA, 2017; pp. 33–63.

56. Frías, J.; Vidal-Valverde, C.; Sotomayor, C.; Díaz-Pollan, C.; Urbano, G. Influence of processing on available carbohydrate content and antinutritional factors of chickpeas. *Eur. Food Res. Technol.* **2000**, *210*, 340–345. [CrossRef]

57. Krokida, M.K.; Marinos-Kouris, D. Rehydration kinetics of dehydrated products. *J. Food Eng.* **2003**, *57*, 1–7. [CrossRef]

58. Monteiro, R.L.; Domschke, N.N.; Tribuzi, G.; Teleken, J.T.; Carciofi, B.A.M.; Laurindo, J.B. Producing crispy chickpea snacks by air, freeze, and microwave multi-flash drying. *LWT-Food Sci. Technol.* **2021**, *140*, 110781. [CrossRef]

59. Uzogara, S.; Morton, I.; Daniel, J. Influence of various salts in the cooking water on pectin losses and cooked texture of cowpeas (*Vigna unguiculata*). *J. Food Biochem.* **2007**, *14*, 283–291. [CrossRef]

60. Aravindakshan, S.; Nguyen, T.H.A.; Kyomugasho, C.; Buvé, C.; Dewettinck, K.; Van Loey, A.; Hendrickx, M.E. The impact of drying and rehydration on the structural properties and quality attributes of pre-cooked dried beans. *Foods* **2021**, *10*, 1665. [CrossRef]

61. Ulloa, J.A.; Ibarra-Zavala, S.J.; Ramírez-Salas, S.P.; Rosas-Ulloa, P.; Ramírez-Ramírez, J.C.; Ulloa-Rangel, B.E. Chemical, physico-chemical, nutritional, microbiological, sensory and rehydration characteristics of instant whole beans (*Phaseolus vulgaris*). *Food Technol. Biotechnol.* **2015**, *53*, 48–56. [CrossRef]

62. Chen, K.; Zhang, M.; Bhandari, B.; Sun, J.; Chen, J. Novel freeze drying based technologies for production and development of healthy snacks and meal replacement products with special nutrition and function: A review. *Dry. Technol.* **2022**, *40*, 1582–1597. [CrossRef]

63. Gupta, S.; Liu, C.; Sathe, S.K. Quality of a chickpea-based high protein snack. *J. Food Sci.* **2019**, *84*, 1621–1630. [CrossRef]

64. Jayasena, D.D.; Ahn, D.U.; Nam, K.C.; Jo, C. Flavour chemistry of chicken meat: A review. *Asian-Australas. J. Anim. Sci.* **2013**, *26*, 732–742. [CrossRef]

65. Kerth, C.R.; Miller, R.K. Beef flavor: A review from chemistry to consumer. *J. Sci. Food Agric.* **2015**, *95*, 2783–2798. [CrossRef]

66. Juárez, M.; Aldai, N.; López-Campos, Ó.; Dugan, M.; Uttaro, B.; Aalhus, J. Beef texture and juiciness. In *Handbook of Meat and Meat Processing*, 2nd ed.; Hui, Y.H., Ed.; CCR Press: Boca Ratón, FL, USA, 2012; pp. 177–206. [CrossRef]

67. Caliskan, G.; Dirim, S.N. Drying characteristics of pumpkin (*Cucurbita moschata*) slices in convective and freeze dryer. *Heat Mass Transf./Waerme-Und Stoffuebertragung* **2017**, *53*, 2129–2141. [CrossRef]

68. Jukniene, I.; Zaborskiene, G.; Jankauskiene, A.; Kabašinskiene, A.; Zakariene, G.; Bliznikas, S. Effect of lyophilization process on nutritional value of meat by-products. *Appl. Sci.* **2022**, *12*, 12984. [CrossRef]

69. Etherington, D.J.; Sims, T.J. Detection and estimation of collagen. *J. Sci. Food Agric.* **1981**, *32*, 539–546. [CrossRef]

70. Sims, J.S.; Bailey, J.B. Connective tissue. In *Developments in Meat Science*, 1st ed.; Lawrie, R.A., Ed.; Applied Science: London, UK, 1981; Volume 2, pp. 29–59.

71. Álvarez, M.D.; Canet, W.; Tortosa, M.E. Kinetics of thermal softening of potato tissue (cv. Monalisa) by water heating. *Eur. Food Res. Technol.* **2001**, *212*, 588–596. [CrossRef]

72. Bordoloi, A.; Singh, J.; Kaur, L. In vitro digestibility of starch in cooked potatoes as affected by guar gum: Microstructural and rheological characteristics. *Food Chem.* **2012**, *133*, 1206–1213. [CrossRef]

73. Paulus, K.; Saguy, I. Effect of heat treatment on the quality of cooked carrots. *J. Food Sci.* **1980**, *45*, 239–241. [CrossRef]

74. Paciulli, M.; Ganino, T.; Carini, E.; Pellegrini, N.; Pugliese, A.; Chiavaro, E. Effect of different cooking methods on structure and quality of industrially frozen carrots. *J. Food Sci. Technol.* **2016**, *53*, 2443–2451. [CrossRef] [PubMed]

75. Kaur, L.; Singh, N.; Singh Sodhi, N.; Singh Gujral, H. Some properties of potatoes and their starches I. Cooking, textural and rheological properties of potatoes. *Food Chem.* **2002**, *79*, 177–181. [CrossRef]

76. Wang, L.; Qiao, K.; Duan, W.; Zhang, Y.; Xiao, J.; Huang, Y. Comparison of taste components in stewed beef broth under different conditions by means of chemical analyzed. *Food Sci. Nutr.* **2020**, *8*, 955–964. [CrossRef]

77. Pereira-Lima, C.I.; Ordóñez, J.A.; García de Fernando, G.D.; Cambero, M.I. Influence of heat treatment on carnosine, anserine and free amino acid composition of beef broth and its role in flavour development. *Eur. Food Res. Technol.* **2000**, *210*, 165–172. [CrossRef]

78. Pérez-Palacios, T.; Eusebio, J.; Palma, S.F.; Carvalho, M.J.; Mir-Bel, J.; Antequera, T. Taste compounds and consumer acceptance of chicken soups as affected by cooking conditions. *Int. J. Food Prop.* **2017**, *20*, S154–S165. [CrossRef]

79. Zhang, L.; Hao, Z.; Zhao, C.; Zhang, Y.; Li, J.; Sun, B.; Tang, Y.; Yao, M. Taste compounds, affecting factors, and methods used to evaluate chicken soup: A review. *Food Sci. Nutr.* **2021**, *9*, 5833–5853. [CrossRef]

80. Dekker, M.; Hennig, K.; Verkerk, R. Differences in thermal stability of glucosinolates in five brassica vegetables. *Czech J. Food Sci.* **2009**, *27*, S85–S88. [CrossRef]

81. Renz, M.; Dekker, M.; Rohn, S.; Hanschen, F.S. Plant matrix concentration and redox status influence thermal glucosinolate stability and formation of nitriles in selected *Brassica* vegetable broths. *Food Chem.* **2023**, *404*, 134594. [CrossRef]

82. Toldrá, F.; Reig, M. Sausages. In *Handbook of Food Products Manufacturing*, 1st ed.; Hui, Y.H., Ed.; John Wiley & Sons: Hoboken, NJ, USA, 2007; Volume 2, pp. 251–264. [CrossRef]

83. Lorenzo, J.M.; Franco Ruiz, D.; Carballo, J. Fat content of dry-cured sausages and its effect on chemical, physical, textural and sensory properties. In *Fermented Meat Products*, 1st ed.; Zdolec, N., Ed.; CRC Press: Boca Ratón, FL, USA, 2016; Volume 1, pp. 474–487.

84. Paris, J.M.G.; Falkenberg, T.; Nöthlings, U.; Heinzel, C.; Borgemeister, C.; Escobar, N. Changing dietary patterns is necessary to improve the sustainability of Western diets from a One Health perspective. *Sci. Total Environ.* **2022**, *811*, 151437. [CrossRef]

Review

Pomegranate Husk Scald Browning during Storage: A Review on Factors Involved, Their Modes of Action, and Its Association to Postharvest Treatments

Mahshad Maghoumi [1,*], **Maria Luisa Amodio** [1], **Danial Fatchurrahman** [1], **Luis Cisneros-Zevallos** [2] **and Giancarlo Colelli** [1]

[1] Dipartimento di Scienze Agrarie, Degli Alimenti e Dell'ambiente, Università di Foggia, Via Napoli 25, 71122 Foggia, Italy

[2] Department of Horticultural Sciences, Texas A&M University, College Station, TX 77843, USA

* Correspondence: mahshad.maghoumi@unifg.it

Abstract: The pomegranate (*Punica granatum* L.), which contains high levels of health-promoting compounds, has received much attention in recent decades. Fruit storage potential ranges from 3 to 4 months in air and from 4 to 6 months in Controlled Atmospheres (CA) with 3–5% oxygen and 10–15% carbon dioxide. Storage life is limited by decay, chilling injury, weight loss (WL), and husk scald. In particular, husk scald (HS) limits pomegranate long-term storage at favorable temperatures. HS appears as skin browning which expands from stem end towards the blossom end during handling or long-term storage (10–12 weeks) at 6–10 °C. Even though HS symptoms are limited to external appearance, it may still significantly reduce pomegranate fruit marketability. A number of postharvest treatments have been proposed to prevent husk scald, including atmospheric modifications, intermittent warming, coatings, and exposure to 1-MCP. Long-term storage may induce phenolic compounds accumulation, affect organelles membranes, and activate browning enzymes such as polyphenol oxidases (PPO) and peroxidases (POD). Due to oxidation of tannins and phenolics, scalding becomes visible. There is no complete understanding of the etiology and biochemistry of HS. This review discusses the hypothesized mechanism of HS based on recent research, its association to postharvest treatments, and their possible targets.

Keywords: pomegranate; browning; oxidative stress; long term storage; husk scald; polyphenol oxidase; postharvest treatments

Citation: Maghoumi, M.; Amodio, M.L.; Fatchurrahman, D.; Cisneros-Zevallos, L.; Colelli, G. Pomegranate Husk Scald Browning during Storage: A Review on Factors Involved, Their Modes of Action, and Its Association to Postharvest Treatments. *Foods* **2022**, *11*, 3365. https://doi.org/10.3390/foods11213365

Academic Editor: Jinhe Bai

Received: 20 September 2022
Accepted: 19 October 2022
Published: 26 October 2022

1. Introduction

One of the most well-known fruits in the world, the pomegranate (*Punica granatum* L.), is likely native from Iran and grown in areas with a Mediterranean environment [1]. The fruit is made up of an albedo, septa, membranes, a hard, leathery outer layer, and many arils, where a translucent sac that holds 80% juice and 20% seed envelops each edible aril [2]. This fruit is considered as a non-climacteric fruit [3] and is characterized by low rate of postharvest respiration and ethylene production. Pomegranate is a rich source of organic acids, micro- and macro-nutrients, as well as anti-mutagenic, anti-inflammatory, anti-hypertension, and antioxidant compounds [4], which has influenced the increase in the demand for pomegranate fruit and of their products due to the fruit's high-valued health advantages and the public's growing awareness of functional foods [5]. Pomegranate harvest season is in the fall and lasts for fewer than three months, although it may be kept for several months in cold storage with a high humidity level. The fruit nevertheless experiences both qualitative and quantitative losses [6].

Due to the slowing down of cell metabolism at low temperatures and the delay in senescence, refrigerated storage of fruits and vegetables allows for the preservation of their quality after harvest [7–9]. However, some tropical and subtropical fruit, such as pomegranates, are not ideal for cold storage preservation since it results in physiological damage.

The main factors that, in general, may limit pomegranate ability to be stored and accelerate their quality decline include WL, shriveling, decay, chilling injury, husk scald and decrease flavor acceptability. Fruit transpiration, that causes WL and hardening and browning of the husk, is a significant storage challenge for pomegranate fruit [10,11] while additional factors affecting pomegranate storage life include decay, that is frequently brought on by the presence of a fungal inoculum in the fruit's blossom end [6] and this condition is worse at temperatures above 5 °C [12]. Chilling injury (CI) is a physiological disorder that takes place at low storage temperature (<5 °C) in pomegranates, and it manifests as internal discoloration of arils and albedo as well as brown spots and pitting in the skin, while on the other hand, husk scald (HS) typically originates from the stem end and manifests as a superficial browning of the skin and happens during storage temperatures above 5 °C. Differing from CI, HS does not damage the arils or the white locular septa of the fruit, and its severity can reach up to 60% of the skin [13–15] Figure 1.

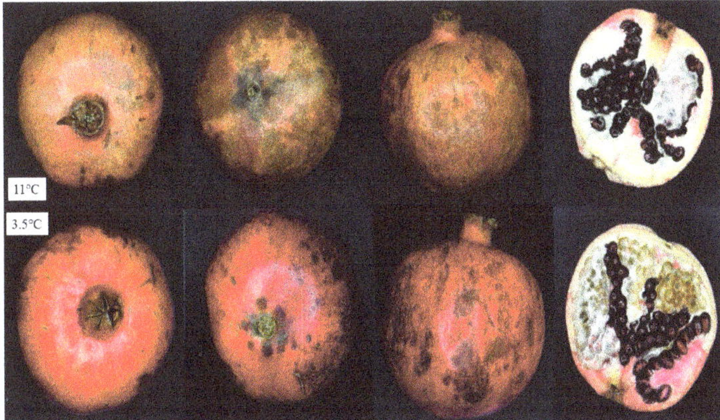

Figure 1. Husk scald and chilling injury in pomegranate "Wonderful" fruit skin after 120 days of storage at 11 and 3.5 °C and ≥95% RH, respectively.

When fruit is stored at 6–10 °C to avoid CI, the effects of superficial scald (husk scald) are typically more severe, and so it can also happen at storage temperatures above 10 °C and these scalded portions of the fruit may become prone to fungal deterioration in advanced stages [16]. Table 1 provides a summary of the differences between husk scald and chilling injury. It is assumed that CI and HS do not appear in the same fruit since they are mutually exclusive due to differences in storage temperatures, however, it has been suggested recently that both disorders can appear simultaneously at temperatures of 7 °C [17]. In general, scalding decreases the marketability of fresh pomegranate even when the internal quality remains in good conditions, because the quality of fruit is determined by both internal and external attributes [5,15,18]. It is thought that skin browning is the result of the oxidation of phenolic compounds [15,19].

Table 1. Difference between husk scald and chilling injury in pomegranate fruit during storage.

Husk Scald (HS)	Chilling Injury (CI)
Long-term storage > 10–12 weeks	Short-term storage (starting from 6–8 weeks)
Storage between 6–10 °C or above	Storage < 5 °C
Loss of red pigment and brownish appearance on the stem end that spreads towards the blossom end and peel hardening	Dark brown region scatters in whole fruit skin, pitting
No injury to the arils or locular septa	Affects internal parts and arils

Although there is a lack of information on scald disorder, the review's goal is to address factors that impact pomegranate husk scald development, as well as the potential mode of action involved.

2. Evidence for Husk Scald Incidence Mechanism

Weight loss, microbiological contamination, oxidative stresses, and browning are important variables that might decrease the storage life of fruit and vegetables during cold storage and handling. Lower or mild degrees of stress tend to enhance the levels of cellular antioxidant compounds, whereas severe stress may result in a decrease in such levels and the development of postharvest physiological disorders [20–22]. Reactive oxygen species (ROS) are scavenged by a sophisticated antioxidant defense system in plants that include both non-enzymatic and enzymatic components. Non-enzymatic antioxidant compounds include the main cellular buffers as well as a diversity of compounds including phenolic compounds, carotenoids and alkaloids [23]. Phenolics are plant secondary metabolites that are generated and accumulate in the plant, for example, by the activation of phenylalanine ammonia-lyase (PAL) [24]. These substances could be categorized as tannins, flavonoids, hydroxycinnamate esters and lignin among others. On the other hand, superoxide dismutase (SOD), catalase (CAT), ascorbate peroxidase (APX), and other enzymes are examples of enzymatic antioxidants [25,26]. Pomegranates are rich in phenolic compounds, and various studies have suggested that oxidative stress that occurs in the fruit skin while stored at temperatures over 5 °C, may play a significant role in husk scald disorder [27]. Enzymatic browning is the second leading cause of quality loss in fruits and vegetables, after microbial infection, and the food sector has prioritized preservation against oxidation during storage and handling [28].

Polyphenol oxidase (PPO) and peroxidase (POD) are the main enzymes that cause browning in plants. PPO is a type of enzyme known as an oxidoreductase. It catalyzes the conversion of monohydroxy phenols (e.g., phenol, tyrosine, and p-cresol) into *o*-dihydroxy phenols (e.g., catechol, dopamine, and adrenalin), as well as the dehydrogenation of *o*-dihydroxy phenols into o-quinones. Figure 2 illustrates how the formation of melanin and the oxidation of phenolic chemicals to quinones give foods their dark color [29].

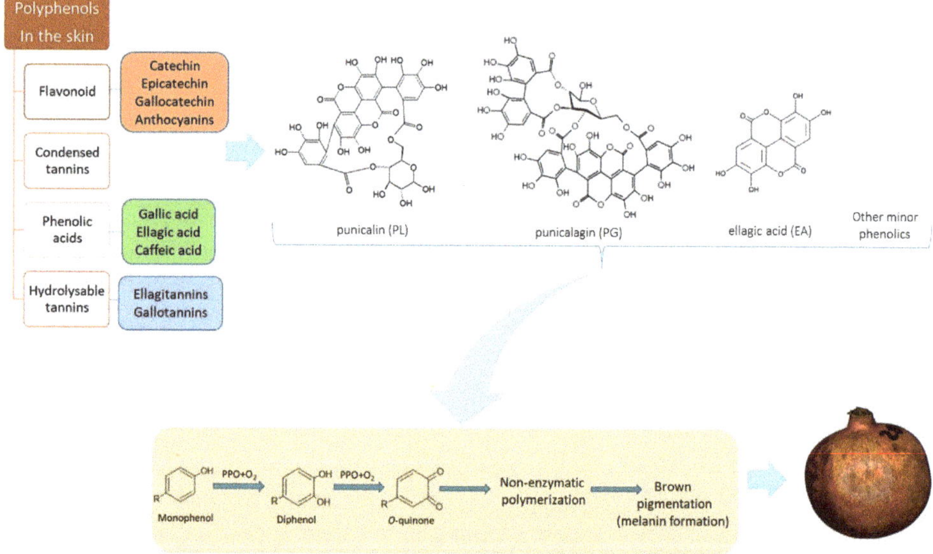

Figure 2. Pomegranate polyphenols in the skin, chemical structures of punicalin (PL), punicalagin (PG) and ellagic acid (EA), and brown pigment (melanin) formation from phenolic compounds [29–31].

Peroxidases (POD), which perform mono-electron oxidation on a wide range of composites in the presence of hydrogen peroxide, are additional enzymes which have polyphenol oxidation activity [32]. POD is a common enzyme in plants, but it still has an unclear mechanism of action. It has been suggested that the function of this enzyme depends on the availability of hydrogen [33].

As illustrated in Figure 2, the principal polyphenols in pomegranate peel are ellagitannins (hydrolysable tannins), phenolic acids (mostly hydroxybenzoic acids), and flavonoids (anthocyanins and other complex flavonoids) [34]. Punicalagin (PG) ($C_{48}H_{28}O_{30}$) is the most abundant polyphenol in pomegranate skin [35], which are hexahydroxydiphenic acid (HHDP) esters coupled to polyols [36]. The PG content in the peel can reach up to 66% of the total polyphenols, it is soluble in water and can be found naturally as reversible anomers, alpha and beta; nonetheless, they are frequently referred to as punicalagin [37]. By releasing one HHDP, PG can be hydrolyzed to punicalin (PL), another ellagitannin, and both PG and PL can be hydrolyzed to Ellagic acid (EA) under particular conditions [38]. Although the structural units of PG, PL, and EA are similar (Figure 2), they differ in terms of molecular size, polarity, and solubility [30]. Moreover, flavonoids of pomegranate skin include catechin, epicatechin, quercetin, anthocyanins and procyanidins [39].

It was discovered that in scalded pomegranate fruits, PPO and POD activity increased while total phenolic and total tannin components decreased [15]. Additionally, there was a correlation between husk scald injury and the quantity of extractable *o*-dihydroxy phenols detected in the skin affected regions. Although the amount of *o*-dihydroxy phenols in pomegranate skin is quite little, it has been demonstrated that during storage, husk scald can occur as result of the enzyme-mediated oxidation of phenols present in the skin [18]. Scalded fruits had decreased antioxidant activity and total phenolic components, which was followed by a high expression ratio of PPO to PAL [17].

Temperature has a considerable impact on the rate of enzymatic browning. Fruit and vegetables do not become brown when stored at low temperature, and PPO's enzymatic activity ceases at temperatures below 7 °C even though it is not deactivated [29].

In addition, CAT activity in pomegranate peel negatively correlated with the browning index while ascorbic acid oxidase (AAO), PPO, and POD favorably correlated with the skin browning index [19]. Accordingly, it seems that the accumulation of H_2O_2 with the decline of CAT activity made the tannins oxidize to browning substances. In the peel of scalded fruit, it was shown membrane instability, increased lipid peroxidation, and enhanced carotenoid production, which may suggest a response to oxidative stress [18].

Punicalin levels were strongly and negatively connected with husk scald, but punicalagin had no significant correlation to the incidence of scald. This finding raises the possibility that high punicalin levels could prevent the occurrence of HS in pomegranate fruit [40].

The symptoms and occurrence of pomegranate and apple scald are similar, however, current research indicates that the biochemical reasons and control mechanisms of the two disorders are different. Diphenylamine (DPA) treatment as an antioxidant ingredient, for instance, can decrease apple scald, but it has no effect on pomegranate scald [18,41]. Although DPA does not affect scald development in pomegranates, researchers did not eliminate the possibility that an oxidation process is involved in symptom development [18].

Anthocyanins are water-soluble polyphenolic pigments that are responsible for the appealing red-violet-blue hues of many fruits, especially pomegranate arils and husks [42]. Anthocyanins also have significant antioxidant action [43]. However, no links between anthocyanins and the signs of husk scald were discovered [40].

The findings imply that neither the total soluble solids content nor the titratable acidity or husk color are connected to the development of husk disorders [40].

The study on the anatomical changes in HS of pomegranate skin showed cracks in the epidermis and cuticle as well as cell degradation only in the very outer layers of the skin, while the inner layers seemed intact [17]. HS fruit also showed skin hardness as a result of transpiration and water loss [44]. Additionally, the monitoring of changes in tissue thickness was conducted to determine the trends of water loss on pomegranate fruit in relation to the peel tissues and location. During fruit storage, waxy cuticles fragmented more rapidly and microcracks widened. Compared to the stem end of the fruit, the calyx-end and equatorial region of the fruit exhibited a significantly higher number of lenticels, a larger lenticel size, and a generally thinner peel. The calyx end of the fruit was more susceptible to water loss than the equatorial- and stem-end regions of the fruit [45].

Gene *Pgr023188* encoding uncharacterized protein upregulated in HS fruits, this gene has 50% similarity to gene *BVC80_1667g87* that encodes for glycine rich protein (GRPs). These genes are involved in increasing tolerance during water stress [46]. Moreover, genes associated with stress (*Pgr000486* and *Pgr006284*), defense (*Pgr011016*), oxidative stress (*Pgr006284*) and Glycosyltransferase (GTs) (*Pgr007593*) which is crucial for the biosynthesis of secondary metabolites were highly presented in HS skin [44]. However, no comprehensive model of mode of action of HS development involving these genes have been reported so far in the literature.

3. Hypothetical Model of Pomegranate Husk Scald Incidence

A possible mechanism of husk scald development is summarized in Figure 3. It is suggested that when water loss is present in pomegranate fruit browning of the skin is enhanced [47,48]. In addition, it has been reported that oxidative stress is induced by water stress, which leads to the synthesis of nicotinamide adenine dinucleotide phosphate (NADPH) oxidase and trigger the oxidative burst [49,50]. The increase in ROS, in addition to promoting phenolic biosynthesis mediated by PAL activity, also induces phenolic oxidation due to an increase in PPO activity and loss of membrane compartmentalization in a similar fashion as a wound-like effect [51].

Losing selective permeability in the presence of oxygen may lead to membrane leakage and interaction between enzymes (PPO, POD) which are located in the cytoplasm with substrate (phenolic compounds such as hydroxy cinnamic acid and tannins including punicalin, punigalagin in the cells rind vacuole). Higher ratio of PPO/PAL activity is associated with oxidation of phenolic compounds to quinones which creates colored brown products [44]. Eventually, the accumulation of these browning agents in the skin become visible and considered as husk scald.

To avoid this phenomenon of HS in pomegranate during storage various methods were developed in the past as postharvest treatments without knowing the mode of action of HS development. However, we can infer according to the proposed model herein (Figure 3) that the role of these methods is either to inactivate PPO (through limitation of oxygen exposure such as CA, modified atmosphere packaging (MAP), coating or heat) or to avoid contact between the enzyme and its substrate by increasing cell integrity using intermittent warming.

Investigating postharvest treatments preventing husk scald disorder, is helpful in order to confirm the possible mechanism of HS presented herein. Numerous methods and strategies for postharvest storage of pomegranate avoiding husk scald will be discussed in this review paper and their possible targets.

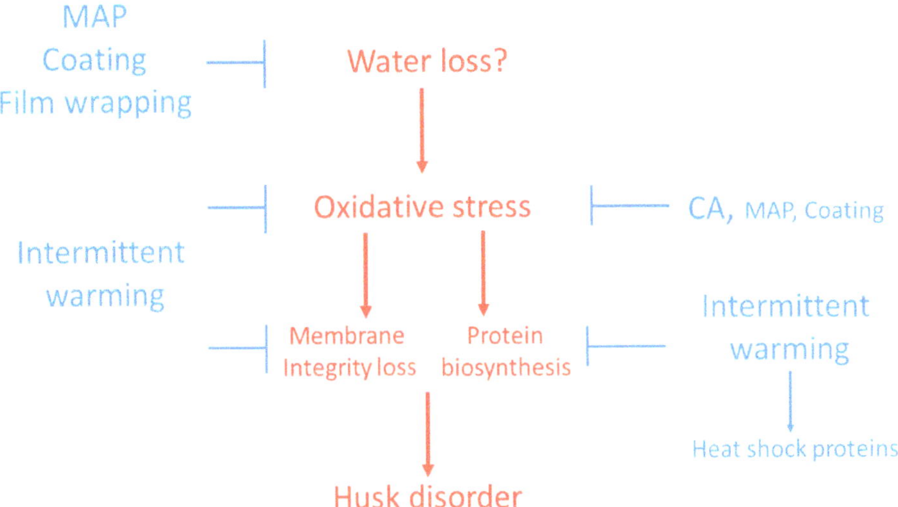

Figure 3. Hypothesis of pomegranate husk scald incidence during storage and the possible targets of postharvest treatments. Accordingly, MAP, coating and film wrapping systems would work mainly by preventing water loss, while CA would have a mode of action involving a decrease in oxidative stress by reducing the presence of oxygen. Furthermore, MAPs and coatings may also play in part this role of reducing oxidative stress. On the other hand, intermittent warming may exert its effects by inducing heat shock-like proteins in a hierarchical stress response model or alternatively by reducing oxidative stress and preserving membrane fluidity.

4. Factors Affecting Husk Scald Incidence

4.1. Pre-Harvest Factors

Adverse environmental conditions can disrupt normal homeostasis leading to cell death in plant cells and consequently physiological disorder may appear [52]. The tolerance of plant to regulate and maintain cell homeostasis depends on the interaction between genetics and environmental factors. In fact, genetics are a determining factor of plant resistance to physiological disorders [53–55]. The information regarding pre-harvest factors affecting husk scald development in pomegranates during storage is limited in the literature.

For instance, the results of a research on seven pomegranate accessions showed that higher antioxidant capacity, total phenolic compound, and punicalin caused lower WL, husk scald incidence and fungi decay in fruit. On the contrary, the amount of punicalagin had not significant correlation with the scald incidence. In addition, anthocyanin content did not show any significant correlation with HS incidence. However, light colored fruit were more sensitive to chilling injury and husk scald, regardless of having high antioxidant activity [40]. Furthermore, neither total soluble solids or total acidity had an effect on HS incidence [40].

There was a correlation between catechol content and peel browning in pomegranate fruit harvested in late season, and there was a stable catechol level in less browned peel during storage [56]. In addition, it has been suggested that pomegranates having a high antioxidant capacity and high total phenolics content in their husks are more tolerant to husk scald and may retard the formation of scalding [15,16,18,40,57–59].

Research by [18] showed that late-season harvested pomegranate are more susceptible to HS than mid-season harvested fruits, which is in contrast with the result reported by [13] that indicated that a delay in harvest reduces the incidence of HS. In general, total antioxidant, the level of hydrolysable tannins, such as punicalagin, punicalin and gallagic acid, and total phenols are high in pomegranate skin during the development stage, however, they decrease during the maturation stage and decrease further during storage [59,60]. This might in part explain the reason of higher susceptibility to HS of late-harvested pomegranates.

In Figure 4, we present an integral model based on the above data linking pre-harvest factors and its effects on HS development during storage. This integrative approach may be used as a reference for identifying and introducing future studies on pre-harvest factors and to dissect their effects on different pomegranate fruit quality attributes and physiological disorders besides HS (e.g., chilling injury effects).

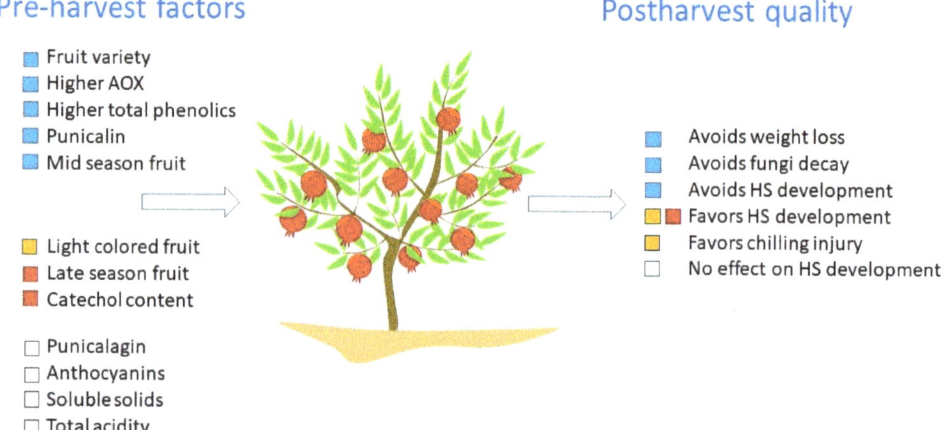

Figure 4. Integral model linking pre-harvest factors and its effects on HS development during postharvest storage. This integrative approach based on reported studied in the literature (see Section 4.1) can be used as a reference for identifying and introducing future studies on pre-harvest factors and also dissect their effects on different pomegranate fruit quality attributes and physiological disorders besides HS (e.g., chilling injury effects).

4.2. Postharvest Treatments for Husk Scald Control

In Table 2, different postharvest treatments and their effects on controlling husk scald incidence in pomegranate fruit are summarized, including the use of CA, MAP, film wrapping, intermittent warming, coatings, exposure to 1-MCP, and the use of exogenous putrescine treatment. In addition, in Figure 3, we present the possible targets of these postharvest treatments based on a model of HS development proposed herein.

4.2.1. Effect of CA

Controlled atmosphere (CA) and modified atmosphere (MA) storage include altering the storage gas environment, specifically the CO_2 and O_2 concentrations surrounding the commodities, and can often help to avoid CI or delay the onset of symptoms [25].

Even though pomegranates do not have climacteric respiratory systems, a suitable O_2 to CO_2 ratio can prevent peel browning [13]. The ratio has a higher impact on browning avoidance when combined with the proper storage temperature. Previous studies have shown that 2% O_2, without CO_2, reduced husk scald and postponed the onset of chilling injury symptoms in Israeli "Wonderful" pomegranate at 2 °C, but husk scald persisted after removal from storage. "Mollar" pomegranates were kept for up to eight weeks at 5 °C and at or above 95% relative humidity (RH) in air and in various combinations of controlled atmospheres, including 10% O_2 + 5% CO_2, 5% O_2 + 5% CO_2, 5% O_2 + 10% CO_2, and 5% O_2 + 0% CO_2 ethylene-free. Except at 10% O_2 + 5% CO_2, controlled atmospheric storage decreased WL, the risk of decay, and the severity of husk scald.

In contrast to the level previously suggested for the "Wonderful" cv., the "Mollar" cv. appears to be more vulnerable to husk scald than other types. Different cultivars may exhibit varying sensitivity to this disorder [47]. Higher CO_2 content had a stronger effect in preventing the rate and severity of husk scald after 5 months of storage at 6 °C [48]. The use of CA storage, which combines 5% O_2 + 15% CO_2, has the highest success rate for decreasing decay and husk scald problems of all of these methods, it has been shown to prolong the postharvest life of pomegranates for up to 5 months at 7 °C [6]. The increased accumulation of acetaldehyde, ethyl-acetate, and ethanol brought on by anaerobic respiration during CA storage is a drawback because these compounds alter the fruit's flavor [6,13,18]. Furthermore, CA with less oxygen may induce the formation of ethanol reduce the fruit's marketability [13]. On the other hand, reduced anthocyanin content, which affects the color of the arils, is a disadvantage of storing with elevated CO_2 levels [61].

Browning reduction may relate to increased membrane integrity and decreased peroxidase (POD) and polyphenol oxidase (PPO) enzyme activity [62]. Enzymatic browning occurs when various enzymes such as POD and PPO and their substrates such as phenolics, decompartmentalize due to a decrease in membrane integrity [63].

It was shown that loss of phospholipid content in pomegranate skin cells that were stored in air was lower than CA-stored fruit. However, lipophilic conjugates of *p*-coumaric acid (LPCAC) isolated from the neutral lipids fraction of peel tissue lipid extracts in air-stored fruit increased nearly three-fold than CA-stored fruit [18]. LPCAC are known as cutin- or suberin-like oligomers which increase in response of oxidative stress or water loss [64]. Furthermore, hydroxycinnamic acid synthesis may contribute to the browning events involved in scald formation [13].

In general, these results are in agreement with the proposed model of HS development presented herein (Figure 3), where water stress and oxidative stress are key factors in HS development in pomegranate fruit. It is likely that CA reduces oxidative stress when exposed to low levels of oxygen thus reducing HS development.

4.2.2. Effect of MAP

It has been shown that modified atmosphere packaging (MAP) can increase storage life by three months or more and prevent water loss, visible shriveling symptoms, husk scald, and rot in 'Mollar de Elche' [10], 'Ganesh' [65], 'Primosole' [2], 'Wonderful' [66], 'Hicrannar' and 'Hicaznar' [67–69] pomegranates.

MAP started to be used frequently for pomegranate shipping and storage. The effect of MAP pomegranate was studied using unperforated polypropylene (UPP) film of of 25 µm thickness and Perforated polypropylene (PPP) film of 20 µm thickness in 2 °C and 5 °C storage.

PPP at 5 °C was the best treatment for maintaining red skin-color of the arils at the end of storage, however, husk-scald development was higher at 5 °C possibly because of an increase in polyphenol oxidase activity. Additionally, only unpackaged control at 5 °C fruits showed a moderate (commercially objectionable) level of pitting and husk-scald severity index [10].

Although there are several studies on the effectiveness of MAP for prevention of HS, there is not any information about antioxidants, phenolic compounds or anthocyanin alteration in the fruit skin during storage under MAP treatment. However, in litchi fruit, MAP reduced skin browning through a reduction in PPO and POD activities [70]. Browning of the skin in litchi during storage is the major issue that limits the marketability, the loss of membrane integrity and oxidative stress, which induces skin browning [71]. Higher CO_2 concentration in MAP has been found effective in maintaining higher antioxidant activity enzymes including CAT, SOD and APX [72], on the other hand due to lower O_2 concentration, POD and PPO do not catalyze phenols oxidation [70]. Based on our assumption, MAP may have the same effect in pomegranate skin as it has in litchi fruit skin. Furthermore, it has been reported that litchi pericarp laccases play important roles in browning [73] and since laccase and PPO may share common substrate, such as catechol, studies are suggested to determine if laccase activity plays any role in pomegranate fruit HS development.

It is likely that MAP systems may operate by different modes of action against HS, likely through reduction in water stress and the corresponding oxidative stress signals, and simultaneously by reducing oxygen levels that reduces oxidative stress within the cells (Figure 3).

4.2.3. Film Wrapping

Pomegranates were preserved at 8 °C for 6 or 12 weeks, and a combination of film wrapping and 600 mgL^{-1} fludioxonil was utilized [74]. The results demonstrated that wrapping could increase marketability, almost entirely prevent husk scald or browning discoloration, and maintain fruit freshness during the entire storage period. Despite the fact that wrapped fruits had higher respiration rates than controls, film wrapping had a significant impact in reducing water loss. According to the model presented herein, this film wrapping likely reduced water loss and did not allow the accumulation of oxidative stress signaling preventing HS development (Figure 3).

4.2.4. Effect of Intermittent Warming

Intermittent warming is the interruption of low-temperature storage with one or more short periods of warm temperature for various periods of time. It has been shown to be beneficial in reducing CI and improving in keeping quality of several horticultural crops, presumably by allowing the continuation of aspects of ripening that had been halted by the low temperature storage [75]. Intermittent warming alleviates husk scald in pomegranates [76]. In chilling injury, intermittent warming has been associated with an increase in unsaturated fatty acid content of membranes increasing fluidity and to an increase in antioxidant levels due to an increment of polyamines that may protect membranes through antioxidant effects [77].

Ref. [78] compared pomegranate intermittent warming cycles of 1 day at 20 °C every 6 days, while the fruit was stored at 0 °C or 5 °C. The results showed that storage at 0 °C with intermittent warming avoided decay although increased the risk of chilling injuries such us pitting and husk scald; on the other hand, 5 °C storage with intermittent warming reduced these injuries but fungal attacks were not inhibited.

The application of intermittent warming for 24 h at 15 ± 0.5 °C every 5 days for pomegranate fruit, stored at 5.0 ± 0.5 °C and 5.0% CO$_2$ + 8.0% O$_2$ could delay browning, with the browning index being 0.15 after 120 days of storage. The results showed that the activities of the browning index of peel were correlated positively with ascorbic acid oxidase (AAO), PPO and POD, while correlated negatively with catalase (CAT) activity. The application of intermittent warming and pomegranate fruit storage under the conditions of 5% CO$_2$ + 8% O$_2$ gas component could delay peel browning [19].

Although the mode of action of intermittent warming is not clear in HS, it is possible that different mode of actions might be taking place; on the one hand, changes in membrane fluidity and reduced oxidative stress may be favored [77] or alternatively the proposed mechanism of heat shock treatments might take place involving a hierarchical response to stresses, where heat shock-proteins synthesis is favored to a heat stress compared to a subsequent water stress exposure [79]. Thus, according to the proposed model herein, intermittent warming will favor membrane fluidity an antioxidant protection, or may affect proteins synthesized and HS development would not be favored (Figure 3). More research is needed to test this hypothesis of redirecting protein synthesis of key proteins and enzymes due to the application of heat treatments or intermittent warming.

4.2.5. Effect of Coating

Fruit and vegetables are coated to extend their shelf life. Coatings can be made of protein, lipid, polysaccharide, resin, or any combination of, as well as originate from both plant and animal sources. During processing, handling, and storage, coating creates a barrier for gases and moisture and also serve as natural delivery systems for postharvest chemicals [80–87]. Fruit and vegetables can be treated with soy lecithin-based compounds to delay ripening, maintain firmness, reduce physiological problems, and enhance their appearance and marketability [88,89]. The application of commercial formulation of soy lecithin coating on pomegranates significantly reduced fruit WL as well as incidence and severity of husk scald in 'Primosole' cultivar. Lecithin slightly decreased the transpiration rate and, therefore, its effect on WL was quite low. It is assumed that the effect of lecithin is because of the antioxidant properties of soy lecithin [75]. The application of MAP and coating with chitosan (CH) alone or in combination decreased HS incidence especially in MAP + CH treatment. MAP and CH + MAP treatments reduced WL by about 5-fold, compared to control treatment while CH coating alone was not effective as much as these treatments in reducing WL during storage and shelf life. Interestingly, CH treatment alone was not as effective as MAP or MAP + CH in controlling HS symptoms [90]. Retention of the husk color in the pomegranate fruit subjected to coating treatments could be attributed to their ability to modify surrounding atmosphere of the fruit, resulting in reduced respiration rate, WL, and inhibited the activity of enzymes that is associated with husk discoloration [91,92]. In general, according to the model presented herein, it is likely that coatings reduce HS incidence by decreasing water loss and avoiding oxidative stress signals (Figure 3).

4.2.6. Effect of 1-MCP

It has been reported that 1-MCP can minimize superficial apple and pear skin scalding and improve the storage and shelf life of fruit [93,94].

Pomegranate is a non-climacteric fruit; studies have shown that some non-climacteric fruits can benefit somewhat from 1-MCP [95–98].

Ref. [18] compared the application of diphenylamine (DPA) and 1-MCP alone and in combination; results showed neither diphenylamine at 1100 or 2200 μLL^{-1}, nor 1-methylcyclopropene at 1 μLL^{-1}, alone or together reduced scald incidence and severity in pomegranates. On the contrary, it has been reported that the peel browning of 'Dahongpao' pomegranates was significantly reduced using 0.25, 0.5, and 1 $\mu L/l$ 1-MCP concentrations, possibly related to a reduction in PPO activity [99]. Moreover, [100] examined an experiment using 1-MCP in combination with CA (2% O_2 + 5% CO_2) and air on "Wonderful" pomegranate. The results showed that 1-MCP can prevent husk scald in pomegranates stored in air storage. The partial control or delay of scald symptoms by 1-MCP shows that the biochemical mechanism(s) causing scald are activated, but not entirely by ethylene [101,102]. Furthermore, 1-MCP and DPA do not effectively prevent superficial scald in pomegranates as they do in apples [41,103].

Even though non-climacteric fruit do not even actually ripen, the low levels of ethylene they release are likely involved in the fruit's senescence, therefore treating them with 1-MCP may have some advantages. Pomegranates may benefit from lower husk scald and superior aril appearance and taste due to an inhibitory influence on the fruit's metabolic activity, which leads to senescence. More work has to be conducted to elucidate the role of ethylene in HS development before promoting 1-MCP to control HS development.

4.2.7. Exogenous Putrescine Treatment

Polyamines (PAs) are compounds naturally synthesized in plants and have roles in development, ripening and senescence processes. Putrescine (PUT), spermine (SPE), and spermidine (SPD) are common PAs applied externally to increase the shelf-life of fruit [104]. The application of 2 and 3 mmol/L putrescine before 4 months storage at 5C and 95% RH was effective for husk scald prevention for the first 3 months storage. However, all of the fruit developed scald at the end of 4th month of storage. Moreover, a higher PUT concentration showed a significant decrease in WL of fruit due to maintenance of cell integrity and permeability. Despite the fact that PUT is considered to have antioxidant properties, in this study the treatment was not applied as a coating and thus, there was enough oxygen supply for enzymatic oxidation of phenolic compounds during long-term storage [105]. In general, more work is needed to elucidate any possible role of polyamines in reducing HS incidence and their mode of action.

Table 2. Summary of postharvest treatments applied to prevent husk scald (HS) in pomegranate fruit during storage.

Reference	Key Findings	Treatment Description	Fruit Treatment
[13]	Reduced HS	$2\% \ O_2 + 0\% CO_2$	CA
[47]	The more O_2 available the more HS occurs	$10\% \ O_2 + 5\% \ CO_2$ $5\% \ O_2 + 5\% \ CO_2$ $5\% \ O_2 + 0\% \ CO_2$	CA
[6]	The best combination controlling HS	$5\% \ O_2 + 15\% \ CO_2$	CA
[48]	The higher CO_2 concentration The better HS control	3% and $6\% \ CO_2$	CA
[10]	PPP film and storage at 5 °C the best for controlling HS	Unperforated polypropylene (UPP) film and Perforated polypropylene (PPP) film	MAP
[2]	Inhibited HS	Film wrapping in combination with $600 \ mgL^{-1}$ fludioxonil	MAP
[78]	Decreased HS	Intermittent warming cycles of 1 day at 20 °C every 6 days	Intermittent warming
[19]	Delay HS incidence	Intermittent warming for 1 day at 15 °C every 5 days and $5\% \ CO_2 + 8\% \ O_2$	Intermittent warming
[75]	Reduce severity of HS	Soy lecithin	Coating
[90]	Successfully controlled HS	Chitosan treatment 1% with MAP	Coating
[18]	No effect on controlling HS	1-MCP alone or in combination with diphenylamine (DPA)	1-MCP
[99]	Peel browning index decreased in all treatments, up to 35% at 0.5 µL/L concentration	1-MCP (0.25, 0.5, and 1 µL/L)	1-MCP
[100]	1-MCP partially prevents HS specially in air storage	1-MCP combined with controlled atmosphere ($2\% \ O_2 + 5\% \ CO_2$) and air	1-MCP
[105]	Effective for husk scald prevention for the first 3 months storage, afterwards not effective	2 and 3 mmol/L before storage at 5 °C	Putrescine

5. Conclusions

Pomegranate is an exotic fruit with high valuable nutritional benefits. This fruit even has the potential to be considered a functional or medicinal food that recently has received much attention among consumers. Due to increased consumptions and demand, extending the marketing season, creating the opportunity for the sustainable consumption of fresh pomegranate and/or its availability to the processing industry by prolonging the storage life of pomegranates is crucial. HS, on the other hand, has a significant negative effect on pomegranate marketability. The results of this review paper and the hypothetical model described herein suggest that storage duration more likely triggers a cascade of oxidative changes that eventually cause a wound-like effect within skin cells and leads to skin enzymatic browning. This process of HS development apparently starts with water loss and a sequential oxidative stress mechanism. Further studies are needed to confirm water loss as the triggering mechanism of HS development. It is interesting that all of the successful postharvest treatments proposed so far in the literature to halt HS development can be explained by the proposed model of HS development described herein, either through control of water loss, control of oxidative stress, maintaining membrane integrity or through redirecting of protein synthesis of key proteins/enzymes involved in HS development. Furthermore, studies are needed to determine the specific phenolics [39] involved in HS development, since the type of phenolic has been shown to influence the hue of the browning polymerization process [106].

This review paper on HS development and mode of action can be the basis to revisit known postharvest treatments and to find and develop new techniques that can prevents HS development. For instance, the combination of treatments using a hurdle approach based on mode of actions of individual treatments may allow us to obtain additive or synergistic effects on the reduction in HS incidence. In spite of the fact that there are a number of studies controlling HS in the literature, proposing a model describing scald etiology and its prevention or even prediction seems essential for developing novel approaches to reduce HS.

Author Contributions: Conceptualization, M.M., L.C.-Z. and G.C.; investigation, M.M.; resources, L.C.-Z. and G.C.; writing—original draft preparation, M.M. and L.C.-Z.; review and editing, L.C.-Z., G.C., D.F. and M.L.A.; supervision, L.C.-Z. and G.C. All authors have read and agreed to the published version of the manuscript.

Funding: This research was funded by "Masseria Frutti Rossi srl": grant number 657/2017 UNIFG-CLE n. 0022936 III/19 08/09/2017.

Data Availability Statement: The data presented in this study are available on request from the corresponding author.

Conflicts of Interest: The authors declare no conflict of interest.

References

1. Holland, D.; Hatib, K.; Bar-Ya'akov, I. Pomegranate: Botany, Horticulture, Breeding. *Hortic. Rev.* **2009**, *35*, 127–191.
2. D'Aquino, S.; Palma, A.; Schirra, M.; Continella, A.; Tribulato, E.; La Malfa, S. Influence of film wrapping and fludioxonil application on quality of pomegranate fruit. *Postharvest Biol. Technol.* **2010**, *55*, 121–128. [CrossRef]
3. Elyatem, S.M.; Kader, A.A. Post-Harvest Physiology and Storage Behaviour of Pomegranate Fruits. *Sci. Hortic.* **1984**, *24*, 287–298. [CrossRef]
4. Opara, L.U.; Al-Ani, M.R.; Al-Shuaibi, Y.S. Physico-chemical Properties, Vitamin C Content, and Antimicrobial Properties of Pomegranate Fruit (*Punica granatum* L.). *Food Bioprocess Technol.* **2009**, *2*, 315–321. [CrossRef]
5. Erkan, M.; Dogan, A. Pomegranate. In *Postharvest Physiological Disorders in Fruits and Vegetables*; Freitas, S.T.D., Pareek, S., Eds.; CRC Press Taylor & Francis Group: Boca Raton, FL, USA, 2019; pp. 529–550.
6. Hess-Pierce, B.; Kader, A.A. Responses of 'Wonderful' pomegranates to controlled atmospheres. *Acta Hort.* **2003**, *600*, 751–757. [CrossRef]
7. McGlasson, W.B.; Scott, K.J.; Mendoza, D.B. The refrigerated storage of tropical and subtropical products. *Int. J. Refrig.* **1979**, *2*, 199–206. [CrossRef]
8. Hardenburg, R.E.; Watada, A.E.; Wang, C.Y. *The Commercial Storage of Fruits, Vegetables, and Florist and Nursery Stocks*; US Department of Agriculture: Washington, DC, USA, 1986; p. 130.
9. Makule, E.; Dimoso, N.; Tassou, S.A. Precooling and Cold Storage Methods for Fruits and Vegetables in Sub-Saharan Africa & mdash; A Review. *Horticulturae* **2022**, *8*, 776.
10. Artes, F.; Villaescusa, R.; Tudela, J.A. Modified Atmosphere Packaging of Pomegranate. *J. Food Sci.* **2000**, *65*, 1112–1116. [CrossRef]
11. Caleb, O.J.; Mahajan, P.V.; Opara, U.L.; Witthuhn, C.R. Modelling the respiration rates of pomegranate fruit and arils. *Postharvest Biol. Technol.* **2012**, *64*, 49–54. [CrossRef]
12. Roy, S.K.; Waskar, D.P. Pomegranate. In *Postharvest Physiology and Storage of Tropical and Subtropical Fruits*; Mitra, S., Ed.; Cab International: Wallingford, UK, 1997.
13. Ben-Arie, R.; Or, E. The development and control of husk scald on 'Wonderful' pomegranate fruit during storage. *J. Am. Soc. Hort. Sci. Hort.* **1986**, *111*, 395–399. [CrossRef]
14. Erkan, M.; Kader, A.A. Pomegranate (*Punica granatum* L.). In *Postharvest Biology and Technology of Tropical and Subtropical Fruits*; Yahia, E.M., Ed.; Woodhead: New York, NY, USA, 2011; pp. 287–311.
15. Arendse, E. Determining Optimum Storage Conditions for Pomegranate Fruit (cv. Wonderful). Ph.D. Thesis, University of Stellenbosch, Stellenbosch, South Africa, 2014.
16. Kader, A.A. Postharvest biology and technology of pomegranates. In *Pomegranates: Ancient Roots to Modern Medicine*; Seeram, N.P., Schulman, R.N., Heber, D., Eds.; CRC Press: Boca Raton, FL, USA, 2006; pp. 211–220.
17. Baghel, R.S.; Keren-Keiserman, A.; Ginzberg, I. Metabolic changes in pomegranate fruit skin following cold storage promote chilling injury of the peel. *Sci. Rep.* **2021**, *11*, 9141. [CrossRef]
18. Defilippi, B.G.; Whitaker, B.D.; Hess-Pierce, B.M.; Kader, A.A. Development and control of scald on wonderful pomegranates during long-term storage. *Postharvest Biol. Technol.* **2006**, *41*, 234–243. [CrossRef]

19. Zhang, Y.L.; Zhang, R.G. Study on the mechanism of browning of pomegranate (*Punica granatum* L. cv. "Ganesh") peel in different storage conditions. *Agric. Sci. China* **2008**, *7*, 65–73. [CrossRef]

20. Zhanyuan, D.; William, J.B. Peroxidative Activity of Apple Peel in Relation to Development of Poststorage Disorders. *HortScience* **1995**, *30*, 205–209. [CrossRef]

21. Toivonen, P. Postharvest Storage Procedures and Oxidative Stress. *HortScience* **2004**, *39*, 938–942. [CrossRef]

22. Cisneros-Zevallos, L.; Jacobo-Velázquez, D.A. Controlled Abiotic Stresses Revisited: From Homeostasis through Hormesis to Extreme Stresses and the Impact on Nutraceuticals and Quality during Pre- and Postharvest Applications in Horticultural Crops. *J. Agric. Food Chem.* **2020**, *68*, 11877–11879. [CrossRef]

23. Sharma, P.; Jha, A.; Dubey, R.; Pessarakli, M. Reactive Oxygen Species, Oxidative Damage, and Antioxidative Defense Mechanism in Plants under Stressful Conditions. *J. Bot.* **2012**, *2012*, 217037. [CrossRef]

24. Dixon, R.A.; Paiva, N.L. Stress-induced phenylpropanoid metabolism. *Plant Cell* **1995**, *7*, 1085. [CrossRef]

25. Wang, C.Y. Approaches to Reduce Chilling Injury of Fruits and Vegetables. *Hortic. Rev.* **2010**, *15*, 63–95. [CrossRef]

26. Mondal, K.; Sharma, N.; Malhotra, S.; Dhawan, K.; Singh, R. Antioxidant systems in ripening tomato fruits. *Biol. Plant.* **2004**, *48*, 49–53. [CrossRef]

27. Caleb, O.J.; Opara, U.L.; Witthuhn, C.R. Modified atmosphere packaging of pomegranate fruit and arils. *Food Bioprocess Technol.* **2012**, *5*, 15–30. [CrossRef]

28. Ioannou, I.; Ghoul, M. Prevention of enzymatic browning in fruit and vegetables. *Eur. Sci. J.* **2013**, *9*, 310–341.

29. Singh, B.; Suri, K.; Shevkani, K.; Kaur, A.; Kaur, A.; Singh, N. Enzymatic Browning of Fruit and Vegetables: A Review. *Enzyme. Food Technol.* **2018**, 63–78. [CrossRef]

30. Sun, Y.-q.; Tao, X.; Men, X.-m.; Xu, Z.-w.; Wang, T. In vitro and in vivo antioxidant activities of three major polyphenolic compounds in pomegranate peel: Ellagic acid, punicalin, and punicalagin. *J. Integr. Agric.* **2017**, *16*, 1808–1818. [CrossRef]

31. Kandylis, P.; Kokkinomagoulos, E. Food Applications and Potential Health Benefits of Pomegranate and its Derivatives. *Foods* **2020**, *9*, 122. [CrossRef]

32. Dunford, H.; Stillman, J. On the function and mechanism of action of peroxidases. *Coord. Chem. Rev.* **1976**, *19*, 187–251. [CrossRef]

33. Amiot, M.J.; Tacchini, M.; Aubert, S.; Nicolas, J. Phenolic composition and browning susceptibility of various apple cultivars at maturity. *J. Food Sci.* **1992**, *57*, 958–962.

34. Çam, M.; Hışıl, Y. Pressurised water extraction of polyphenols from pomegranate peels. *Food Chem.* **2010**, *123*, 878–885. [CrossRef]

35. Fischer, U.A.; Jaksch, A.V.; Carle, R.; Kammerer, D.R. Determination of Lignans in Edible and Nonedible Parts of Pomegranate (*Punica granatum* L.) and Products Derived Therefrom, Particularly Focusing on the Quantitation of Isolariciresinol Using HPLC-DAD-ESI/MSn. *J. Agric. Food Chem.* **2012**, *60*, 283–292. [CrossRef]

36. Landete, J. Ellagitannins, ellagic acid and their derived metabolites: A review about source, metabolism, functions and health. *Food Res. Int.* **2011**, *44*, 1150–1160. [CrossRef]

37. Li, J.; Li, G.; Zhao, Y.; Yu, C. Composition of pomegranate peel polyphenols and its antioxidant activities. *Sci. Agric. Sin.* **2009**, *42*, 4035–4041.

38. Cerdá, B.; Llorach, R.; Cerón, J.J.; Espín, J.C.; Tomás-Barberán, F.A. Evaluation of the bioavailability and metabolism in the rat of punicalagin, an antioxidant polyphenol from pomegranate juice. *Eur. J. Nutr.* **2003**, *42*, 18–28. [CrossRef]

39. Singh, B.; Singh, J.P.; Kaur, A.; Singh, N. Phenolic compounds as beneficial phytochemicals in pomegranate (*Punica granatum* L.) peel: A review. *Food Chem.* **2018**, *261*, 75–86. [CrossRef]

40. Matityahu, I.; Glazer, I.; Holland, D.; Bar-Ya'akov, I.; Ben-Arie, R.; Amir, R. Total Antioxidative Capacity and Total Phenolic Levels in Pomegranate Husks Correlate to Several Postharvest Fruit Quality Parameters. *Food Bioprocess Technol.* **2013**, *7*, 1938–1949. [CrossRef]

41. Whitaker, B. Oxidative Stress and Superficial Scald of Apple Fruit. *HortScience* **2004**, *39*, 933–937. [CrossRef]

42. Gil, M.I.; García-Viguera, C.; Artés, F.; Tomás-Barberán, F.A. Changes in pomegranate juice pigmentation during ripening. *J. Sci. Food Agric.* **1995**, *68*, 77–81. [CrossRef]

43. Seeram, N.P.; Aviram, M.; Zhang, Y.; Henning, S.M.; Feng, L.; Dreher, M.; Heber, D. Comparison of antioxidant potency of commonly consumed polyphenol-rich beverages in the United States. *J. Agric. Food Chem.* **2008**, *56*, 1415–1422. [CrossRef]

44. Belay, Z.A.; Caleb, O.J.; Vorster, A.; van Heerden, C.; Opara, U.L. Transcriptomic changes associated with husk scald incidence on pomegranate fruit peel during cold storage. *Food Res. Int.* **2020**, *135*, 109285. [CrossRef]

45. Lufu, R.; Ambaw, A.; Opara, U.L. Functional characterisation of lenticels, micro-cracks, wax patterns, peel tissue fractions and water loss of pomegranate fruit (cv. Wonderful) during storage. *Postharvest Biol. Technol.* **2021**, *178*, 111539. [CrossRef]

46. Giarola, V.; Krey, S.; von den Driesch, B.; Bartels, D. The *Craterostigma plantagineum* glycine-rich protein CpGRP1 interacts with a cell wall-associated protein kinase 1 (CpWAK1) and accumulates in leaf cell walls during dehydration. *New Phytol.* **2016**, *210*, 535–550. [CrossRef]

47. Artes, F.; Maren, J.G.; Martenez, J.A. Controlled atmosphere storage of pomegranate. *Z. Lebensm.-Unters. Forsch.* **1996**, *203*, 33–37. [CrossRef]

48. Nerya, O.; Gizis, A.; Tsyilling, A.; Gemarasni, D.; Sharabi-Nov, A.; Ben-Arie, R. Controlled Atmosphere Storage of Pomegranate. *Acta Hortic.* **2006**, *712*, 655–660. [CrossRef]

49. Low, P.S.; Merida, J.R. The oxidative burst in plant defense: Function and signal transduction. *Physiol. Plant* **1996**, *96*, 533–542. [CrossRef]

50. Van Breusegem, F.; Vranová, E.; Dat, J.F.; Inzé, D. The role of active oxygen species in plant signal transduction. *Plant Sci.* **2001**, *161*, 405–414. [CrossRef]
51. Cisneros-Zevallos, L.; Jacobo-Velázquez, D.; Pech, J.-C.; Koiwa, H. Signaling Molecules Involved in the Postharvest Stress Response of Plants. In *Handbook of Plant and Crop Physiology*; CRC Press: Boca Raton, FL, USA, 2014; pp. 259–276. [CrossRef]
52. Pareek, S. *Postharvest Physiological Disorders in Fruit and Vegetables*; CRC Press: Boca Raton, FL, USA, 2019; pp. 3–14.
53. Volz, R.K.; Alspach, P.A.; Fletcher, D.J.; Ferguson, I.B. Genetic variation in bitter pit and fruit calcium concentrations within a diverse apple germplasm collection. *Euphytica* **2006**, *149*, 1. [CrossRef]
54. De Freitas, S.T.; Amarante, C.V.T.d.; Labavitch, J.M.; Mitcham, E.J. Cellular approach to understand bitter pit development in apple fruit. *Postharvest Biol. Technol.* **2010**, *57*, 6–13. [CrossRef]
55. Juan Pablo, F.-T.; Javier, O.; Juan Antonio, M.; Antonio Luis, A.; Iban, E.; Pere, A.; Antonio José, M. Mapping Fruit Susceptibility to Postharvest Physiological Disorders and Decay Using a Collection of Near-isogenic Lines of Melon. *J. Amer. Soc. Hort. Sci.* **2007**, *132*, 739–748. [CrossRef]
56. Gao, M.Y.; Shi, F.H.; Chang, N.J. Experiment of pomegranate storage. *Deciduous Fruit Tree* **1987**, *3*, 28–30. (In Chinese)
57. Matityahu, I.; Marciano, P.; Holland, D.; Ben-Arie, R.; Amir, R. Differential effects of regular and controlled atmosphere storage on the quality of three cultivars of pomegranate (*Punica granatum* L.). *Postharvest Biol. Technol.* **2016**, *115*, 132–141. [CrossRef]
58. Tzulker, R.; Glazer, I.; Bar-Ilan, I.; Holland, D.; Aviram, M.; Amir, R. Antioxidant activity, polyphenol content, and related compounds in different fruit juices and homogenates prepared from 29 different pomegranate accessions. *J. Agric. Food Chem.* **2007**, *55*, 9559–9570. [CrossRef]
59. Shwartz, E.; Glazer, I.; Bar-Ya'akov, I.; Matityahu, I.; Bar-Ilan, I.; Holland, D.; Amir, R. Changes in chemical constituents during the maturation and ripening of two commercially important pomegranate accessions. *Food Chem.* **2009**, *115*, 965–973. [CrossRef]
60. Liu, C.; Zhang, Z.; Dang, Z.; Xu, J.; Ren, X. New insights on phenolic compound metabolism in pomegranate fruit during storage. *Sci. Hortic.* **2021**, *285*, 110138. [CrossRef]
61. Deirdre, M.H.; Maria, I.G.; Adel, A.K. Effect of Carbon Dioxide on Anthocyanins, Phenylalanine Ammonia Lyase and Glucosyltransferase in the Arils of Stored Pomegranates. *J. Am. Soc. Hortic. Sci. Jashs* **1998**, *123*, 136–140. [CrossRef]
62. Zauberman, G.; Ronen, R.; Akerman, M.; Weksler, A.; Rot, I.; Fuchs, Y. Post-harvest retention of the red colour of litchi fruit pericarp. *Sci. Hortic.* **1991**, *47*, 89–97. [CrossRef]
63. Jiang, Y.; Li, Y.; Li, J. Browning control, shelf life extension and quality maintenance of frozen litchi fruit by hydrochloric acid. *J. Food Eng.* **2004**, *63*, 147–151. [CrossRef]
64. Moire, L.; Schmutz, A.; Buchala, A.; Yan, B.; Stark, R.E.; Ryser, U. Glycerol Is a Suberin Monomer. New Experimental Evidence for an Old Hypothesis1. *Plant Physiol.* **1999**, *119*, 1137–1146. [CrossRef]
65. Nanda, S.; Sudhakar Rao, D.V.; Krishnamurthy, S. Effects of shrink film wrapping and storage temperature on the shelf life and quality of pomegranate fruits cv. Ganesh. *Postharvest Biol. Technol.* **2001**, *22*, 61–69. [CrossRef]
66. Porat, R.; Kosto, I.; Daus, A. Bulk storage of "Wonderful" pomegranate fruit using modified atmosphere bags. *Isr. J. Plant Sci.* **2016**, *63*, 45. [CrossRef]
67. Selcuk, N.; Erkan, M. Changes in antioxidant activity and postharvest quality of sweet pomegranates cv. Hicrannar under modified atmosphere packaging. *Postharvest Biol. Technol.* **2014**, *92*, 29–36. [CrossRef]
68. Selcuk, N.; Erkan, M. Changes in phenolic compounds and antioxidant activity of sour–sweet pomegranates cv. 'Hicaznar' during long-term storage under modified atmosphere packaging. *Postharvest Biol. Technol.* **2015**, *109*, 30–39. [CrossRef]
69. Candir, E.; Özdemir, A.E.; Aksoy, M.C. Effects of modified atmosphere packaging on the storage and shelf life of Hicaznar pomegranate fruits. *Turk. J. Agric. For.* **2019**, *43*, 241–253. [CrossRef]
70. Ali, S.; Sattar Khan, A.; Ullah Malik, A.; Anjum, M.A.; Nawaz, A.; Shoaib Shah, H.M. Modified atmosphere packaging delays enzymatic browning and maintains quality of harvested litchi fruit during low temperature storage. *Sci. Hortic.* **2019**, *254*, 14–20. [CrossRef]
71. Ali, S.; Khan, A.S.; Malik, A.U.; Shaheen, T.; Shahid, M. Pre-storage methionine treatment inhibits postharvest enzymatic browning of cold stored 'Gola' litchi fruit. *Postharvest Biol. Technol.* **2018**, *140*, 100–106. [CrossRef]
72. Song, L.; Chen, H.; Gao, H.; Fang, X.; Mu, H.; Yuan, Y.; Yang, Q.; Jiang, Y. Combined modified atmosphere packaging and low temperature storage delay lignification and improve the defense response of minimally processed water bamboo shoot. *Chem. Cent. J.* **2013**, *7*, 147. [CrossRef] [PubMed]
73. Fang, F.; Zhang, X.-L.; Luo, H.-H.; Zhou, J.-J.; Gong, Y.-H.; Li, W.-J.; Shi, Z.-W.; He, Q.; Wu, Q.; Li, L. An intracellular laccase is responsible for epicatechin-mediated anthocyanin degradation in litchi fruit pericarp. *Plant Physiol.* **2015**, *169*, 2391–2408. [CrossRef]
74. D'Aquino, S.; Schirra, M.; Gentile, A.; Tribulato, E.; La Malfa, S.; Palma, A. Postharvest Lecithin Application Improves Storability of 'Primosole' Pomegranates. *Acta Hortic.* **2012**, *934*, 733–739. [CrossRef]
75. Biswas, P.; East, A.R.; Hewett, E.W.; Heyes, J.A. Intermittent Warming in Alleviating Chilling Injury—A Potential Technique with Commercial Constraint. *Food Bioprocess Technol.* **2016**, *9*, 1–15. [CrossRef]
76. Artés, F.; Tudela, J.A.; Villaescusa, R. Thermal postharvest treatments for improving pomegranate quality and shelf life. *Postharvest Biol. Technol.* **2000**, *18*, 245–251. [CrossRef]
77. Taghipour, L.; Rahemi, M.; Assar, P.; Ramezanian, A.; Mirdehghan, S.H. Alleviating Chilling Injury in Stored Pomegranate Using a Single Intermittent Warming Cycle: Fatty Acid and Polyamine Modifications. *Int. J. Food Sci.* **2021**, *2021*, 2931353. [CrossRef]

78. Artés, F.; Tudela, J.A.; Gil, M.I. Improving the keeping quality of pomegranate fruit by intermittent warming. *Z. Lebensm.-Forsch. A* **1998**, *207*, 316–321. [CrossRef]

79. Saltveit, M.E. Wound induced changes in phenolic metabolism and tissue browning are altered by heat shock. *Postharvest Biol. Technol.* **2000**, *21*, 61–69. [CrossRef]

80. Sharma, H.P.; Chaudhary, V.; Kumar, M. Importance of edible coating on fruits and vegetables: A review. *J. Pharmacogn. Phytochem.* **2019**, *8*, 4104–4110.

81. Gennadios, A.; McHugh, T.H.; Krochta, J.M. Edible Coatings and Films. In *Edible Coatings and Films to Improve Food Quality*; Technomic Publ. Co.: Lancaster, PA, USA, 1994; p. 201.

82. Cisneros-Zevallos, L.; Krochta, J.M. Internal Modified Atmospheres of Coated Fresh Fruits and Vegetables: Understanding Relative Humidity Effects. *J. Food Sci.* **2002**, *67*, 2792–2797. [CrossRef]

83. Cisneros-Zevallos, L.; Krochta, J.M. Whey Protein Coatings for Fresh Fruits and Relative Humidity Effects. *J. Food Sci.* **2003**, *68*, 176–181. [CrossRef]

84. Cisneros-Zevallos, L.; Krochta, J.M. Dependence of Coating Thickness on Viscosity of Coating Solution Applied to Fruits and Vegetables by Dipping Method. *J. Food Sci.* **2003**, *68*, 503–510. [CrossRef]

85. Amin, U.; Khan, M.K.I.; Khan, M.U.; Ehtasham Akram, M.; Pateiro, M.; Lorenzo, J.M.; Maan, A.A. Improvement of the Performance of Chitosan—Aloe vera Coatings by Adding Beeswax on Postharvest Quality of Mango Fruit. *Foods* **2021**, *10*, 2240. [CrossRef]

86. Lu, J.; Li, T.; Ma, L.; Li, S.; Jiang, W.; Qin, W.; Li, S.; Li, Q.; Zhang, Z.; Wu, H. Optimization of heat-sealing properties for antimicrobial soybean protein isolate film incorporating diatomite/thymol complex and its application on blueberry packaging. *Food Packag. Shelf Life* **2021**, *29*, 100690. [CrossRef]

87. Wu, X.; Hu, Q.; Liang, X.; Fang, S. Fabrication of colloidal stable gliadin-casein nanoparticles for the encapsulation of natamycin: Molecular interactions and antifungal application on cherry tomato. *Food Chem.* **2022**, *391*, 133288. [CrossRef]

88. Ozgen, M.; Palta, J.P. Use of Lysophosphatidylethanolamine (Lpe), A Natural Lipid, to Accelerate Ripening and Enhance Shelf Life of Cranberry Fruit. *Acta Hortic.* **2003**, *628*, 141–146. [CrossRef]

89. Hoa, T.T.; Ducamp, M.-N. Effects of different coatings on biochemical changes of 'cat Hoa loc' mangoes in storage. *Postharvest Biol. Technol.* **2008**, *48*, 150–152. [CrossRef]

90. Candir, E.; Ozdemir, A.E.; Aksoy, M.C. Effects of chitosan coating and modified atmosphere packaging on postharvest quality and bioactive compounds of pomegranate fruit cv. 'Hicaznar'. *Sci. Hortic.* **2018**, *235*, 235–243. [CrossRef]

91. Ghaouth, A.; Arul, J.; Ponnampalam, R.; Boulet, M. Chitosan Coating Effect on Storability and Quality of Fresh Strawberries. *J. Food Sci.* **1991**, *56*, 1618–1620. [CrossRef]

92. Zhang, D.; Quantick, P.C. Effects of chitosan coating on enzymatic browning and decay during postharvest storage of litchi (*Litchi chinensis* Sonn.) fruit. *Postharvest Biol. Technol.* **1997**, *12*, 195–202. [CrossRef]

93. Fan, X.; Mattheis, J.P.; Blankenship, S. Development of apple superficial scald, soft scald, core flush, and greasiness is reduced by MCP. *J. Agric. Food Chem.* **1999**, *47*, 3063–3068. [CrossRef]

94. Zohar, S.; Amnon, L.; Susan, L. Effect of Heat or 1-Methylcyclopropene on Antioxidative Enzyme Activities and Antioxidants in Apples in Relation to Superficial Scald Development. *J. Am. Soc. Hortic. Sci. Jashs* **2003**, *128*, 761–766. [CrossRef]

95. Balogh, A.; Koncz, T.; Tisza, V.; Kiss, E.; Heszky, L. The Effect of 1-MCP on the Expression of Several Ripening-related Genes in Strawberries. *HortScience* **2005**, *40*, 2088–2090. [CrossRef]

96. Cai, C.; Chen, K.; Xu, W.; Zhang, W.; Li, X.; Ferguson, I. Effect of 1-MCP on postharvest quality of Loquat fruit. *Postharvest Biol. Technol.* **2006**, *40*, 155–162. [CrossRef]

97. Cao, S.; Yang, Z.; Zheng, Y. Effect of 1-methylcyclopene on senescence and quality maintenance of green bell pepper fruit during storage at 20 °C. *Postharvest Biol. Technol.* **2012**, *70*, 1–6. [CrossRef]

98. Donald, J.H. Suppression of Ethylene Responses Through Application of 1-Methylcyclopropene: A Powerful Tool for Elucidating Ripening and Senescence Mechanisms in Climacteric and Nonclimacteric Fruits and Vegetables. *HortScience Horts* **2008**, *43*, 106–111. [CrossRef]

99. Zhang, L.H.; Zhang, Y.H.; Li, L.L.; Li, Y.X. Effects of 1-MCP on peel browning of pomegranates. *Acta Hortic.* **2008**, *774*, 275–282. [CrossRef]

100. Gamrasni, D.; Gadban, H.; Tsvilling, A.; Goldberg, T.; Neria, O.; Ben-Arie, R.; Wolff, T.; Stern, Y. 1-Mcp Improves The Quality of Stored 'Wonderful' Pomegranates. *Acta Hortic.* **2015**, *1079*, 229–234. [CrossRef]

101. Watkins, C.B.; Barden, C.L.; Bramlage, W.J. Relationships between Alpha-Farnesene, Ethylene Production and Superficial Scald Development of Apples. *Acta Hortic.* **1993**, *343*, 155–160. [CrossRef]

102. Watkins, C.B.; Nock, J.F.; Whitaker, B.D. Responses of early, mid and late season apple cultivars to postharvest application of 1-methylcyclopropene (1-MCP) under air and controlled atmosphere storage conditions. *Postharvest Biol. Technol.* **2000**, *19*, 17–32. [CrossRef]

103. Zanella, A. Control of apple superficial scald and ripening—A comparison between 1-methylcyclopropene and diphenylamine postharvest treatments, initial low oxygen stress and ultra low oxygen storage. *Postharvest Biol. Technol.* **2003**, *27*, 69–78. [CrossRef]

104. Sharma, S.; Pareek, S.; Sagar, N.A.; Valero, D.; Serrano, M. Modulatory Effects of Exogenously Applied Polyamines on Postharvest Physiology, Antioxidant System and Shelf Life of Fruits: A Review. *Int. J. Mol. Sci.* **2017**, *18*, 1789. [CrossRef]

105. Fawole, O.A.; Atukuri, J.; Arendse, E.; Opara, U.O. Postharvest physiological responses of pomegranate fruit (cv. Wonderful) to exogenous putrescine treatment and effects on physico-chemical and phytochemical properties. *Food Sci. Hum. Wellness* **2020**, *9*, 146–161. [CrossRef]

106. Gomes, M.H.; Vieira, T.; Fundo, J.F.; Almeida, D.P. Polyphenoloxidase activity and browning in fresh-cut 'Rocha' pear as affected by pH, phenolic substrates, and antibrowning additives. *Postharvest Biol. Technol.* **2014**, *91*, 32–38. [CrossRef]

Article

Postharvest Quality Improvement of Tomato (*Solanum lycopersicum* L.) Fruit Using a Nanomultilayer Coating Containing *Aloe vera*

María L. Flores-López [1,2,†], Jorge M. Vieira [1], Cristina M. R. Rocha [1], José M. Lagarón [3], Miguel A. Cerqueira [4], Diana Jasso de Rodríguez [5] and António A. Vicente [1,*]

[1] Centre of Biological Engineering, Universidade do Minho, Campus de Gualtar, 4710-057 Braga, Portugal; lilianaflores@uadec.edu.mx (M.L.F.-L.); d8778@ceb.uminho.pt (J.M.V.); cmrocha@ceb.uminho.pt (C.M.R.R.)

[2] Centro de Investigación e Innovación Científica y Tecnológica, Universidad Autónoma de Coahuila, Saltillo 25070, Coahuila, Mexico

[3] Novel Materials and Nanotechnology Group, Institute of Agrochemistry and Food Technology (IATA), Spanish Council for Scientific Research (CSIC), Calle Catedrático Agustín Escardino Benlloch 7, 46980 Paterna, Spain; lagaron@iata.csic.es

[4] International Iberian Nanotechnology Laboratory, Av. Mestre José Veiga s/n, 4715-330 Braga, Portugal; miguel.cerqueira@inl.int

[5] Plant Breeding Department, Universidad Autónoma Agraria Antonio Narro, Calzada Antonio Narro No. 1923, Colonia Buenavista, Saltillo 25315, Coahuila, Mexico; diana.jasso@uaaan.edu.mx

* Correspondence: avicente@deb.uminho.pt

† This article is part of the PhD thesis of María L. Flores-López.

Citation: Flores-López, M.L.; Vieira, J.M.; Rocha, C.M.R.; Lagarón, J.M.; Cerqueira, M.A.; Jasso de Rodríguez, D.; Vicente, A.A. Postharvest Quality Improvement of Tomato (*Solanum lycopersicum* L.) Fruit Using a Nanomultilayer Coating Containing *Aloe vera*. Foods **2024**, 13, 83. https://doi.org/10.3390/foods13010083

Academic Editor: Malco Cruz-Romero

Received: 7 June 2023
Revised: 19 December 2023
Accepted: 21 December 2023
Published: 26 December 2023

Abstract: The effectiveness of an alginate/chitosan nanomultilayer coating without (NM) and with *Aloe vera* liquid fraction (NM+Av) was evaluated on the postharvest quality of tomato fruit at 20 °C and 85% relative humidity (RH) to simulate direct consumption. Both nanomultilayer coatings had comparable effects on firmness and pH values. However, the NM+Av coating significantly reduced weight loss (4.5 ± 0.2%) and molds and yeasts (3.5–4.0 log CFU g^{-1}) compared to uncoated fruit (16.2 ± 1.2% and 8.0 ± 0.0 log CFU g^{-1}, respectively). It notably lowered O_2 consumption by 70% and a 52% decrease in CO_2 production, inhibiting ethylene synthesis. Visual evaluation confirmed NM+Av's efficacy in preserving the postharvest quality of tomato. The preservation of color, indicated by the Minolta color (a*/b*) values, demonstrated NM+Av's ability to keep the light red stage compared to uncoated fruit. The favorable effects of NM+Av coating on enhancing postharvest quality indicates it as a potential alternative for large-scale tomato fruit preservation.

Keywords: nanomultilayer coatings; tomato; *Aloe vera* liquid fraction; gas barrier properties; postharvest quality

1. Introduction

Tomato (*Solanum lycopersicum* L.) is the second most important crop in the world after potatoes, with a production of approximately 11.6 billion tons of fresh weight in 2020 [1]. It is classified as a climacteric fruit, which means its ripening process is accompanied by an increase in respiration rate and ethylene production, leading to a relatively short postharvest shelf life [2]. For example, tomatoes stored at room temperature typically have a shelf life of around 7–11 d, depending on the variety [3,4]. The plant hormone ethylene is essential for normal fruit ripening, as it triggers various physical, physiological, and biochemical changes that enhance the appeal of tomatoes for consumption [5]. During the transition of tomato fruit from the mature green stage to fully ripe, which can occur on the plant or after harvesting, various quality parameters such as color, texture, and flavor undergo significant changes [6]. Additionally, the respiratory process brings about physiological consequences that are less desirable, including senescence, decay, chlorophyll degradation, and subsequent deterioration [7]. Transpiration, on the other hand, leads to shrinkage and weight loss of the produce by facilitating the movement of water vapor from the surface to the surrounding air [8].

Furthermore, tomatoes are susceptible to fungal attacks, such as *Rhizopus stolonifer* and *Penicillium expansum*, which can colonize injured fruits during harvesting and handling, rapidly spreading to adjacent fruits and causing significant losses [9]. Bacterial infections, including *Escherichia coli*, can also pose a risk to human health when contaminated tomatoes are consumed [10].

In general, it is crucial to control the processes of respiration and transpiration, as well as microbial contamination to extend the shelf life and maintain the quality of tomatoes. The conventional method used to delay and/or reduce ethylene production is storage at low temperatures, but this approach may cause chilling injury [11]. Pesticides and sanitizers are often employed to reduce pathogen levels; however, their application can result in residues that may exceed the maximum allowable limits, posing a serious problem for human health [12]. To address these challenges, recent research has focused on exploring improved and more efficient postharvest processing and preservation techniques. Among these techniques, the application of edible coatings has emerged as an alternative for extending the postharvest life of tomato fruit [2,13,14].

Edible coatings can modify the internal atmosphere of coated produce, creating a semi-permeable barrier against O_2, CO_2, and moisture. As a result, they can reduce respiration, water loss, and oxidation reactions [15]. Polysaccharide-based coatings are the most used materials to extend the shelf life of fruits and vegetables. Examples of these materials include sodium alginate [16], chitosan [17], galactomannans [18], pullulan [19,20], pectin [21], among others.

Recent studies have suggested that these coatings can exhibit improved functionality when used at nanoscale and applied in the form of a nanomultilayer on produce. Nanomultilayer coatings can be constructed by alternating the deposition of polyelectrolytes with opposite charges using the layer-by-layer (LbL) deposition technique [22]. Examples of this technique include chitosan/sodium alginate and chitosan/pectin coatings [15,23–26]. The objective is to combine their bioactivity and barrier properties, resulting in enhanced efficiency and an improved gas barrier compared to conventional coatings. Additionally, nanomultilayer coatings can serve as carriers for bioactive compounds, either directly or encapsulated, enabling controlled release of the active agents and prolonged bioactivity over time.

Recent studies have emphasized *Aloe vera*'s and its fractions' potential in food preservation coatings. Tarangini et al. [27] combined *A. vera*'s bioactivity with sericin, chitosan, and glycerol, extending tomato shelf life by reducing deterioration for up to 21 d at 25 °C. *A. vera* gel-based edible coatings (60–80% gel) maintained higher levels of lycopene, ascorbic acid, sugar, carotenoids, flavonoids, and pectin, while reducing microbial counts on tomatoes stored at 10 °C for 30 d [28]. Additionally, *A. vera* liquid fraction displayed potent antioxidant and antifungal properties against key fungi during tomato postharvest [29,30].

However, there are no studies incorporating *A. vera* into nanomultilayer coatings. Therefore, the objective of this study was to assess the impact of incorporating *A. vera* liquid fraction into an alginate/chitosan nanomultilayer coating on the physicochemical parameters associated with the postharvest quality of tomatoes during storage at room temperatures (20 °C/85% RH, respectively). Additionally, the study aimed to investigate the role of the *A. vera* liquid fraction in controlling microbial spoilage and the respiration process of the tomato fruit.

2. Materials and Methods

2.1. Materials

Sodium alginate was obtained from Manutex RSX (Kelco International, Ltd., Portugal), and chitosan (91.23% deacetylation degree and high molecular weight) was purchased from Golden-Shell Biochemical Co., Ltd. (Taizhou, Zhejiang, China). Lactic acid of 90% purity and oxalic acid dehydrate were purchased from Merck (Darmstadt, Germany). Tween 80 was purchased from Acros Organics (Geel, Belgium). Sodium hydroxide was obtained from Riedel-de Haën (Seelze, Germany), and ascorbic acid was obtained from VWR (Radnor, PA, USA). Dichloran-rose Bengal-chloramphenicol (DRBC), glycerol, sodium chloride, and phenolphthalein were supplied by Panreac (Barcelona, Spain). The dye 2,6-dichlorophenol-indophenol (DCPIP) was obtained from Sigma (St. Louis, MO, USA). Plate count agar (PCA) and peptone bacteriological were purchased from HiMedia Laboratories (Mumbai, India).

The liquid fraction of *A. vera* (Aloe Vera Ecológico, Alicante, Spain) was obtained using the method described by Flores-López et al. [29].

Tomatoes (*Solanum lycopersicum* L.) of the round variety were selected at the turning-pink stage of ripening (average weight = 192 g), following the USDA standard tomato color classification chart (USDA, 1991), were purchased from a local supermarket in Braga, Portugal. The fruits were visually selected for uniformity in size, color, and absence of fungal infection, and were kept at 6 °C until use. Before the treatments were applied, the tomatoes were washed with a solution of sodium hypochlorite (0.05% *v/v*) for 3 min, then rinsing with distilled water, and air-dried at room temperature.

2.2. Experimental Design

The shelf-life analyses were performed at 20 °C and 85% RH (representing commercial storage conditions). Three different treatments were evaluated: uncoated tomatoes (control), nanomultilayer coating (NM), and nanomultilayer coating with *A. vera* liquid fraction (NM+Av), as presented in Table 1. The physicochemical and microbiological analyses were conducted at regular intervals (0, 3, 6, 9, 12, and 15 d) and respiration rate was evaluated daily for 8 d.

Table 1. Treatments applied to tomato fruit.

Treatment	1st Layer	2nd Layer	3rd Layer	4th Layer	5th Layer
Uncoated					
Nanomultilayer coating (NM)	Alg	Ch	Alg	Ch	Alg
Nanomultilayer coating + *A. vera* liquid fraction (NM+Av)	Alg/Av *	Ch/Av **	Alg/Av *	Ch/Av **	Alg/Av *

Liquid fraction of *A. vera* (Av) at concentration of 0.2% (*w/v*) for alginate (Alg) * and 0.6% (*w/v*) for chitosan (Ch) **.

2.3. Coating Preparation

Polyelectrolyte solutions based on sodium alginate (Alg) and chitosan (Ch) were prepared according to the method described by Fabra et al. [22]. The concentrations of each polysaccharide, surfactant (Tween 80), and plasticizer (glycerol) were determined based on the spreading coefficient (Ws) studies on the tomato surface, as described by Casariego et al. [31]. Briefly, a 0.2% (*w/v*) Alg solution was prepared by dissolving Alg in distilled water and stirring at room temperature until complete dissolution. Glycerol (0.05%, *w/v*) and Tween 80 (0.05%, *w/v*) were added as a plasticizer and surfactant, respectively. The pH of the solution was adjusted to 7.0 using a 1 mol L^{-1} sodium hydroxide solution.

The Ch solution (0.6%, *w/v*) was dissolved in a solution of lactic acid (1.0%, *v/v*) and stirred until Ch was completely dissolved. Glycerol (0.1%, *w/v*) and Tween 80 (0.1%, *w/v*) were added, and the pH of the Ch solution was adjusted to 3.0 using a 1 mol L^{-1} lactic acid solution.

Subsequently, the *A. vera* liquid fraction was added to both the Ch (Ch-Av) and Alg (Alg-Av) coating solutions to achieve a final concentration of 0.6% or 0.2% (*w/v*), respectively, and mixed for 2 h at room temperature until homogenization. The concentrations of the *A. vera* liquid fraction were selected based on its reported range of bioactivity (antifungal and antioxidant) by Flores-López et al. [29], and considering the amounts of Alg and Ch used in the solutions (0.2% and 0.6% w/v, respectively), to obtain a polysaccharide: *A. vera* liquid fraction ratio of 1:1.

Zeta Potential

The zeta potential (ζ-potential) of each Alg and Ch coating solution was determined using a particle micro-electrophoresis instrument (Zetasizer Nano ZS-90, Malvern Instruments, Malvern, UK). Each sample was loaded into disposable capillary cells (DTS 1060, Malvern Instruments) and assessed at room temperature [2]. Additionally, the effect of adding *A. vera* liquid fraction to the coating solutions was evaluated in triplicate.

2.4. Nanomultilayer Coating Application on Tomato Fruit

The coatings were applied to the test groups using the LbL deposition technique, as shown in Table 1. No coating was applied to the control group (uncoated). Briefly, the tomatoes were immersed in a 0.2% (w/v) Alg solution at pH 7.0 for 10 s, and then rinsed with distilled water at the same pH (7.0). The samples were dried at 30 °C for 20 min in an oven with air circulation (Binder KBF, GmbH, Tuttlingen, Germany). The process was repeated using a Ch solution (0.6% w/v) at pH 3.0, followed by rinsing with distilled water at the same pH (3.0). This process was repeated with alternate deposition of a total of five layers (Alg-Ch-Alg-Ch-Alg). The immersion time, drying time between layers, and drying temperature were established based on preliminary tests. These conditions were selected to facilitate future scale-up of the process and application at the industrial level.

For each treatment, three replicates of five tomatoes (n = 15) were placed in trays and placed inside a controlled temperature and humidity chamber (Binder GmbH, Tuttlingen, Germany) at 20 °C and 85% RH. Temperature and RH during the shelf life and respiration tests were recorded using an iButton data logger (Thermochron, Dallas, TX, USA). Photographic documentation was utilized to capture the visual changes in the appearance of tomatoes by comparing the initial images of the treatments and the control over the storage period.

2.5. Physicochemical Analyses
2.5.1. Weight Loss

The weight loss of five tomatoes per treatment was evaluated by weighing all samples using a precision balance (METTLER AE200, Mettler-Toledo, Giesen, Germany) at the beginning of storage (0 d) and during the experimental storage period. The percentage (%) of weight loss was determined using the following equation:

$$\text{Weight loss}(\%) = \frac{W_i - W_f}{W_i} \times 100 \tag{1}$$

where W_i is the initial sample weight and W_f is the final sample weight.

2.5.2. Titratable Acidity (TA), pH, Soluble Solid Content (SSC)

At consistent intervals of 3 d during 15 d, three tomatoes from each treatment were analyzed. The samples were cut into small pieces, subsequently 50 g of each treatment was ground in a blender and filtered through Whatman filter paper no. 1 under vacuum. Titratable acidity (TA) was measured by utilizing 10 mL of previously obtained juice. Two drops of 1% (w/v) phenolphthalein were added, followed by titration using NaOH (0.1 mol L^{-1}) using the 942.15 AOAC method. The results were expressed as a percentage (%) of citric acid. The pH of each treatment was determined using a pH meter (Hanna Instruments Inc., Bucharest, Romania) by directly submerging the electrode into the homogenized sample.

The juice from both the treatments and control groups was used to determine the soluble solid content (SSC) following the 932.12 AOAC method [32]. Briefly, a drop of the tomato juice was applied to a refractometer's surface (HI 96801, Hanna Instruments Inc., Bucharest, Romania) calibrated with distilled water to measure the refractive index. Results were expressed as percentage (%). For all physicochemical tests, three samples per treatment were analyzed at each sampling time.

2.5.3. Ascorbic Acid (AA) Determination

The ascorbic acid (AA) content was estimated using the DCPIP titration method of Ranggana [33] with some modifications. Briefly, the juice obtained from the fruit was centrifuged (Sigma 4K15, Sartorius, Göttingen, Germany) for 5 min at $12,000\times g$ at room temperature. The supernatant (2.0 mL) was mixed with 5.0 mL of oxalic acid (4.0% w/v) and 2.0 mL of distilled water. The volume required to cause a color change in the DCPIP solution (24.0 g L^{-1} in distilled water) was recorded. A standard solution of ascorbic acid at a concentration of 20 g L^{-1} in distilled water was used as a reference. The results were expressed as mg kg^{-1} of AA per fresh weight (FW). All determinations were performed in triplicate.

2.6. Color

The color of the tomato skin was evaluated by measuring it with a Minolta colorimeter (CR 400; Minolta, Osaka, Japan). Average readings were taken at three points on the circumference of each fruit. The instrument was calibrated using a standard white color plate (Y = 93.5, x = 0.3114, y = 0.3190). The results were reported as Minolta color values according to the scale proposed by Batu [34] for tomato fruit, that indicates a direct correlation between the Minolta a*/b* ratio and the USDA ripening stages (Table 2), which were calculated using the following equation:

$$\text{Minolta color} = \frac{a^*}{b^*} \tag{2}$$

where a* value corresponds to the degree of redness and the b* value represents yellowness in the Minolta colorimeter.

Table 2. Classification of USDA mature stages of tomato fruit according to Minolta color values.

Minolta Color Values (a*/b*)	USDA Tomato Color Stages
−0.59 to −0.47	Green
−0.47 to −0.27	Breaker
−0.27 to 0.08	Turning
0.08 to 0.60	Pink
0.60 to 0.95	Light red
0.96 to 1.21	Red

Adapted from Batu [34].

2.7. Firmness

Fruit firmness was determined using a texture analyzer (TA.XT, Stable Micro Systems, Godalming, UK). The tomato fruit was positioned at the center of the platform, and the force (N) required to penetrate 2.0 cm into the fruit was measured at the break point using a 6 mm flat-head stainless steel cylindrical probe. The test speed was set at 5.0 mm s^{-1}. Firmness measurements were taken at the beginning and end of each test, and the results were reported as the mean $\pm$ SE (n = 10) and expressed in Newtons (N).

2.8. Microbiological Analyses

Microbiological analyses were conducted to count the total aerobic mesophilic microorganisms and molds and yeasts during the storage conditions, following the method described by Olivas et al. [35]. A sample weighing 10 g was aseptically collected from tomato surfaces and placed in a sterilized filter stomacher bag (VWR Scientific, West Chester, PA, USA) containing 90 mL of sterilized peptone water (0.1% w/v). The mixture was blended for 120 s using a Stomacher blender (3500, Seward Medical, London, UK). Serial decimal dilutions of the filtrate in 0.1% peptone water were pour-plated in duplicate on PCA agar and incubated at 37 °C for 2 d to count aerobic mesophilic microorganisms. Simultaneously, the same decimal dilutions were spread-plated on DRBC agar, a selective medium for the isolation and quantification of molds and yeasts. These plates were then incubated for 5 d at 25 °C.

The results were quantified and expressed as log colony-forming units per gram (log CFU g^{-1}), encompassing the total count of aerobic mesophilic microorganisms as well as the combined count of molds and yeast observed on the selective DRBC agar plates. All analyses were performed with two replicates.

2.9. Gas Transfer Rate and Ethylene Production

The closed system method was used to measure the gas exchange (O_2 and CO_2) and ethylene (C_2H_4) production of tomato fruit. Acrylic air-tight cylindrical containers with a top lid fitted with a septum for gas sampling were used for each fruit and measured daily. A whole intact fruit sample was placed within each container, which was then placed in a controlled temperature and humidity chamber (Binder, Binder GmbH, Tuttlingen, Germany) to maintain the storage conditions. Evaluations were made daily for 8 d. Temperature and RH were recorded using an iButton data logger (Thermochron, Dallas, TX, USA) placed inside the container.

The O_2 and CO_2 contents were determined using a gas chromatograph (Bruker Scion 456, Markham, ON, Canada) equipped with two thermal conductivity detectors (TCD). The gas chromatograph had two columns: SS MolSieve 13 $\times$ (80/100), 2 m $\times$ 2 mm $\times$ 1/8″ for O_2 determination and BR Q PLOT, 30 m $\times$ 0.53 mm for CO_2 measurement. Argon and helium were used as carrier gases, respectively. Calibration was performed using a mixture containing 10% CO_2, 20% O2, and 70% N_2. The C_2H_4 production was evaluated using a Varian Star 3400 CX (Palo Alto, CA, USA) gas chromatograph, coupled with a flame ionization detector (FID). The chromatograph was equipped with a vf-5 ms 30 m $\times$ 0.25 mm, 0.25 µm column. Helium, nitrogen, air, and hydrogen were used as carrier gases. Ethylene at 500 mg L^{-1} (Calgaz, Staffordshire, UK) was used as a standard for calibration.

The determination of gas transfer rate was performed with three replicates for each group of samples. Three full replicates were performed for both the control and the coated fruit groups. The O_2 consumption and CO_2 and C_2H_4 production rates were determined as described by Cerqueira et al. [36] with some modifications.

2.10. Statistical Analyses

The data analyses were conducted using FAUANL software v. 2015 [37] and Statistica software (release 7, edition 2004, Statsoft, Tulsa, OK, USA). Analysis of variance (ANOVA) was performed to determine significant differences. Mean values that were significantly different ($p < 0.05$) were separated using the Tukey test for a randomized block experimental design.

3. Results and Discussion

3.1. Physicochemical Analyses

3.1.1. Weight Loss

A weight loss above 5% is considered a limiting factor for the postharvest life of fruit crops, as there is a known relationship between this parameter, temperature, and storage time [38]. Figure 1 presents the weight loss of tomato fruit during storage at 20 °C and 85% RH. There was a significant difference between tomatoes coated with NM+Av coating and those coated with NM coating or uncoated tomatoes. This difference was more pronounced ($p < 0.05$) on the 15th d, with weight loss values of 5.2 $\pm$ 1.2% for tomatoes coated with NM+Av, and 8.5 $\pm$ 0.6% and 16.2 $\pm$ 1.2% for those coated with NM coating and uncoated tomatoes, respectively. The improved water barrier provided by the nanomultilayer coatings compared to uncoated tomatoes can be attributed to the electrostatic interactions between the Alg and Ch layers. This was corroborated by the ζ-potential values of the Alg solution (-60.40 ± 4.20 mV), which were lower and carried an opposite charge compared to the Ch solution (65.40 ± 3.70 mV). These interactions increase the tortuosity of the system, thereby reducing the diffusion of molecules through the coating materials [15].

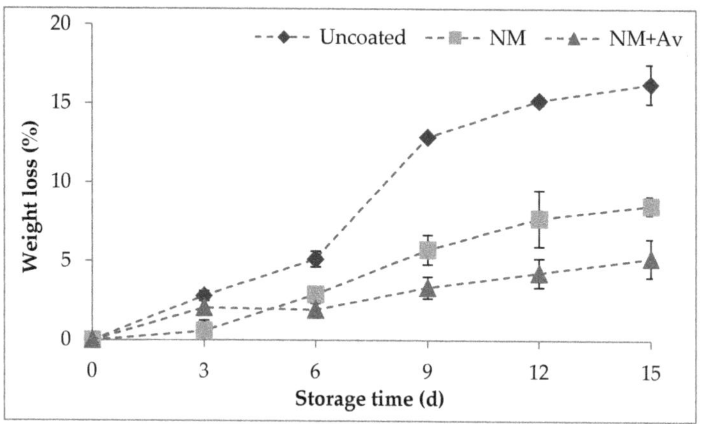

Figure 1. Tomato fruit weight loss (%) during storage at 20 °C/85% RH for 15 d. Values are the mean $\pm$ SE.

The observed reduction in moisture loss and the subsequent improvement of postharvest quality can be attributed to the *A. vera* functionalization, which may be comprehended through its interaction with the hydrophilic groups inherent in Ch. The incorporation of *A. vera* liquid fraction into the Alg solution resulted in a significant increase in the charge (-45.50 ± 3.30 mV), potentially associated with the partial neutralization of Alg's carboxylic groups by positively charged components of *A. vera* (e.g., proteins). However, there were no significant differences observed in the Ch coating solution upon the incorporation of the *A. vera* liquid fraction (72.20 ± 4.50 mV). These conditions guarantee the occurrence of electrostatic interaction between these polysaccharides even when functionalized. *A. vera*, known for its bioactive compounds, likely forms hydrogen bonds or other molecular interactions with Ch's hydrophilic sites. This interaction alters the Ch-water dynamics, possibly by forming a protective barrier or modifying the surface properties of the Ch-based coating. Consequently, this impedes water permeation or enhances water vapor resistance, subsequently diminishing the rate of moisture loss from the coated fruit [39]. A study conducted by Morad et al. [40] indicated that tomatoes coated with *A. vera* gel also created a physical barrier that reduced the transfer of moisture from the inside of the fruit to the outside, attributed to its hygroscopic nature, the presence of hydrophobic compounds, and a higher polysaccharide content. Vieira et al. [30], also observed a significant decrease in weight loss in a chitosan-based coating containing *A. vera* liquid fraction after being stored at 5 °C and 90% RH for 25 d. This highlights the efficacy of such interactions in preserving produce quality during storage under specified conditions.

3.1.2. Titratable Acidity (TA), pH, Soluble Solid Content (SSC)

The pH, TA, and SSC values for tomatoes under storage at 20 °C and 85% RH are presented in Table 3. The SSC values in tomatoes ranged between 3.6 and 4.6%, which is consistent with the values reported by Zapata et al. [41] for tomato fruit. The SSC values remained stable throughout the 15-d storage period, showing no significant differences between treatments, or compared to day 0. This stability suggests a maintained level of sweetness in the tomatoes, which is consistent with the results reported by Javanmardi and Kubota [42] for tomatoes stored at temperatures ranging from 25 to 27 °C. The pH values remained relatively steady during storage, yet an increase was noted in coated tomatoes by day 15 compared to day 0. This pH rise in coated tomatoes suggests an ongoing maturation, possibly influencing taste development while also being protected against weight loss, as shown in Figure 1.

Table 3. Physicochemical properties of tomato fruit during storage at 20 °C/85% RH for 15 d.

Storage Time (d)		0	3	6	9	12	15
Uncoated	TA	0.3 ± 0.0 [Aa]	0.2 ± 0.0 [Ba]	0.2 ± 0.1 [Aa]	0.3 ± 0.0 [Aa]	0.2 ± 0.1 [Aa]	0.2 ± 0.0 [Aa]
	pH	4.5 ± 0.1 [Aa]	4.5 ± 0.2 [Aa]	4.5 ± 0.2 [Aa]	4.5 ± 0.2 [Aa]	4.5 ± 0.1 [Aa]	4.6 ± 0.1 [Aa]
	SSC	3.9 ± 0.2 [Aa]	4.3 ± 0.2 [Aa]	4.1 ± 0.2 [Aa]	4.0 ± 0.4 [Aa]	4.0 ± 0.3 [Aa]	3.6 ± 0.1 [Aa]
NM	TA	0.3 ± 0.0 [Aa]	0.2 ± 0.0 [Aa]	0.2 ± 0.0 [Ba]	0.2 ± 0.0 [Bb]	0.2 ± 0.0 [Ba]	0.2 ± 0.0 [Ba]
	pH	4.5 ± 0.1 [Aa]	4.5 ± 0.2 [Aa]	4.6 ± 0.1 [Aa]	4.7 ± 0.1 [Aa]	4.6 ± 0.0 [Aa]	4.8 ± 0.0 [Bb]
	SSC	3.9 ± 0.2 [Aa]	4.6 ± 0.0 [Ba]	3.9 ± 0.4 [Aa]	3.9 ± 0.1 [Aa]	4.0 ± 0.2 [Aa]	3.8 ± 0.1 [Aa]
NM+Av	TA	0.3 ± 0.0 [Aa]	0.3 ± 0.0 [Aa]	0.3 ± 0.1 [Aa]	0.2 ± 0.0 [Bb]	0.2 ± 0.0 [Ba]	0.2 ± 0.1 [Aa]
	pH	4.5 ± 0.1 [Aa]	4.5 ± 0.2 [Aa]	4.4 ± 0.2 [Aa]	4.6 ± 0.2 [Aa]	4.7 ± 0.1 [Aa]	4.9 ± 0.1 [Bb]
	SSC	3.9 ± 0.2 [Aa]	4.0 ± 0.3 [Aa]	4.0 ± 0.3 [Aa]	4.0 ± 0.2 [Aa]	3.9 ± 0.1 [Aa]	4.5 ± 0.1 [Ab]

TA = titratable acidity (% citric acid); SSC = soluble solid content (%). Means followed by the same lowercase letters in the columns and uppercase letters in rows did not show a statistically significant difference by Tukey's test ($p < 0.05$).

On the other hand, the acidity of tomatoes significantly contributes to their taste and is intricately linked to the maturation process. However, the acidity does not change linearly over time. Research indicates varying trends, such as decreasing malic acid and increasing citric acid until the turning stage, while contrasting studies show a gradual rise in malic acid throughout maturation [13]. These variations partially explain the results obtained, as the TA values remained stable for all treatments during storage, despite observed changes in other maturation-related parameters such as color. The balance between SSC and TA changes might play a crucial role in taste and quality maintenance during storage.

3.1.3. Ascorbic Acid (AA)

In general, fruits are a natural source of AA, and its levels are reduced during maturation and processing. Due to its sensitivity, AA is used as an indicator of the severity of postharvest fruit damage. Figure 2 illustrates the concentration of AA in uncoated and coated tomatoes. A reduction in AA levels can be observed on the second day of analysis, which aligns with previous findings, and it is attributed to AA being used as a substrate or converted into sugars during ripening [43].

The AA levels were also observed to remain constant throughout the storage period in tomatoes coated with either of the coatings, with differences ($p < 0.05$) only being detected compared to uncoated tomatoes starting from day 12 (Figure 2). The significant reduction in AA content in uncoated tomatoes can be associated with the advanced ripeness of the fruit. This reduction may be attributed to the antioxidant function of AA, where ripening cells absorb higher levels of oxygen due to an increase in respiration rate, which is a characteristic physiological change in climacteric fruits and vegetables at ripeness [44]. The application of nanomultilayer coatings aided in reducing AA loss, although the incorporation of *A. vera* liquid fraction did not influence AA retention, as no significant differences were found between the two coatings. Similarly, multilayer systems such as the Ch-(β-cyclodextrin + trans-cinnamaldehyde complex)-pectin-based multilayer edible coating (with a thickness of 300 ± 1 μm) reported by Brasil et al. [45] demonstrated the ability to retain higher AA values in papaya compared to uncoated papaya during storage at 4 °C.

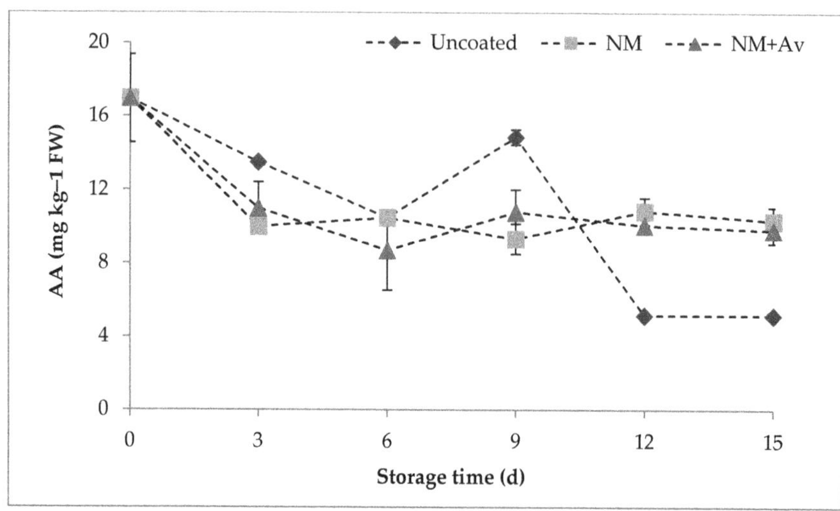

Figure 2. Ascorbic acid (AA) content (mg kg^{-1} FW) in tomato fruit during storage at 20 °C/85% RH for 15 d. Values are the mean ± SE.

3.2. Color and Firmness

Color serves as an important indicator of ripeness and quality in tomatoes [6]. The a* value is used to monitor red color development and the degree of ripening in tomato fruit, while the b* value indicates yellow discoloration. Batu [34] provided a scale of a*/b* values to express the redness and its relationship with the maturation stage of tomatoes (Table 2). Figure 3a demonstrates that the color development, as indicated by an increase in Minolta color a*/b* values, was more pronounced ($p < 0.05$) in uncoated tomatoes throughout the storage period. However, the application of nanomultilayer coatings helped preserve the color attributes of the fruit, with the light red stage of tomatoes being sustained throughout the entire test. The effectiveness of *A. vera*-based coatings in reducing color development in table grapes and mushrooms has been previously reported by other researchers [46,47].

Regarding fruit firmness, the application of NM-Av at day 0 significantly increased the firmness of tomatoes compared to uncoated and NM-coated tomatoes. This increase may be attributed to the greater thickness of NM+Av (500 nm) compared to NM (420 nm) (Figure 3b). Similarly, Athmaselvi et al. [13] reported higher firmness in tomatoes coated with an *A. vera*-based coating. A significant decrease in firmness occurred at the end of the storage period, with uncoated tomatoes exhibiting the lowest firmness. The coated tomatoes maintained higher firmness, but no significant differences were found between the two types of nanomultilayer coatings studied. Consistent with these results, Ali et al. [6] reported lower loss of firmness in tomatoes coated with a gum arabic-based coating during storage at 20 °C and 80–90% RH. Fruit softening is a result of the degradation of cell structure and the internal composition of the cell wall by enzymes (e.g., hydrolases) acting on pectin and starch. These actions are closely linked to the progress of fruit ripening. The effect of coatings on delaying fruit softening is associated with their ability to act as a barrier for O_2 uptake. The results of the gas transfer rate showed that coated tomatoes exhibited significantly lower O_2 consumption than uncoated tomatoes, thereby slowing down metabolic activity and, consequently, the ripening process [48].

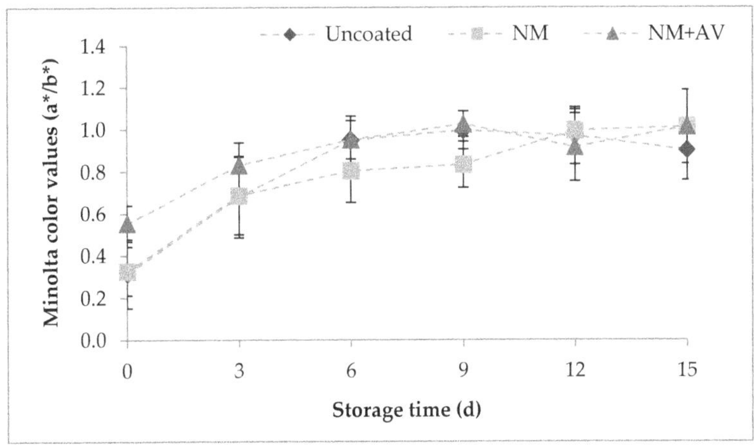

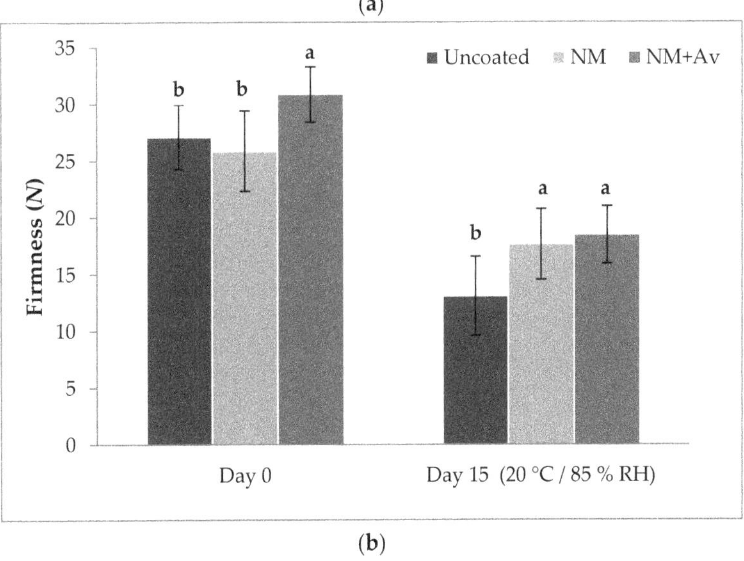

(b)

Figure 3. (**a**) Minolta color (a*/b*) values and (**b**) firmness of tomato fruit during storage at 20 °C/85% RH for 15 d. Values are the mean ± SE. For firmness, different letters on the same day indicate statistical differences ($p < 0.05$).

3.3. Microbial Analyses

During the storage period, NM+Av exhibited better inhibition until day 9, and after day 12, no significant differences were found between tomatoes coated with both nano-multilayer coatings (Figure 4a). However, these treatments showed reduced ($p < 0.05$) populations of molds and yeasts (4.0 ± 0.0 log CFU g^{-1}) compared to uncoated tomatoes (8.0 ± 0.0 log CFU g^{-1}).

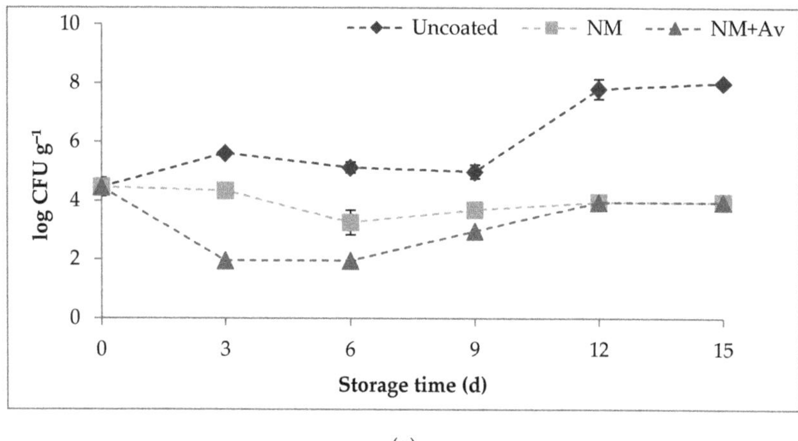

(**a**)

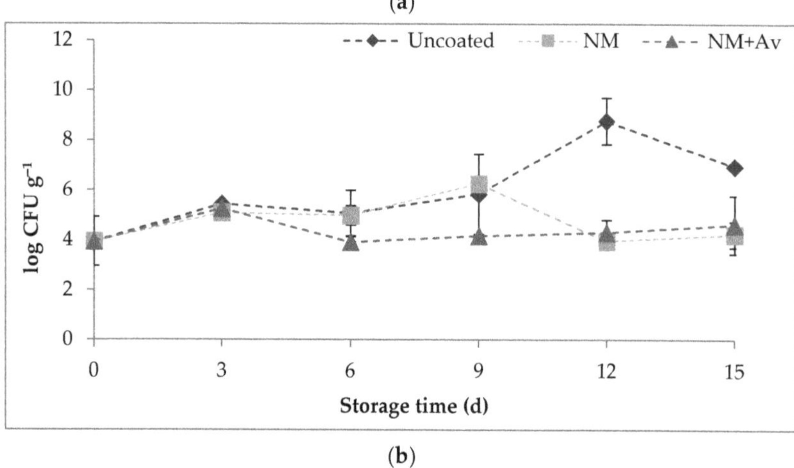

(**b**)

Figure 4. (**a**) Microbiological counting of molds and yeasts and (**b**) aerobic mesophilic microorganisms throughout storage time at 20 °C/85% RH. Values are the mean ± SE.

The antifungal activity of *A. vera* liquid fraction has been associated with the suppression of germination and inhibition of mycelial growth in fungi such as *Rhizoctonia solani*, *Fusarium oxysporum*, *Colletotrichum coccodes*, *B. cinerea*, and *P. expansum* [30,49]. These activities can be attributed to the presence of more than one active compound, although the specific mechanism of action is still unknown [47]. Recently, Vieira et al. [30] reported significantly lower counts of yeasts and molds on blueberry fruit coated with Ch- and Ch-liquid fraction of *A. vera*-based coatings after 25 d of storage at 5 °C and 90% RH. This activity was higher when *A. vera* was incorporated into the coating, although the authors also indicated a combination of the effects of Ch and *A. vera*.

The initial count of mesophilic microorganisms was 4.0 ± 0.7 log CFU g^{-1}, and it increased during storage (Figure 4b). However, no significant differences in mesophilic counts were found between coated and uncoated tomatoes, but a slight reduction ($p < 0.05$) was observed at day 12 compared to uncoated samples.

The effectiveness of the evaluated nanomultilayer coatings was higher for yeasts and molds than for mesophilic microorganisms. Valverde et al. [47] also found the same behavior on table grapes coated with *A. vera* gel. It is supposed that the antimicrobial activity of *A. vera* cannot be sustained throughout storage, probably due to the stability of its bioactive compounds, mainly phenolic compounds and organic acids, which are responsible for its antimicrobial activity, as reported by Flores-López et al. [29]. Visual evaluation confirmed that the uncoated tomatoes had extensive spoilage on the surface after 15 d of storage (Figure 5).

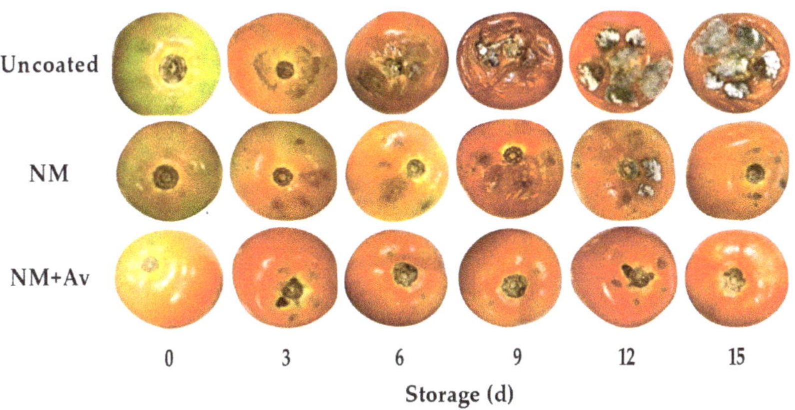

Figure 5. Tomato fruit's gradual change throughout a 15-d storage period at 20 °C/85% RH.

3.4. Gas Transfer Rate and Ethylene Production

Innovation in the design of gas barrier materials plays a key role in the agro-food industry, as they can extend the shelf life of produce. Figure 6a shows the effect of coated treatments on CO_2 production, in which a statistically significant reduction in the respiration rate was observed at day 8 for tomatoes coated with NM+Av compared to uncoated and NM-coated tomatoes.

NM+Av exhibited a lower gas transfer rate, with a 70% lower O_2 consumption and 52% lower CO_2 production after storage, compared to uncoated samples. Recent studies have demonstrated the effectiveness of κ-carrageenan and lysozyme nanomultilayer coatings in reducing O_2 and CO_2 exchange in fresh-cut and whole pears, as well as their role as a barrier to water loss [24]. Additionally, Medeiros et al. [23] associated the reduction in gas flow with the extension of shelf life in mangoes coated with a pectin and Ch-based nanomultilayer coating.

The ethylene production rate at the end of the experiment (day 8) showed differences between treatments (Figure 6b). These results indicate that the NM+Av coating did not significantly alter the gas balance in the tomato fruit, as it primarily reduced the respiration rates [50]. The observed reduction in ethylene production rates in tomatoes coated with NM+Av suggests influence on the fruit's physiology. While the incorporation of *A. vera* into the nanomultilayer coating notably decreased ethylene production, it appears that this effect is not only attributable to altering gas balance. Instead, it is likely a combined outcome of the reduced respiration rates and enhanced gas barrier properties conferred by the NM+Av coating.

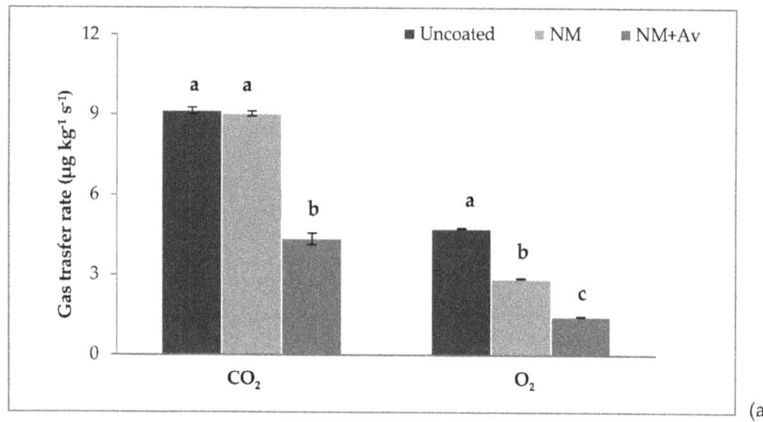

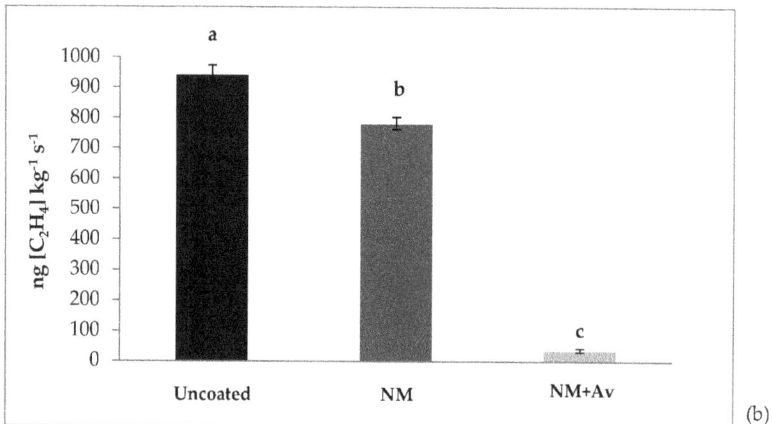

Figure 6. (**a**) CO_2 and O_2 transfer rates and (**b**) ethylene production during storage at 20 °C/85% RH for 8 d. Values are the mean ± SE. Different letters (a–c) indicate statistical differences between treatments ($p < 0.05$).

This reduction can be attributed to the fact that the application of the coating on the tomato's surface restricts the permeation of respiratory gases. Additionally, the lower counts of yeasts and molds may influence the lower ethylene production in tomato fruit coated with NM+Av, suggesting that the accumulated ethylene in uncoated fruit is produced by fungi rather than the tomato fruit itself. The reduction in ethylene synthesis and the gas barrier properties of *A. vera* gel have been previously reported for climacteric fruits such as peaches, plums, nectarines [51], apples [52], and tomatoes [14].

Visual evaluation of the tomato fruit at the end of the storage conditions (20 °C/85% RH) also confirmed the significant effect of the NM+Av coating on maintaining quality and appearance (Figure 5). It provided a barrier against ethylene production and gas exchange between the inner and outer environments, resulting in a delay in the maturation process and an extension of the shelf life of the tomato fruit. Further work is required to scale up this technology to an industrial level to make its application feasible and accessible for producers.

4. Conclusions

The effectiveness of an alginate/chitosan nanomultilayer coating containing *A. vera* liquid fraction in extending the postharvest life of tomato fruit was evaluated. The application of nanomultilayer coatings (with and without *A. vera*) regulated maturation during storage conditions at 20 °C and 85% RH for 15 d, as it resulted in a reduction in the gas transfer rate in the coated tomato fruit. Among the coatings, NM+Av exhibited superior protective properties against weight loss, reduction in gas transfer rates, and ethylene production. Additionally, it effectively decreased microbial spoilage, thereby improving the overall quality of the tomato fruit. These beneficial properties of NM+Av make it a novel alternative for extending the postharvest quality of tomato fruit, which holds significant commercial value for both producers and consumers.

Author Contributions: Investigation and methodology, writing—original draft, writing—reviewing and editing, M.L.F.-L.; investigation and methodology and validation, J.M.V. and C.M.R.R.; supervision, project administration, J.M.L.; formal analysis, project administration and data curation, supervision, writing—reviewing and editing M.A.C.; conceptualization and supervision, D.J.d.R.; conceptualization, investigation, resources, and supervision, A.A.V. All authors have read and agreed to the published version of the manuscript.

Funding: This research received no external funding.

Institutional Review Board Statement: Not applicable.

Informed Consent Statement: Not applicable.

Data Availability Statement: Data is contained within the article.

Acknowledgments: María L. Flores-López thanks the Mexican Science and Technology Council (CONAHCYT, Mexico) for PhD fellowship support (CONAHCYT grant number: 215499/310847). This study was supported by the Portuguese Foundation for Science and Technology (FCT) under the scope of the strategic funding of the UID/BIO/04469/2020 unit and COMPETE 2020 (POCI-01-0145-FEDER-006684) and of the Project RECI/BBB-EBI/0179/2012 (FCOMP-01-0124-FEDER-027462). The authors would also like to thank Eng. Luis Fernando Camacho Guerra of CIICyT (Centro de Investigación e Innovación Científica y Tecnológica, Universidad Autónoma de Coahuila, Mexico) for his assistance during this study. The authors are thankful to Madalena Vieira for her valuable suggestions during this study.

Conflicts of Interest: The authors declare no conflicts of interest.

References

1. FAO. Available online: https://www.fao.org/ (accessed on 6 June 2023).
2. Salas-Méndez, E.D.J.; Vicente, A.; Pinheiro, A.C.; Ballesteros, L.F.; Silva, P.; Rodríguez-García, R.; Hernández-Castillo, F.D.; Díaz-Jiménez, M.D.L.V.; Flores-López, M.L.; Villarreal-Quintanilla, J.Á.; et al. Application of edible nanolaminate coatings with antimicrobial extract of Flourensia cernua to extend the shelf-life of tomato (*Solanum lycopersicum* L.) fruit. *Postharvest Biol. Technol.* **2019**, *150*, 19–27. [CrossRef]
3. Garuba, T.; Mustapha, O.T.; Oyeyiola, G.P. Shelf life and proximate composition of tomato (*Solanum lycopersicum* L.) fruits as influenced by storage methods. *Ceylon J. Sci.* **2018**, *47*, 387. [CrossRef]
4. Mama, S.; Yemer, J.; Woelore, W. Effect of hot water treatments on shelf life of tomato (*Lycopersicon esculentum* Mill). *J. Nat. Sci. Res.* **2016**, *6*, 69–77.
5. Deng, H.; Pirrello, J.; Chen, Y.; Li, N.; Zhu, S.; Chirinos, X.; Bouzayen, M.; Liu, Y.; Liu, M. A novel tomato F-box protein, SlEBF3, is involved in tuning ethylene signaling during plant development and climacteric fruit ripening. *Plant J.* **2018**, *95*, 648–658. [CrossRef] [PubMed]
6. Ali, A.; Maqbool, M.; Ramachandran, S.; Alderson, P.G. Gum arabic as a novel edible coating for enhancing shelf-life and improving postharvest quality of tomato (*Solanum lycopersicum* L.) fruit. *Postharvest Biol. Technol.* **2010**, *58*, 42–47. [CrossRef]
7. Kandasamy, P.; Moitra, R.; Mukherjee, S. Measurement and modeling of respiration rate of tomato (Cultivar Roma) for modified atmosphere storage. *Recent Pat. Food Nutr. Agric.* **2015**, *7*, 62–69. [CrossRef] [PubMed]
8. Damdam, A.; Al-Zahrani, A.; Salah, L.; Salama, K.N. Effect of Combining UV-C Irradiation and Vacuum Sealing on the Shelf Life of Fresh Strawberries and Tomatoes. *J. Food Sci.* **2023**, *88*, 595–607. [CrossRef] [PubMed]
9. Bautista-Baños, S.; Velázquez-del Valle, M.G.; Hernández-Lauzardo, A.N.; Ait Barka, E. The Rhizopus stolonifer e tomato interaction. In *Plant Microbe Interaction*; Barka, E.A., Clément, C., Eds.; Research Signpost: Kerala, India, 2008; pp. 269–289.
10. Ramos-García, M.; Bosquez-Molina, E.; Hernández-Romano, J.; Zavala-Padilla, G.; Terrés-Rojas, E.; Alia-Tejacal, I.; Barrera-Necha, L.; Hernández-López, M.; Bautista-Baños, S. Use of chitosan-based edible coatings in combination with other natural compounds, to control *Rhizopus stolonifer* and *Escherichia coli* DH5α in fresh tomatoes. *Crop Prot.* **2012**, *38*, 1–6. [CrossRef]

11. Park, M.H.; Sangwanangkul, P.; Choi, J.W. Reduced chilling injury and delayed fruit ripening in tomatoes with modified atmosphere and humidity packaging. *Sci. Hortic. (Amst.)* **2018**, *231*, 66–72. [CrossRef]

12. Mishra, R.K.; Bohra, A.; Kamaal, N.; Kumar, K.; Gandhi, K.; GK, S.; Saabale, P.R.; SJ, S.N.; Sarma, B.K.; Kumar, D.; et al. Utilization of Biopesticides as Sustainable Solutions for Management of Pests in Legume Crops: Achievements and Prospects. *Egypt. J. Biol. Pest Control* **2018**, *28*, 1–11. [CrossRef]

13. Athmaselvi, K.A.; Sumitha, P.; Revathy, B. Development of *Aloe vera* based edible coating for tomato. *Int. Agrophysics* **2013**, *27*, 369–375. [CrossRef]

14. Chauhan, O.P.; Nanjappa, C.; Ashok, N.; Ravi, N.; Roopa, N.; Raju, P.S. Shellac and *Aloe vera* gel based surface coating for shelf life extension of tomatoes. *J. Food Sci. Technol.* **2013**, *52*, 1200–1205. [CrossRef] [PubMed]

15. Acevedo-Fani, A.; Salvia-Trujillo, L.; Rojas-Graü, M.A.; Martín-Belloso, O. Edible Films from Essential-Oil-Loaded Nanoemulsions: Physicochemical Characterization and Antimicrobial Properties. *Food Hydrocoll.* **2015**, *47*, 168–177. [CrossRef]

16. Zhu, D.; Guo, R.; Li, W.; Song, J.; Cheng, F. Improved postharvest preservation effects of *Pholiota nameko* mushroom by sodium alginate–based edible Composite Coating. *Food Bioprocess Technol.* **2019**, *12*, 587–598. [CrossRef]

17. Pinzon, M.I.; Sanchez, L.T.; Garcia, O.R.; Gutierrez, R.; Luna, J.C.; Villa, C.C. Increasing shelf life of strawberries (*Fragaria* ssp) by using a banana starch-chitosan-*Aloe vera* gel composite edible coating. *Int. J. Food Sci. Technol.* **2020**, *55*, 92–98. [CrossRef]

18. Ainee, A.; Hussain, S.; Nadeem, M.; Al-Hilphy, A.R.; Siddeeg, A. Extraction, purification, optimization, and application of galactomannan-based edible coating formulations for guava using response surface methodology. *J. Food Qual.* **2022**, *2022*, 5613046. [CrossRef]

19. Li, L.; Sun, J.; Gao, H.; Shen, Y.; Li, C.; Yi, P.; He, X.; Ling, D.; Sheng, J.; Li, J.; et al. Effects of Polysaccharide-Based Edible Coatings on Quality and Antioxidant Enzyme System of Strawberry during Cold Storage. *Int. J. Polym. Sci.* **2017**, *2017*, 1–8. [CrossRef]

20. Pobiega, K.; Igielska, M.; Włodarczyk, P.; Gniewosz, M. The Use of Pullulan Coatings with Propolis Extract to Extend the Shelf Life of Blueberry (*Vaccinium corymbosum*) Fruit. *Int. J. Food Sci. Technol.* **2021**, *56*, 1013–1020. [CrossRef]

21. Aguilar-Veloz, L.M.; Calderón-Santoyo, M.; Carvajal-Millan, E.; Martínez-Robinson, K.; Ragazzo-Sánchez, J.A. *Artocarpus heterophyllus* Lam. Leaf Extracts Added to Pectin-Based Edible Coating for *Alternaria* Sp. Control in Tomato. *Lwt* **2022**, *156*, 1–9. [CrossRef]

22. Fabra, M.J.; Flores-López, M.L.; Cerqueira, M.A.; Jasso de Rodríguez, D.; Lagaron, J.M.; Vicente, A.A. Layer-by-Layer technique to developing functional nanolaminate films with antifungal activity. *Food Bioprocess Technol.* **2016**, *9*, 471–480. [CrossRef]

23. Medeiros, B.G.D.S.; Pinheiro, A.C.; Carneiro-Da-Cunha, M.G.; Vicente, A.A. Development and characterization of a nanomulti-layer coating of pectin and chitosan—Evaluation of its gas barrier properties and application on "Tommy Atkins" mangoes. *J. Food Eng.* **2012**, *110*, 457–464. [CrossRef]

24. Medeiros, B.G.D.S.; Pinheiro, A.C.; Teixeira, J.A.; Vicente, A.A.; Carneiro-da-Cunha, M.G. Polysaccharide/protein nanomultilayer coatings: Construction, characterization and evaluation of their effect on "Rocha" Pear (*Pyrus communis* L.) shelf-life. *Food Bioprocess Technol.* **2012**, *5*, 2435–2445. [CrossRef]

25. Wei, F.; Ye, F.; Li, S.; Wang, L.; Li, J.; Zhao, G. Layer-by-Layer Coating of Chitosan/Pectin Effectively Improves the Hydration Capacity, Water Suspendability and Tofu Gel Compatibility of Okara Powder. *Food Hydrocoll.* **2018**, *77*, 465–473. [CrossRef]

26. Li, K.; Zhu, J.; Guan, G.; Wu, H. Preparation of Chitosan-Sodium Alginate Films through Layer-by-Layer Assembly and Ferulic Acid Crosslinking: Film Properties, Characterization, and Formation Mechanism. *Int. J. Biol. Macromol.* **2019**, *122*, 485–492. [CrossRef] [PubMed]

27. Tarangini, K.; Kavi, P.; Jagajjanani Rao, K. Application of sericin-based edible coating material for postharvest shelf-life extension and preservation of tomatoes. *eFood* **2022**, *3*, e36. [CrossRef]

28. Firdous, N.; Khan, M.R.; Butt, M.S.; Ali, M.; Asim Shabbir, M.; Din, A.; Hussain, A.; Siddeeg, A.; Manzoor, M.F. Effect of *Aloe vera* gel-based edible coating on microbiological safety and quality of tomato. *CYTA—J. Food* **2022**, *20*, 355–365. [CrossRef]

29. Flores-López, M.L.; Romaní, A.; Cerqueira, M.A.; Rodríguez-García, R.; Jasso de Rodríguez, D.; Vicente, A.A. Compositional features and bioactive properties of whole fraction from *Aloe vera* processing. *Ind. Crops Prod.* **2016**, *91*, 179–185. [CrossRef]

30. Vieira, J.M.; Flores-López, M.L.; Jasso de Rodríguez, D.; Sousa, M.C.; Vicente, A.A.; Martins, J.T. Effect of chitosan–*Aloe vera* coating on postharvest quality of blueberry (*Vaccinium corymbosum*) fruit. *Postharvest Biol. Technol.* **2016**, *116*, 88–97. [CrossRef]

31. Casariego, A.; Souza, B.W.S.; Vicente, A.A.; Teixeira, J.A.; Cruz, L.; Díaz, R. Chitosan coating surface properties as affected by plasticizer, surfactant and polymer concentrations in relation to the surface properties of tomato and carrot. *Food Hydrocoll.* **2008**, *22*, 1452–1459. [CrossRef]

32. AOAC. *Official Methods of Analysis of the Association of Official Analytical Chemists (AOAC)*, 16th ed.; AOAC International, W., Ed.; Association of Official Analytical Chemists: Washington, DC, USA, 1997.

33. Ranggana, S. *Manual Analysis of Fruit and Vegetable Products*; Ltd., T.M., Ed.; Hill Publish. Comp. Ltd.: New Delhi, India, 1977.

34. Batu, A. Determination of acceptable firmness and colour values of tomatoes. *J. Food Eng.* **2004**, *61*, 471–475. [CrossRef]

35. Olivas, G.I.; Mattinson, D.S.; Barbosa-Cánovas, G.V. Alginate coatings for preservation of minimally processed 'Gala' apples. *Postharvest Biol. Technol.* **2007**, *45*, 89–96. [CrossRef]

36. Cerqueira, M.A.; Lima, Á.M.; Souza, B.W.S.; Teixeira, J.A.; Moreira, R.A.; Vicente, A.A. Functional polysaccharides as edible coatings for cheese. *J. Agric. Food Chem.* **2009**, *57*, 1456–1462. [CrossRef] [PubMed]

37. Olivares, S.E. *Paquete de Diseños Experimentales*; Autonomous University of Nuevo: León, Mexico, 2015.

38. Aktas, H.; Bayindir, D.; Dilmaçünal, T.; Koyuncu, M.A. The effects of minerals, ascorbic acid, and salicylic acid on the bunch quality of tomatoes (*Solanum lycopersicum*) at high and low temperatures. *HortScience* **2012**, *47*, 1478–1483. [CrossRef]
39. Khoshgozaran-Abras, S.; Azizi, M.H.; Hamidy, Z.; Bagheripoor-Fallah, N. Mechanical, physicochemical and color properties of chitosan based-films as a function of *Aloe vera* gel incorporation. *Carbohydr. Polym.* **2012**, *87*, 2058–2062. [CrossRef]
40. Morad, M.M.; Abou EL -Yazied, A.; Abd El-Gawad, H.G.; Osman, H.S.; Abou-Elwafa, S.M.; Metwally, A.A. Extending storage life and maintaining the quality of tomato fruits by using postharvest exogenous edible coating of *Aloe vera* gel, starch, and casein. *J. Pharm. Negat. Results* **2022**, *13*, 1708–1727. [CrossRef]
41. Zapata, P.J.; Guillén, F.; Martínez-Romero, D.; Salvador Castillo, D.V.; Serrano, M. Use of alginate or zein as edible coatings to delay postharvest ripening process and to maintain tomato (*Solanum lycopersicon* Mill) quality. *J. Sci. Food Agric.* **2008**, *88*, 1287–1293. [CrossRef]
42. Javanmardi, J.; Kubota, C. Variation of lycopene, antioxidant activity, total soluble solids and weight loss of tomato during postharvest storage. *Postharvest Biol. Technol.* **2006**, *41*, 151–155. [CrossRef]
43. Domínguez, I.; Lafuente, M.T.; Hernández-Muñoz, P.; Gavara, R. Influence of modified atmosphere and ethylene levels on quality attributes of fresh tomatoes (*Lycopersicon esculentum* Mill.). *Food Chem.* **2016**, *209*, 211–219. [CrossRef]
44. Abushita, A.A.; Hebshi, E.A.; Daood, H.G.; Biacs, P.A. Determination of antioxidant vitamins in tomatoes. *Food Chem.* **1997**, *60*, 207–212. [CrossRef]
45. Brasil, I.M.; Gomes, C.; Puerta-Gomez, A.; Castell-Perez, M.E.; Moreira, R.G. Polysaccharide-based multilayered antimicrobial edible coating enhances quality of fresh-cut papaya. *LWT—Food Sci. Technol.* **2012**, *47*, 39–45. [CrossRef]
46. Mohebbi, M.; Ansarifar, E.; Hasanpour, N.; Amiryousefi, M.R. Suitability of *Aloe vera* and gum tragacanth as edible coatings for extending the shelf Life of button mushroom. *Food Bioprocess Technol.* **2012**, *5*, 3193–3202. [CrossRef]
47. Valverde, J.M.; Valero, D.; Martínez-Romero, D.; Guillén, F.; Castillo, S.; Serrano, M. Novel edible coating based on Aloe vera gel to maintain table grape quality and safety. *J. Agric. Food Chem.* **2005**, *53*, 7807–7813. [CrossRef] [PubMed]
48. Sogvar, O.B.; Koushesh Saba, M.; Emamifar, A. *Aloe vera* and ascorbic acid coatings maintain postharvest quality and reduce microbial load of strawberry fruit. *Postharvest Biol. Technol.* **2016**, *114*, 29–35. [CrossRef]
49. Jasso de Rodríguez, D.; Hernández-Castillo, D.; Rodríguez-García, R.; Angulo-Sánchez, J.L. Antifungal activity in vitro of *Aloe vera* pulp and liquid fraction against plant pathogenic fungi. *Ind. Crops Prod.* **2005**, *21*, 81–87. [CrossRef]
50. Chrysargyris, A.; Nikou, A.; Tzortzakis, N. Effectiveness of Aloe Vera Gel Coating for Maintaining Tomato Fruit Quality. *New Zeal. J. Crop Hortic. Sci.* **2016**, *44*, 203–217. [CrossRef]
51. Paladines, D.; Valero, D.; Valverde, J.M.; Díaz-Mula, H.; Serrano, M.; Martínez-Romero, D. The addition of rosehip oil improves the beneficial effect of *Aloe vera* gel on delaying ripening and maintaining postharvest quality of several stonefruit. *Postharvest Biol. Technol.* **2014**, *92*, 23–28. [CrossRef]
52. Chauhan, O.P.; Raju, P.S.; Singh, A.; Bawa, A.S. Shellac and Aloe-Gel-Based Surface Coatings for Maintaining Keeping Quality of Apple Slices. *Food Chem.* **2011**, *126*, 961–966. [CrossRef]

 *foods*

Article

Effects of Frozen Storage Temperature on Water-Holding Capacity and Physicochemical Properties of Muscles in Different Parts of Bluefin Tuna

Jinfeng Wang [1,2,3,4], Hanwen Zhang [1,2,3,4], Jing Xie [1,2,3,4,*], Wenhui Yu [1,2,3,4] and Yuyao Sun [1,2,3,4]

1   College of Food Science and Technology, Shanghai Ocean University, Shanghai 201306, China; jfwang@shou.edu.cn (J.W.); zhw1045993510@163.com (H.Z.); strfor@163.com (W.Y.); bj2012_0225@126.com (Y.S.)
2   Shanghai Professional Technology Service Platform on Cold Chain Equipment Performance and Energy Saving Evaluation, Shanghai 201306, China
3   Shanghai Engineering Research Center of Aquatic Product Processing & Preservation, Shanghai 201306, China
4   National Experimental Teaching Demonstration Center for Food Science and Engineering, Shanghai Ocean University, Shanghai 201306, China
*   Correspondence: jxie@shou.edu.cn; Tel.: +86-156-9216-5513

Abstract: The effects of different freezing temperatures on the water-holding capacity and physicochemical properties of bluefin tuna were studied. The naked body, big belly and middle belly parts of bluefin tuna were stored at −18 °C and −55 °C for 180 days. The tuna was evaluated by determining the water-holding capacity, color difference, malondialdehyde (MDA), salt-soluble protein content, free amino acid (FAA), endogenous fluorescent proteins and water distribution and migration. The salt-soluble protein content was measured by the Bradford method. The color difference was measured by a CR-400 color difference meter. The water distribution and migration were analyzed by the low-field nuclear magnetic resonance (LF-NMR). The results showed little quality change during short-term frozen storage, but the frozen storage temperature of −55 °C significantly improved the quality of tuna compared with the frozen storage temperature of −18 °C. There were great differences in the salt-soluble protein content, water-holding capacity and water content the different parts of the tuna. The water-holding capacity and the protein content were the highest, and the water distribution of the naked body part was the most uniform of the three different parts. Because of the high fat content in the big belly and the middle belly, the MDA content and the odor of amino acid increased rapidly and the quality seriously decreased during the frozen storage.

Keywords: bluefin tuna; frozen storage; physicochemical properties; water-holding capacity; color difference

Citation: Wang, J.; Zhang, H.; Xie, J.; Yu, W.; Sun, Y. Effects of Frozen Storage Temperature on Water-Holding Capacity and Physicochemical Properties of Muscles in Different Parts of Bluefin Tuna. *Foods* **2022**, *11*, 2315. https://doi.org/10.3390/foods11152315

Academic Editors: Sidonia Martinez and Javier Carballo

Received: 8 July 2022
Accepted: 29 July 2022
Published: 3 August 2022

1. Introduction

As is well-known, tuna is highly perishable, and therefore, storage time can influence the quality of tuna. Generally, frozen storage is chosen as the main technique for extending the shelf life of tuna [1]. Compared with ice storage and cold storage, frozen storage uses a lower storage temperature. The effect of beetroot-peel dip treatment on steak [2] was studied by Hafsa et al. who found that cellular enzymatic activity and biochemical reactions were more inhibited when the frozen storage was at a relatively low temperature. Similarly, frozen storage was found to be an essential technology in meat processing for a long shelf life and good quality [3].

Tuna is a good source of nutrients such as protein, fat, vitamins, amino acids, unsaturated fatty acids, and minerals [4]. Bluefin tuna is suitable as a raw material for the preparation of high quality food [5]. Based on their research on the effects of ultra-low temperature and freshness on the quality of frozen tuna meat, Naho et al. [6] found that the frozen storage temperature of tuna should be −50 °C or lower. During the freezing process,

---

ice crystals are formed in muscle cells, and Teuteberg et al. [7] investigated the effects of frozen storage time and temperature on the quality of thawed pork and found that ice crystals can destroy muscle cells and change the cell structure, resulting in water loss and affecting the meat quality. The water-holding capacity is affected by the ice crystals [8]. For bluefin tuna, different parts contain different nutrients [9]. Freezing different parts of the tuna can maximize its economic value.

The quality of frozen storage products may be affected by several factors, such as the frozen storage mode [10], frozen storage temperature [11], freezing rate [12], etc. According to the research, the frozen storage temperature has a significant impact on the quality of frozen tuna [13], and the water-holding capacity is an important quality parameter that characterizes the sensory characteristics of tuna, thus affecting the quality of tuna.

However, there are few studies on the impact of frozen storage temperature on bluefin tuna's physicochemical properties, as well as the water-holding capacity. There is no public research on the frozen storage of different parts of bluefin tuna. Research on the frozen storage of different parts will help to improve the economic benefits of bluefin tuna. Therefore, the purpose of this study was to explore the effects of different frozen storage temperatures on the water-holding capacity and physicochemical properties of bluefin tuna and to provide a theoretical foundation for improving the fresh-keeping quality of different parts of bluefin tuna.

2. Materials and Methods

2.1. Materials

The tuna used in the study was purchased from Dalian Chenyang Technology Development Co., Ltd. It was frozen immediately after fishing, cut according to the naked body, big belly and middle belly, and then it was sent to the laboratory for frozen storage.

2.2. Main Test Equipment

LF-NMR measurements were carried out by using a Niumag Benchtop Pulsed NMR Analyzer (Niumag Electric Corporation, Shanghai, China) to analyze the water distribution and migration. A color difference meter (CR-400, Konica Minolta, Tokyo, Japan), was used for measuring color changes in the sample. A UV-V spectrophotometer (UV-1102, Shanghai Tianmei Instrument Co., Ltd., Shanghai, China) was used for measuring FAA. High-speed centrifuge was used to centrifuge the mixed solutions. A fluorescence spectrophotometer was used for measuring the endogenous fluorescence of protein.

2.3. Experimental Method

Tuna samples were grouped according to the naked body, middle belly and big belly. Then, 6 parallel tests were conducted in one experiment, and the tuna samples were stored in refrigerators at −55 °C and −18 °C for 6 months. After the tuna arrived in the laboratory, the fresh, initial values including water-holding capacity, color difference, MDA, salt-soluble protein content and endogenous fluorescent proteins were measured and then the experiment was carried out every month.

2.4. Thawing Method

Tuna was thawed in warm salt-water with a concentration of 3% and the probe of the thermometer was inserted into the sample to measure the central temperature. When the temperature reached 5 °C, it was considered that the thawing was complete.

2.5. Water-Holding Capacity

After thawing the tuna meat, the water on the surface was wiped off with filter paper. The meat was weighed, and 2 g was wrapped with two layers of paper, and put in the centrifugal tube. The meat was centrifuged at 4 °C for 10 min, at 5500 R/min, and then the mass of the meat was accurately weighed after centrifugation. The formula for the water-holding capacity is as follows:

$$W\ (\%) = (M_a / M_b) \times 100\% \tag{1}$$

where W is the water-holding capacity, M_a is the quality of meat after centrifugation, and M_b is the quality of meat before centrifugation.

2.6. LF-NMR Measurements

The prepared tuna sample was packed with fresh-keeping film and put into a nuclear magnetic detection tube with a diameter of 70 mm. The coil temperature was 32 °C, and the proton resonance frequency was adjusted to 24 MHz. The CPMG sequence was selected and T2 measurement parameters was set as: sampling frequency SW = 100 kHz, analog gain RG1 = 20, P1 = 18.00 µs, digital gain DRG1 = 6, TD = 400 004, PRG = 1, repeat sampling interval TW = 2000 ms, accumulation times NS = 4, P2 = 34.00 µs, echo time TE = 0.550, and serial number of the echo = 8000. According to the CPMG exponential decay curve, the transverse relaxation time T2 map was obtained by iterative inversion with analysis software.

2.7. MDA

One gram of tuna meat was added to 9 mL of normal saline, and centrifuged (7550 R/min, 10 min), After centrifugation, the supernatant was determined by visible spectrophotometry and by using a malondialdehyde kit (Nanjing Jiancheng Bioengineering Institute, Nanjing, China). Each group of samples was measured three times and the average value was taken.

2.8. Salt-Soluble Protein Assay

Next, 20 mL of a 0.02 mol/L PBS solution was added to 2 g of chopped tuna. The resultant precipitate was then collected after the mixed solution had been homogenized at high speed (6000 R/min, 3 min) and centrifuged at a low temperature (10,000× g, 4 °C, 20 min). After this procedure, 40 mL NaCl solution was prepared. Then, we added 12 mL 3% NaCl solution, homogenized the mixture at a high speed (6000 R/min, 3 min), centrifuged the mixture at a low temperature (10,000× g, 4 °C, 10 min), took the supernatant for storage, repeated the experiment (a total of 3 times), and combined the supernatant. At this time, the supernatant contained myogenic protein, and then a 20 µL sarcoplasmic protein solution was drained into a well-plate (dry in advance). The above draining process was repeated 3 times and the solution was placed in 3 orifice plates, respectively, as a parallel experiment. Then, 200 µL of Bradford solution was added to these 3 orifice plates and was quickly mixed at a room temperature of 25–30 °C. After a 5 min reaction, the microplate reader was inserted and connected to the computer for testing.

2.9. FAA

After homogenizing, drying, and grinding the sample, 0.2 g of dry sample was placed in a 55 mL volumetric flask, and 0.02 mol/L hydrochloric acid was added to a constant volume; it stood for 1 h, and was then filtered, purified, and passed through the membrane. The AccQ-Tag method was used to develop a reverse phase separation, which was carried out with UV 248 nm detection, and the instrument used was a high performance liquid chromatograph.

2.10. Endogenous Fluorescent Protein

Two grams of tuna meat was weighed and 18 mL of distilled water was added, it was homogenized (14,000 R/min, duration 1 min), and then the solution was centrifuged for 10 min (10,000 R/min, 4 °C), filtered to obtain precipitation. Then, 18 mL of NaCl solution with 3% concentration was added, it was homogenized, centrifuged and the supernatant was taken.

The extracted myofibrin solution was diluted to 0.05 mg/mL in 0.6 mol/L KCL solution, and it was set according to the following parameters: laser wavelength 295 nm, excitation, and emission slit width 2.5 nm. The wavelength scanning surface was divided into 300–400 nm, the scanning speed was 12,000 nm/min, and each sample was measured in parallel 5 times.

2.11. Color Difference

The thawed tuna was cut into 20 mm × 20 mm × 15 mm, a CR-400 color difference meter was used to measure the L*, a* value, and ΔE of the tuna block, and whiteboard correction was finished before sample determination. Each group of samples was measured in parallel 3 times.

2.12. Data Analysis

The data for all indexes were processed by Excel 2016. The measured indexes were measured three times with parallel measurement results, drawn by origin 8.5 software (OriginLab, Northampton, MA, USA), and analyzed by SPSS 19.0 software (SPSS Inc., Chicago, IL, USA) ($p < 0.05$ is considered a significant difference).

3. Results

3.1. Changes of Muscle Water-Holding Capacity in Different Parts of Tuna

The water-holding capacity is defined as the physical binding force of a tuna sample to water under the action of external forces [14]. Water is the main component of tuna muscle, and water content has an important impact on the taste and quality of tuna muscle [15]. The three-dimensional network structure in the muscle can affect the distribution state and form of water, so the water-holding capacity of tuna meat is also affected by its muscle composition [16]. Figure 1 shows the change in water-holding capacity of tuna sampled from different parts when frozen at −18 °C and −55 °C. It can be seen that during frozen storage time, the water-holding capacity of tuna meat shows a downward trend, and the water-holding capacity of tuna meat stored at −18 °C is relatively lower than that frozen at −55 °C. At the same temperature, the water-holding capacity of the naked body tuna is stronger, and the water-holding capacity of the belly is weaker because the water-holding capacity of the tuna meat is closely related to the content of protein. The water in tuna meat is in the form of free water. One part of the free water exists between myofibrils and connective tissue, and the other part is combined with the carboxyl groups of proteins and sugars. In the process of frozen storage, the hydrophobic/hydrophilic binding bond around the protein is destroyed, and the water bound with the protein becomes free water, resulting in a decrease in the water-holding capacity [17]. In addition, the formation of ice crystals causes certain damage to the tissue structure of myofibrils, which is also an important factor affecting the water-holding capacity of tuna. The size of the ice-crystal was also affected by the freezing temperature; freezing at lower temperatures resulted in smaller ice crystals and less modification of the food's solid structures [18]. The water-holding capacity of the tuna samples from different parts was better at the frozen temperature of −55 °C.

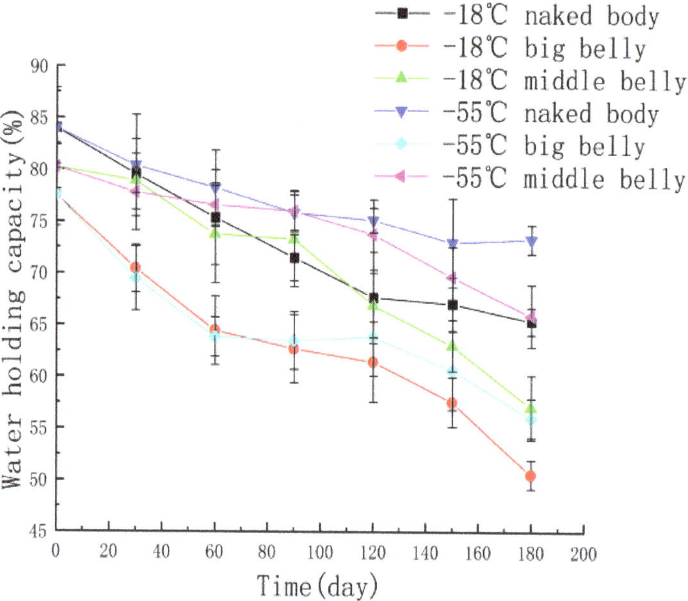

Figure 1. The change in water-holding capacity in different parts of tuna during frozen storage at different temperatures.

3.2. Changes in Muscle Color in Different Parts of Tuna

Biochemical changes such as protein oxidation and fat oxidation in tuna meat will change the color. This is an important index to evaluate the quality of tuna. The color of tuna meat will have an intuitive impact on consumers' purchases [19]. Fresh tuna is bright red, and the color of the meat is expressed by the value of a*. The change in the color of tuna meat is related to the formation of high-iron myoglobin. During frozen storage, ferrous ions bound to myoglobin are easily oxidized to high-iron myoglobin, which results in browning of the flesh [20]. Figure 2 shows that the value of the color difference of the tuna meat had a downward trend over the whole storage period. In the early stage of storage, the a* value decreased rapidly, a large amount of high-iron myoglobin was produced, and the meat color became significantly darker. In the later stage of storage, the decrease in a* slowed down and gradually stabilized. The protein content in the naked body part of tuna is high. During storage, ferrous ions are further oxidized to produce high-iron myoglobin, resulting in the rapid decline in the a* value. Tuna's big belly contains less protein, but it contains more fat, which can prevent tuna from having a bright red color. In addition, during the oxidation process, fats can produce free radicals that oxidize myoglobin and ferrous ions, producing high-iron myoglobin [21]. At the same temperature, the a* value of the big belly part of tuna is always lower than that of the middle belly and naked body part. After 180 days of storage, the naked body part of tuna at the storage temperature of −55 °C had the best a* value.

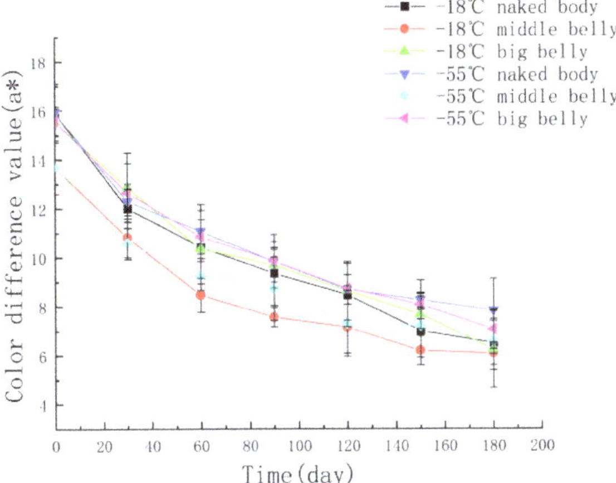

Figure 2. The color change in different parts of tuna during frozen storage at different temperatures.

3.3. Changes in Muscle MDA in Different Parts of Tuna

In the whole process of frozen storage, fat oxidation leads to bad-smelling tuna meat, and the deterioration of the tuna meat, which affects its quality and flavor and reduces the commercial value of the meat. After oxidative degradation, unsaturated fatty acids produce malondialdehyde. The measurement of malondialdehyde can determine the degree of oxidation of tuna fat and characterize the freshness of aquatic products [22]. As shown in Figure 3, during the frozen storage of bluefin tuna naked body, big belly and middle belly at $-18\ ^\circ C$ and $-55\ ^\circ C$, the MDA content continued to rise, which is due to the continuous accumulation of fat oxidation products with the extension of frozen storage time. The content of MDA is significantly affected by temperature and location, and the same location is frozen at different temperatures. The lower the temperature, the slower the increase in MDA content, which is because the low-temperature environment can slow down the reaction rate and reduce the formation of aldehydes, ketones, and carboxylic acids. Under the same temperature, the MDA content of the naked body part was significantly lower than that of the big belly and middle belly, which was consistent with the fat content of the three parts of the tuna. Xu et al. [23] studied the lipid and myoglobin oxidation of bluefin tuna frozen at different temperatures, and obtained similar results.

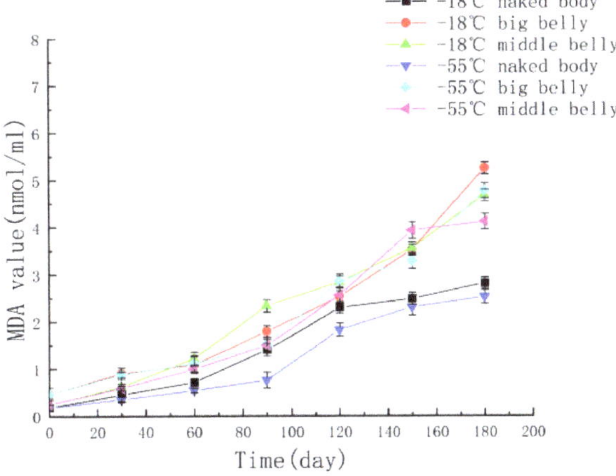

Figure 3. MDA changes in different parts of tuna during frozen storage at different temperatures.

3.4. Changes in Salt-Soluble Protein Content in Different Parts of Tuna Muscle

Protein is the main component of tuna meat. The content of salt-soluble protein in muscle protein accounts for more than 60%. The content of salt-soluble protein can reflect the degree of tuna denaturation [24]. Figure 4 shows the changes in the protein content of tuna meat in different parts at different temperatures. During the process of frozen storage, the content of salt-soluble protein decreased significantly, and the decline rate gradually slowed down, which is due to the over-aggregation and denaturation of hydrogen bonds, disulfide bonds, and hydrophobic bonds in myofibrillar protein molecules [25]. The level of frozen storage temperature has an impact on the content of salt-soluble protein. In the early stage of frozen storage, the protein content of tuna meat frozen at −18 °C decreases rapidly, and the salt-soluble protein content of tuna meat frozen at −55 °C decreases slowly because the low temperature can inhibit the activity of enzymes in organisms. At the end of frozen storage, there is no significant difference in the content of salt-soluble protein of tuna meat frozen at different temperatures, which may be because the mechanical action of ice crystal destroys the myofibril structure, and the protein becomes alkali-soluble protein after freeze denaturation, resulting in the decrease in salt-soluble protein content [26]. After comparing the protein content of three different parts of tuna meat, it was found that the content of salt-soluble protein in the naked body part of tuna meat is the highest and that in the big belly is the lowest, which is due to the different protein content in muscle in different parts of tuna.

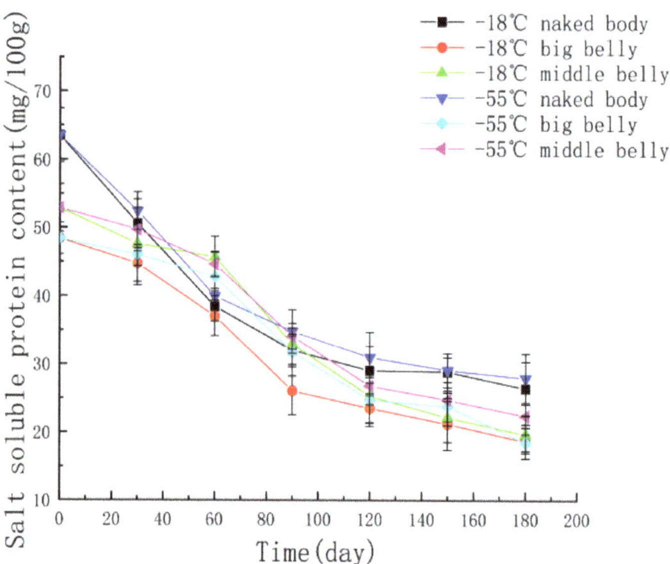

Figure 4. The changes in salt-soluble protein content in different parts of tuna during frozen storage at different temperatures.

3.5. FAA Content in Different Parts of Tuna Muscle

There are many kinds of amino acids in food, mainly including FAAs and bound amino acids. Different kinds of amino acids show different flavors [27]. The composition and content of amino acids in food determine the flavor of food. FAAs come from protein hydrolysis. Aspartic acid, serine, glutamic acid, glycine, alanine, and proline are amino acids that provide flavor, while threonine, valine, methionine, isoleucine, leucine, phenylalanine, and lysine are essential amino acids, which supply human nutrition [28]. Table 1 shows the content of FAAs in different parts of tuna after 180 days of storage at different temperatures. It can be seen that after frozen storage, the FAA content of tuna frozen at $-18\,^{\circ}$C in the same part is significantly lower than that frozen at $-55\,^{\circ}$C, which is consistent with the determination results of salt-soluble protein, indicating that reducing the frozen storage temperature can reduce the degree of protein degradation and the content of FAA. At the same temperature, the content of amino acids in the middle belly of tuna is the highest, and the total content of amino acids in the naked body part is lower. Glycine, histidine, glutamic acid, etc., are odorous amino acids, and the increase in odorous amino acid content will lead to a certain bitterness in tuna, leading to the decline in taste [29]. When frozen at the same temperature, the content of odor acid in naked body parts was the lowest, and the change in the flavor of tuna was the least during frozen storage.

Table 1. Contents of FAAs in different parts of tuna frozen at different temperatures.

Type	$-18\,^{\circ}$C Naked Body	$-18\,^{\circ}$C Middle Belly	$-18\,^{\circ}$C Big Belly	$-55\,^{\circ}$C Naked Body	$-55\,^{\circ}$C Middle Belly	$-55\,^{\circ}$C Big Belly
Aspartic acid	8.45278	28.5659	15.1902	7.82438	11.8069	9.36199
threonine	26.9676	62.6204	42.3936	26.1867	28.8607	24.9122
serine	6.08948	28.302	9.33835	6.92564	10.4408	7.08034
glutamate	29.402	77.3983	56.1695	27.7802	42.5536	44.1216
glycine	43.6959	98.0136	61.2816	39.8779	61.3309	36.0496
alanine	143.353	264.492	108.758	135.117	109.865	73.713
valine	36.3519	69.6088	47.6301	35.8909	32.9729	27.891
methionine	9.01322	38.6347	29.3612	10.7306	9.09695	16.0305
isoleucine	20.7363	36.0292	22.5213	21.1904	15.4014	13.831
leucine	35.9597	61.8645	39.3984	35.2674	27.7481	24.0869
Tyrosine	24.8361	54.2872	32.2359	23.0676	28.4776	21.9596
Phenylalanine	16.8455	45.6567	20.1593	14.4584	19.2527	10.7646
Lysine	264.279	1310.84	141.166	242.389	523.28	98.1079
histidine	8339.41	8802.48	9119.27	7127.83	9926.31	8101.62
Arginine	53.4648	36.7473	32.9904	44.2048	60.5516	20.2973
total	9058.857	11,015.54	9777.864	7798.741	10,907.85	8529.828

3.6. Endogenous Fluorescent Proteins in Different Parts of Tuna Muscle

The internal fluorescence of protein reflects the changes in the tertiary structure of protein molecules. It contains tryptophan, tyrosine, and phenylalanine residues, and can produce fluorescence under the excitation light of 280 nm or 295 nm. By detecting the change in the tryptophan residue microenvironment, the internal structure change in protein can be obtained, and the change in the quality of tuna meat can be determined [30]. Figure 5 shows the changes in endogenous protein fluorescence in different parts of tuna frozen at different temperatures. With the extension of frozen storage time, the endogenous fluorescence intensity of muscle showed a downward trend. This is because low temperature destroys the molecular structure of the myofibrillar protein, exposes tryptophan residues and active groups of protein molecules, and intermolecular aggregation lead to tryptophan residues embedded in protein molecules, resulting in the decrease in endogenous fluorescence intensity [31]. Tuna meat in the same part was frozen at $-55\,^{\circ}$C, and the decline was small, indicating that the temperature can slow down the degree of protein oxidation and prevent the destruction of protein molecular structure. The order of endogenous fluorescence intensity of tuna frozen at the same temperature is as follows: naked body part > middle belly part > big belly part, which is consistent with the protein content of tuna in different parts.

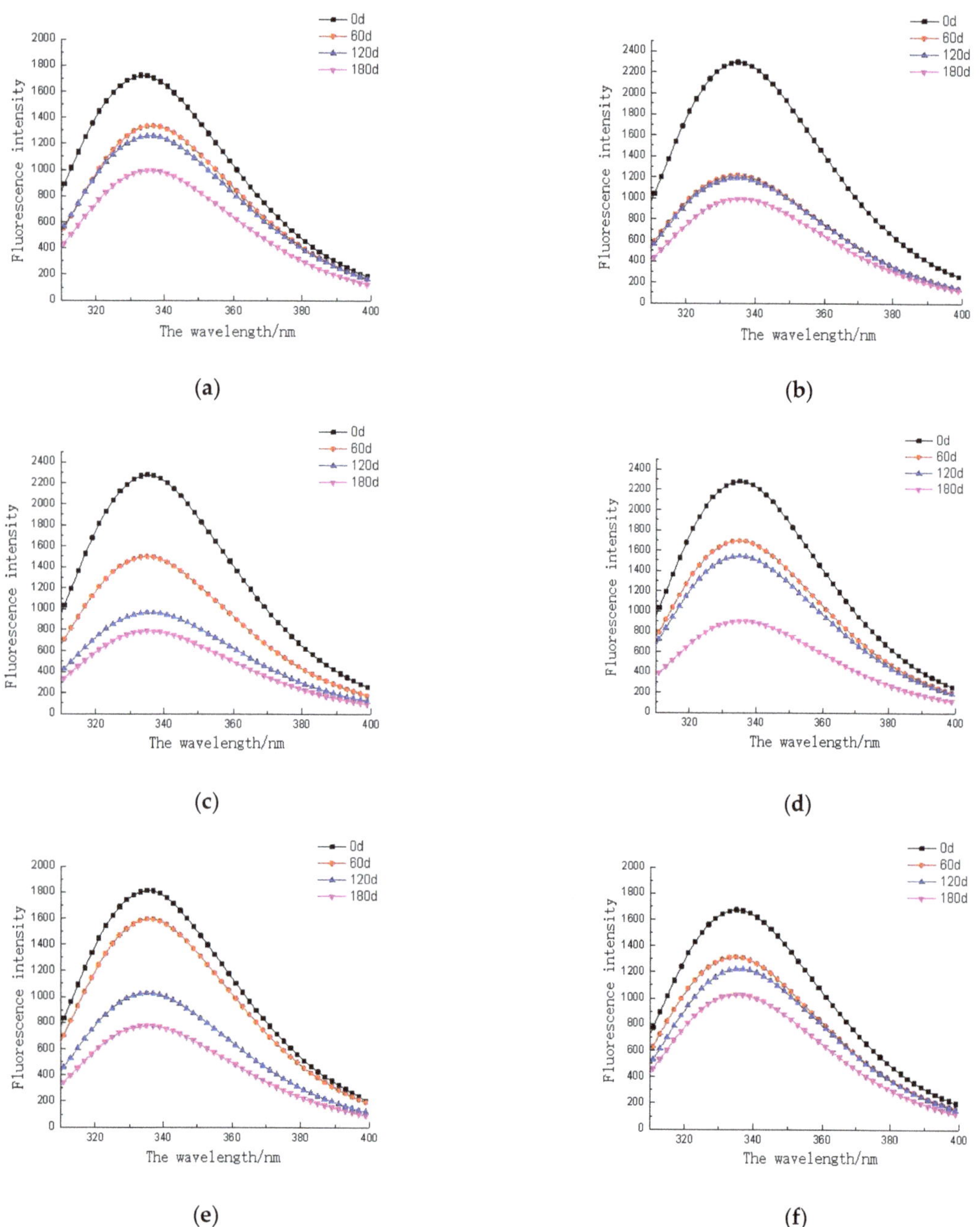

Figure 5. Changes in endogenous fluorescence intensity of tuna in different parts during frozen storage at different temperatures (**a**) frozen storage at −18 °C (**b**) frozen storage at −18 °C (**c**) frozen storage at −18 °C (**d**) frozen storage at −55 °C (**e**) frozen storage at −18 °C (**f**) frozen storage at −55 °C).

3.7. Water Distribution and Migration in Different Parts of Tuna Muscle

LF-NMR uses ^{1}H proton as the medium to determine the water content in the sample by measuring the processing time (relaxation time) of ^{1}H proton from a high-energy state to a low-energy state [32]. The transverse relaxation time is represented by T_2, and the distribution state of T_2 represents the distribution of water, indicating the degree of freedom of water molecules. T_{21} (1~10 ms) represents the combination of water and macromolecules in the form of bound water. T_{22} (10–155 ms) represents water present in the myofibril network and T_{23} (155 ms~) represents water that can flow freely [33]. The changes in the relaxation time of tuna in different parts at different frozen storage temperatures is shown in Figure 6. The peak area of the curve represents the distribution of the water state. There is little difference in the peak area of tuna in all frozen storage experimental groups within 0–10 MS, which indicates that different frozen storage temperatures have little effect on the bound water of tuna in different parts. The water content of T_{22} frozen at −55 °C is significantly higher than that frozen at −18 °C; Harnkarnsujarit et al. [34] obtained a similar result. For frozen tuna, the content of bound water and free water is closely related to the frozen storage temperature; a lower storage temperature results in a lower portion of free water.

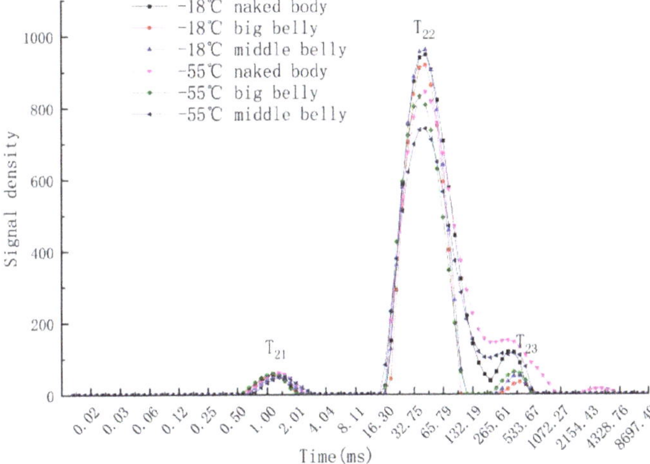

Figure 6. Changes in lateral relaxation time in different parts of tuna under different temperatures.

Table 2 shows the changes in water in different states during the frozen storage time of tuna. The internal water of tuna mainly exists in the form of immobilized water, which accounts for more than 90% of the total water content [35]. For tuna meat frozen at the same temperature, the water content in the naked body part was the highest, and that in the big belly was the lowest. This is because the fat content in the big belly is high, and the tissue structure is fragile. During the process of frozen storage, the water-holding capacity of the myofibril network decreases, and it is not easy to convert immobilized water into free water, resulting in a large loss of water. After 180 days of frozen storage, the naked body part of tuna stored at −55 °C has the highest water content, of which the proportion of immobilized water was as high as 95.7%, indicating that the myofibril structure of tuna was best preserved. Similar studies have also found that the stability of protein structures highly influences the water-holding capacity of seafoods and a higher degree of protein aggregations and denaturation causes low water-holding capacity [36].

Table 2. Proportion of frozen storage peak area in different parts of tuna at different temperatures.

Frozen Storage Mode	Total Peak Area (Water Content)	Bound Water (T21, %)	Immobilized Water (T22, %)	Free Water (T23, %)
−18 °C big belly	7455.455	0.041843	0.941708	0.016449
−18 °C middle belly	8228.852	0.035899	0.941465	0.022635
−18 °C naked body	9767.239	0.032556	0.91532	0.052174
−55 °C big belly	7559.581	0.044403	0.919212	0.036385
−55 °C middle belly	8683.41	0.035597	0.890377	0.074526
−55 °C naked body	10,452.81	0.035959	0.957319	0.006723

LF-NMR can generate pseudo color graphics to display the distribution of water. The stronger the brightness in the image, the higher the signal intensity in this part and the higher the water content in the sample [37]. Figure 7 shows the water distribution of different parts of tuna frozen at different temperatures. The effect of frozen storage temperature on the water distribution of tuna meat is relatively small, and the water distribution in different parts is significantly different. In the naked body part frozen at −55 °C, the tuna meat pattern shows red and yellow, which is evenly distributed, indicating that the signal intensity is high and the water content is the highest. The color of the middle belly and big belly tuna is dim and unevenly distributed, which is due to the high content of fat in the middle belly and big belly, and the distribution of fat affects the water distribution and migration.

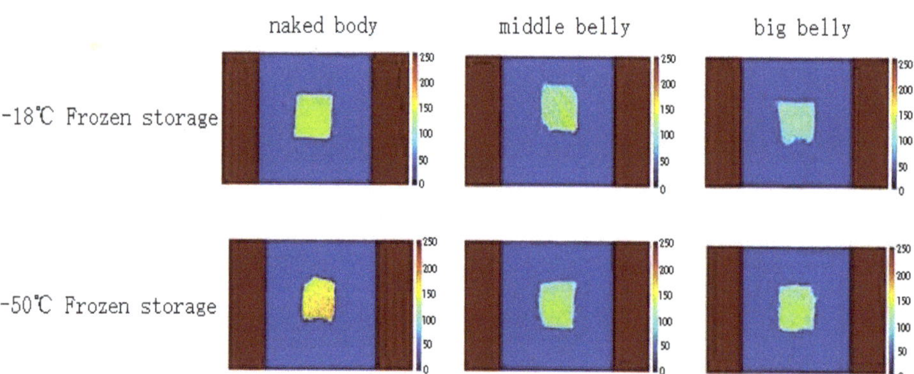

Figure 7. The pseudo color images of different parts of tuna frozen at different temperatures.

The tuna was better maintained after 180 days of frozen storage temperature at −55 °C but the ice crystals formed by frozen storage can damage the salt-soluble protein in the tuna and decrease the hardness of the tuna. Although the shelf life was prolonged, the hardness and taste of tuna meat decreased. Liaomingtao et al. [38] studied the influence of frozen storage temperature on the color change in tuna meat. The quality of the naked body of tuna was better than the quality of the middle belly and big belly parts of tuna when they were stored frozen at −18 °C and −55 °C. The MDA value increased when the storage time increased. The higher the fat content, the more the change in the quality of the tuna; but, for short-term storage for sales, −18 °C was a suitable frozen storage temperature.

4. Conclusions

In this paper, the effects of frozen storage temperatures of −18 °C and −55 °C on the water-holding capacity and physicochemical properties of muscles in different parts of bluefin tuna were studied. The results showed that the water-holding capacity, salt-soluble protein, and water content of tuna decreased as the storage time increased. Compared to the frozen storage temperature of −18 °C, better-quality bluefin tuna was obtained at the frozen storage temperature of $-55\ ^\circ$C. Because there was little change in the quality during short-term frozen storage, the frozen storage temperature of $-18\ ^\circ$C can be suitable for short-term frozen storage for sales. The protein content and the content of salt-soluble protein was highest in the naked body part. There is more fat in the big belly and middle belly, which leads to a rapid rise in MDA in the other two parts. The amino acid content in the middle belly was the highest, and the odor and amino acid content in the naked body part was the lowest. After 180 days of storage, the naked body part of tuna at the storage temperature of $-55\ ^\circ$C has the best a* value, the highest water content and the best water distribution. The results can help to enhance the economic value of bluefin tuna sales. In future research, the big belly and the middle belly of the bluefin tuna in frozen storage requires more attention, and the relationship between the quality of the tuna and the energy consumption of the frozen storage should be considered.

Author Contributions: Methodology and editing, J.W.; writing—original draft preparation, H.Z.; writing—review and editing, J.X.; data curation, W.Y.; supervision, Y.S. All authors have read and agreed to the published version of the manuscript.

Funding: This research was funded by the National Key R&D Program of China (2019YFD0901604).

Institutional Review Board Statement: Not applicable.

Informed Consent Statement: Not applicable.

Data Availability Statement: Authors can confirm that all relevant data are included in the article.

Conflicts of Interest: The authors declare no conflict of interest.

References

1. Chen, X.Q.; Liu, H.Y.; Li, X.X.; Wei, Y.J.; Li, J.R. Effect of ultrasonic-assisted immersion freezing and quick-freezing on quality of sea bass during frozen storage. *LWT* **2022**, *154*, 112737. [CrossRef]
2. Hafsa, M.; Muhammed, P.S.; Zynudheen, A.; Mudassir, A.; Sathish, K. Effect of beetroot peel dip treatment on the quality preservation of Deccan mahseer (*Tor khudree*) steaks during frozen storage ($-18\ ^\circ$C). *LWT* **2021**, *151*, 112222.
3. Astráin-Redín, L.; Abad, J.; Rieder, A.; Kirkhus, B.; Raso, J.; Cebrián, G.; Álvarez, I. Direct contact ultrasound assisted freezing of chicken breast samples. *Ultrason. Sonochem.* **2021**, *70*, 105319. [CrossRef] [PubMed]
4. Bu, Y.; Han, M.L.; Tan, G.Z.; Zhu, W.H.; Li, X.P.; Li, J.R. Changes in quality characteristics of southern bluefin tuna (*Thunnus maccoyii*) during refrigerated storage and their correlation with color stability. *LWT* **2022**, *154*, 112715. [CrossRef]
5. Lin, Y.-C.J.; Theodore, C.B. Embedding food in place and rural development: Insights from the Bluefin Tuna Cultural Festival in Donggang, Taiwan. *J. Rural Stud.* **2020**, *79*, 373–381. [CrossRef]
6. Naho, N.; Ritsuko, W.; Hideto, F.; Ryusuke, T.; Shinji, K.; Emiko, O. Effect of long-term storage, ultra-low temperature, and freshness on the quality characteristics of frozen tuna meat. *LWT* **2022**, *112*, 270–280.
7. Vivien, T.; Ina-Karina, K.; Madeleine, P.; Carsten, K. Effects of duration and temperature of frozen storage on the quality and food safety characteristics of pork after thawing and after storage under modified atmosphere. *Meat Sci.* **2021**, *174*, 108419.
8. Jiang, Q.Q.; Jia, R.; Naho, N.; Hu, Y.Q.; Kazufumi, O.; Emiko, O. Changes in protein properties and tissue histology of tuna meat as affected by salting and subsequent freezing. *J. Rural Stud.* **2019**, *271*, 550–560. [CrossRef]
9. Fatih, P.; Ozlem, S.; Can, A.; Mustafa, S. Some trace elements in front and rear dorsal ordinary muscles of wild and farmed bluefin tuna (*Thunnus thynnus* L. 1758) in the Turkish part of the eastern Mediterranean Sea. *Food Chem. Toxicol.* **2011**, *49*, 1006–1010.
10. Zhu, M.M.; Zhang, J.; Jiao, L.X.; Ma, C.M.; Kang, Z.L.; Ma, H.J. Effects of freezing methods and frozen storage on physicochemical, oxidative properties and protein denaturation of porcine longissimus dorsi. *LWT* **2022**, *153*, 112529. [CrossRef]
11. Leng, D.M.; Zhang, H.N.; Tian, C.Q.; Xu, H.B.; Li, P.R. The effect of magnetic field on the quality of channel catfish under two different freezing temperatures. *Int. J. Refrig.* **2022**, *140*, 49–56. [CrossRef]
12. Qian, S.Y.; Hu, F.F.; Waris, M.; Li, X.; Zhang, C.H.; Christophe, B. The rise of thawing drip: Freezing rate effects on ice crystallization and myowater dynamics changes. *Food Chem.* **2022**, *373*, 131461. [CrossRef] [PubMed]
13. Yang, F.; Jing, D.T.; Yu, D.W.; Xia, W.S.; Jiang, Q.X.; Xu, Y.S.; Yu, P.P. Differential roles of ice crystal, endogenous proteolytic activities and oxidation in softening of obscure pufferfish (*Takifugu obscurus*) fillets during frozen storage. *Food Chem.* **2019**, *278*, 452–459. [CrossRef]
14. Li, X.Y.; Liu, H.X.; Yang, X.S.; Chi, H.; Huang, H.L.; Li, L.Z.; Rao, Y.L. Effects of Temperature on the Quality of Antarctic Krill (*Euphausia superba*) during Frozen Storage. *Mod. Food Sci. Technol.* **2014**, *30*, 191–195.
15. Li, W.N.; Han, J.Z. Study on quality changes of goat meat under cold storage conditions with NMR. *Sci. Technol. Food Ind.* **2010**, *31*, 125–127.

16. Ida, G.A.; Ulf, E.; Emil, V. Water properties and salt uptake in Atlantic salmon fillets as affected by ante-mortem stress, rigor mortis, and brine salting: A low-field 1H NMR and 1H/23Na MRI study. *Food Chem.* **2010**, *120*, 482–489.

17. Bao, J.Q.; Xu, R.; Lu, H. Study on the T.T.T. Curve of the Back Muscle of Frozen Yellow fin Tuna. In Proceedings of the 2007 Academic Annual Meeting of Shanghai Refrigeration Society, Shanghai, China, 10–13 October 2007; pp. 125–127.

18. Harnkarnsujarit, N.; Charoenrein, S. Influence of collapsed structure on stability of β-carotene in freeze-dried mangoes. *Food Res. Int.* **2011**, *44*, 3188–3194. [CrossRef]

19. Jiang, H.Z.; Zhao, F.; Ma, Y.J.; Li, G.D.; Liu, M.; Zhou, D.Q. Effects of Different Freezing Temperatures on Physicochemical Properties and Freshness of Spanish Mackerel. *Sci. Technol. Food Ind.* **2007**, *39*, 258–262.

20. Surendranath, P.S.; Poulson, J. Myoglobin chemistry and meat color. *Annu. Rev. Food Sci. Technol.* **2013**, *4*, 79–99.

21. Cameron, F.; Sun, Q.; Richard, M.; Surendranath, P.S. Myoglobin and lipid oxidation interactions: Mechanistic bases and control. *Meat Sci.* **2010**, *86*, 86–94.

22. Yang, Q.; Xie, J. Effect of ultra high pressure on preservation of frozen hairtail. *Food Ferment. Ind.* **2015**, *41*, 200–206.

23. Xu, K.H.; Liao, M.T.; Lin, S.S.; Zhao, Q.L.; Liu, S.C.; Feng, J.L.; Dai, Z.Y. Oxidation kinetics of lipid and myoglobin in bluefin tuna. *J. Chin. Inst. Food Sci. Technol.* **2015**, *15*, 64–71.

24. Yang, J.S.; Lin, L.; Xia, S.Y.; Xie, C. Study on the mechanism of ultra-low temperature storage on the texture and biochemical characteristics of tuna meat. *Oceanol. Limnol. Sin.* **2015**, *46*, 828–832.

25. Zhou, A.M.; Zeng, Q.X.; Liu, X.; Sun, Y.M. Physicochemical changes of frozen surimi protein in frozen storage and its influencing factors. *Food Sci.* **2003**, *24*, 153–157.

26. Yang, J.S. Study on Cold Storage Technology of Tuna Muscle. Master's Thesis, Zhejiang Ocean University, Zhoushan, China, 2012.

27. Lv, X.J.; Liang, L.L.; Huang, H.J.; Qin, Y.; Ning, Z.X. Effect of free amino acid content on food flavor characteristics. *Food Sci.* **1996**, *17*, 10–12.

28. Jin, Y. Study on the Flavor of Crab Meat. Master's Thesis, Zhejiang Gongshang University, Hangzhou, China, 2011.

29. Chen, J.L.; Shi, W.Z.; Chen, S.S.; Chen, S.H. Effect of different storage temperature on freshness and taste of grass carp meat. *Sci. Technol. Food Ind.* **2013**, *38*, 329–333.

30. Chang, H.X.; Shi, Y.; Wang, H.; Huang, X.Q.; Li, R.P.; Bao, Z.Y.; Tu, Z.C. Effect of Ultrasonic Treatment on Physico-chemical Properties of *Myofifi brillar* Protein from Grass Carp. *Food Sci.* **2015**, *36*, 56–60.

31. Lei, Y.T.; Shi, J.; Gui, P.; Feng, L.G.; Yu, X.P.; Luo, Y.K. Effects of glazing with tea polyphenols on the quality characteristics for Pacific white shrimp (*Litopenaeus vannamei*) during frozen storage. *J. China Agric. Univ.* **2018**, *23*, 92–99.

32. Zhao, H.Q.; Wu, J.X.; Zhang, W.Y.; Lan, W.Q.; Liu, S.C.; Sun, X.H.; Xie, J. Effects of ultra-high pressure treatment on the quality and histological structure of frozen perch slices. *Chin. J. High Press. Phys.* **2017**, *31*, 494–504.

33. Kelly, L.P.; Katja, R.; Henrik, J.A.; David, L.H. Water distribution and mobility in meat during the conversion of muscle to meat and ageing and the impacts on fresh meat quality attributes—A review. *Meat Sci.* **2011**, *89*, 111–124.

34. Harnkarnsujarit, N.; Kawai, K.; Suzuki, T. Effects of Freezing Temperature and Water Activity on Microstructure, Color, and Protein Conformation of Freeze-Dried Bluefin Tuna (*Thunnus orientalis*). *Food Bioprocess Technol.* **2015**, *8*, 619–625. [CrossRef]

35. Kimbuathong, N.; Leelaphiwat, P.; Harnkarnsujarit, N. Inhibition of melanosis and microbial growth in Pacific white shrimp (*Litopenaeus vannamei*) using high CO_2 modified atmosphere packaging. *Food Chem.* **2020**, *312*, 126114. [CrossRef]

36. Yang, H.X.; Liu, D.S.; Li, J.K.; He, Y.; Zhou, P. Effects of changes in protein and tissue structure of herring meat on fish quality under low temperature storage. *Food Ferment. Ind.* **2016**, *42*, 208–213.

37. Miklos, R.; Mora-Gallego, H.; Larsen, F.H.; Serra, X.; Cheong, L.-Z.; Xu, X.; Arnau, J.; Lametsch, R. Influence of lipid type on water and fat mobility in fermented sausages studied by low-field NMR. *Meat Sci.* **2014**, *96*, 617–622. [CrossRef] [PubMed]

38. Liao, M.T. Study on the Color Change of Bluefin Tuna Meat under Frozen Storage. Master's Thesis, Zhejiang Gongshang University, Hangzhou, China, 2013.

 *foods*

Article

Evaluation of Precision and Sensitivity of Back Extrusion Test for Measuring Textural Qualities of Cooked Germinated Brown Rice in Production Process

Kannapot Kaewsorn [1], Pisut Maichoon [2], Pimpen Pornchaloempong [3], Warawut Krusong [4], Panmanas Sirisomboon [2,*], Munehiro Tanaka [5,*] and Takayuki Kojima [5]

[1] Department of Agricultural Engineering, School of Engineering and Innovation, Rajamangala University of Technology Tawan-Ok, Chon Buri 20110, Thailand; kannapot_ka@rmutto.ac.th
[2] Department of Agricultural Engineering, School of Engineering, King Mongkut's Institute of Technology Ladkrabang, Bangkok 10520, Thailand; palmpisut1994@gmail.com
[3] Department of Food Engineering, School of Engineering, King Mongkut's Institute of Technology Ladkrabang, Bangkok 10520, Thailand; pimpen.po@kmitl.ac.th
[4] Division of Fermentation Technology, School of Food Industry, King Mongkut's Institute of Technology Ladkrabang, Bangkok 10520, Thailand; warawut.kr@kmitl.ac.th
[5] Laboratory of Agricultural Production Engineering, Faculty of Agriculture, Saga University, 1 Honjo-machi, Saga 840-8502, Japan; kojimat1733@gmail.com
* Correspondence: panmanas.si@kmitl.ac.th (P.S.); mune@cc.saga-u.ac.jp (M.T.)

Citation: Kaewsorn, K.; Maichoon, P.; Pornchaloempong, P.; Krusong, W.; Sirisomboon, P.; Tanaka, M.; Kojima, T. Evaluation of Precision and Sensitivity of Back Extrusion Test for Measuring Textural Qualities of Cooked Germinated Brown Rice in Production Process. *Foods* **2023**, *12*, 3090. https://doi.org/10.3390/foods12163090

Academic Editors: Sidonia Martinez and Javier Carballo

Received: 17 July 2023
Revised: 4 August 2023
Accepted: 7 August 2023
Published: 17 August 2023

Abstract: The textural qualities of cooked rice may be understood as a dominant property and indicator of eating quality. In this study, we evaluated the precision and sensitivity of a back extrusion (BE) test for the texture of cooked germinated brown rice (GBR) in a production process. BE testing of the textural properties of cooked GBR rice showed a high precision of measurement in hardness, toughness and stickiness tests which indicated by the repeatability and reproductivity test but the sensitivity indicated by coefficient of variation of the texture properties. The findings of our study of the effects on cooked GBR texture of different soaking and incubation durations in the production of Khao Dawk Mali 105 (KDML 105) GBR, as measured by BE testing, confirmed that our original protocol for evaluation of the precision and sensitivity of this texture measurement method. The coefficients of determination (R^2) of hardness, toughness and stickiness tests and the incubation time at after 48 hours of soaking were 0.82, 0.81 and 0.64, respectively. The repeatability and reproducibility of reliable measurements, which have a low standard deviation of the greatest difference between replicates, are considered to indicate high precision. A high coefficient of variation where relatively wide variations in the absolute value of the property can be detected indicates high sensitivity when small resolutions can be detected, and vice versa. The sensitivity of the BE tests for stickiness, toughness and hardness all ranked higher, in that order, than the sensitivity of the method for adhesiveness, which ranked lowest. The coefficients of variation of these texture parameters were 31.26, 20.59, 19.41 and 18.72, respectively. However, the correlation coefficients among the texture properties obtained by BE testing were not related to the precision or sensitivity of the test. By obtaining these results, we verified that our original protocol for the determination of the precision and sensitivity of food texture measurements which was successfully used for GBR texture measurement.

Keywords: germinated brown rice; cooked rice; texture; back extrusion; precision; sensitivity

1. Introduction

Germinated brown rice (GBR) has better texture, nutritional and nutraceutical qualities, compared with brown rice [1–4]. During the GBR soaking process, water is absorbed rapidly and a series of biochemical processes occur, which result in the softening of the texture, degradation of polymers, and enhancement of the synthesis and accumulation of certain phytochemical compounds [1,5–7]. GBR production in a typical small-scale commercial setting may be summarized as follows: First, rough rice is soaked in water at room temperature for 48 h. The water is changed every 4 h and drained at the end of soaking to prevent fermentation of the rice. The rice is kept in a polypropylene sack for incubating at 24 h to obtain the germinated rough rice (GRR). The GRR is then dried using the fluidized bed technique by means of a superheated steam which reduces the moisture content of the rice to around 19%wb. Then, the rice is spread on screens for 3 h at room temperature, and the moisture content declines to about 13–14%wb. Finally, the GRR sample is dehusked to obtain GBR. In addition, we note here the results of a published study, which found that the initial soaking of RD31 rough rice resulted in an increase in moisture content from 12.39–13.42% (dry basis) to 24.95–35.63% (dry basis), although the moisture content only increased with time [8].

Due to its slight flavor, the textural properties of cooked rice—including cooked GBR—are dominant when assessing eating quality. Along with aroma, texture is the most important attribute for the eating quality of cooked rice. Wang et al. [9] reported a large variance of textural attributes, and a total observed number of 39 major volatile organic components, in varieties of Yueguang cooked rice. Pearson correlations showed that the hardness of cooked rice was positively correlated with the content levels of E-2-hexenal, 2-hexanol-monomer, 1-propanol, and E-2-pentenal, while stickiness was positively correlated with 5-methyl-2-furanmethanol and dimethyl trisulfide. Possible mechanisms explaining these relations were discussed in this report [9]. Research findings like these could help the rice industry to develop rice products with the desirable properties of both texture and aroma. Other studies have reported changes in the texture of cooked GBR obtained by different processes. The authors of [10] reported that brown rice with texture that was 32% and 42% softer could be produced by germination and by early harvesting, respectively [10]. Researchers have also reported that, during soaking in water, the germ of GBR produces substances both physiologically and through the action of enzymes which improve the nutritional value and the texture of brown rice [11,12]. Zhu et al. [13] investigated the effects of hydrogen-rich water (HRW) on the germination efficiency and texture of GBR and found that HRW ($1.5\,\mathrm{mg\,L^{-1}}$) treatment significantly increased the germination efficiency of GBR by changing the migration of water during soaking and also improved the texture of GBR due to changes in the ultrastructure of the bran layer and a decrease in insoluble dietary fiber. In addition, subjecting germinated brown rice to microwave precooking and freeze-drying treatments made the brown rice softer, with hardness reduced by about 54.78% after treatment, and also easier to chew, as the chewiness of the precooked brown rice was lowered [14].

Cheevitsopon et al. [15] reported that an Asian Institute of Technology research group developed and used back extrusion (BE) testing [16,17] for measuring the hardness of cooked rice while following the rice cooking method described by Reyes and Jindal [16], Srisawas and Jindal [18], and Parnsakhorn and Noomhorm [19]. A BE sample holder consists of a stainless-steel cylinder of 80 mm length with an internal diameter of 15 mm and a stainless-steel spherical probe of 12.7 mm diameter. There is a 1.15 mm gap between the spherical probe with the wall of the cylinder. In the above-mentioned study, 3 g of cooked rice was placed into the cylinder without any exertion of pressure. Then, the spherical probe was moved downwards and pressed the rice sample in the cylinder at a speed of $1\,\mathrm{mm\,s^{-1}}$ until the spherical probe was 1 mm from the bottom of the cylinder when it stopped moving and returned to its initial position. BE testing followed the methods of Sirisoontaralak and Noomhorm [17], and hardness, adhesiveness and stickiness were all measured. The texture of cooked rice was measured under various measurement conditions using an extrusion test which better predicts values of sensory texture characteristics, as well as tests of Pearson correlations between the maximum force and the gradient value, or the maximum force and the area value, under each measurement condition; these showed high correlations of 0.90 or more [20]. These texture properties were illustrated in the force-time curve obtained by the test. The maximum force indicated hardness, the gradient value indicated firmness, and the area value indicated stickiness. Parnsakhorn and Langkapin [21] used the BE test for measuring the hardness of Zongzi products cooked with white glutinous rice, black sticky rice and riceberry rice as ingredients. Their first sample consisted of 100% white glutinous rice; the second sample was 50% white glutinous rice mixed with 50% black sticky rice; the third sample was 50% white glutinous rice mixed with 50% riceberry rice; and the fourth sample was 50% black glutinous rice mixed with 50% riceberry rice.

Germinated brown rice is more popular for consumption because of its high nutritional value. The texture quality of germinated brown rice is important for consumers and, therefore, the rice industry. Rice incubation is a complicated process, which involves changes in the physical and chemical properties of the rice grain. Starch, protein and lipids are the main rice grain components which affect cooking and eating quality [22]. In addition, rice incubation commences prior to harvest and continues into the postharvest storage period. It involves dramatic changes in the physical and physicochemical properties of the rice grain, including cooking, pasting and thermal properties [23].

The United States Department of Agriculture reported in 2023 that Thailand is a globally important producer and exporter of rice. In the 2020–2021 growing season, Thailand ranked third globally with an 11.9% share of global markets, behind India (a 38.9% share) and Vietnam (12.9%), and followed by Pakistan, the USA and China [24]. Khao Dawk Mali 105 (KDML 105) is the best-known fragrant rice (*Oryza sativa*, L.) variety produced in Thailand. It is exported worldwide and is also the most-used variety for the production of GBR in Thailand. It is included in "Khao Hom Mali" in Thai. This is why we used it as the main variety in this present study.

The objective of this work was to evaluate the effect of different soaking and incubation durations in the production of Khao Dawk Mali 105 (KDML 105) GBR on the texture of cooked GBR, as measured by BE testing. We sought also to determine the precision and sensitivity of the BE test on texture properties of 32 brands of cooked GBR rice in Thailand produced by varieties of rough rice; therefore, an original protocol for the determination of precision and sensitivity of food texture measurement is proposed.

2. Materials and Methods

2.1. Samples

The GBR samples were prepared by a factory of the P.J. Brand in Chonburi Province, Thailand, and purchased from local markets in Thailand. Rough rice of the *Oryza sativa* L. cultivar Khao Dawk Mali 105 (KDML 105) was collected from a field of the P.J. Brand germinated rough rice factory in Chonburi Province, Thailand. The GBR was produced by the method used by the company, which was previously reported by Kaewsorn and Sirisomboon [25]. In brief, rough rice was soaked in water at room temperature for 24 or 48 hours to create the GBR. Every four hours, the water was changed, and it was drained at the end of the soak. To produce the germinated rough rice (GRR), the rice was incubated for seven different incubation times (0, 6, 12, 18, 24, 30 and 36 h). The fluidized-bed method was used to dry the GRR, bringing the rice's moisture content down to about 19%wb. The moisture content of the rice was then reduced to about 13–14%wb after being spread out on a screen where the air could flow through for 3 hours at room temperature. The GRR sample was then dehusked prior to the experiment. This will be referred to as GBR in this paper. In addition, 32 commercial types and brands of germinated brown rice, some with single varieties and others with mixed varieties, as indicated in Kaewsorn and Sirisomboon [25], were purchased from local department stores in Bangkok, Thailand and stored in the laboratory at room temperature. These commercial GBR products were as follows: (m1) Khao Dawk Mali 105; (m2) Khao Dawk Mali 105; (m3) Thai jasmine rice; (m4) Thai jasmine rice; (m5) Khao Dawk Mali 105; (m6) Khao Dawk Mali 105; (m7) Thai jasmine rice; (m8) Khao Dawk Mali 105; (m9) Khao Dawk Mali 105; (m10) Thai jasmine rice; (m11) Thai jasmine rice; (m12) Khao Dawk Mali 105; (m13) Khao Dawk Mali 105; (m14) Khao Dawk Mali 105; (m15) Thai jasmine rice; (m16) Thai jasmine rice; (m17) Thai jasmine rice, Red Hawm Rice; (m18) Thai jasmine rice, Red Hawm Rice; (m19) Thai jasmine rice, Red Hawm Rice; (m20) Thai jasmine rice, Red Hawm Rice; (m21) Khao Dawk Mali 105, Red Hawm Rice, Hom Dang Sukhothai, Sinlek Rice, Homnil, Riceberry, Thai Pathumthani fragrant rice, black sticky rice, RD6; (m22) Khao Dawk Mali 105, Homnil, Red Cargo rice; (m23) species not specified; (m24) Doi rice, Red Doi rice, Kum Doi Rice, Saren Rice, Ubon Ratchathani Hommali Rice; (m25) Khao Dawk Mali 105, Red Hawm Rice, RD6; (m26) Muser Purple Rice, Red Hawm Rice, RD6, Khao Dawk Mali 105; (m27) Thai jasmine rice, Red Hawm Rice, Homnil; (m28) Red Hawm Rice; (m29) Homnil; (m30) Homnil; (m31) Red Cargo rice; (m32) Muser Purple Rice.

2.2. Rice Cooking Method

The rice cooking method followed that described by Sirisomboon et al. [26]. Home electronic rice cookers (RC-10 MM, Toshiba, Thailand) were used to cook 200 g GBR samples using the water-to-rice ratio of 1.6:1 recommended for GBR by rice producers. By such means, cooked rice was obtained with a texture like that of the cooked rice typically consumed by consumers. The cooked GBR was placed into plastic cups containing approximately 5 g of product. In total, 5 cups per sample were prepared at one cooking time.

2.3. Back Extrusion Test

The cooked GBR samples were then subjected to the back extrusion test using 3 g of cooked rice placed into a back extrusion test rig (BE) (Figure 1) which was compressed from the top opening of the rice container by a stainless-steel ball for 99 mm of the total height of 100 mm, with a ball probe speed of $1\,mms^{-1}$. Mean values for each sample were obtained from 5 replicate measurements. The hardness, toughness, stickiness and adhesiveness of cooked GBR were determined by observing the force–time curve and recording the maximum compression force (N) (point H), the area under the curve AHB (Ns which was converted to Nmm), the negative force (N) (point C), and the area BCD (Ns which was converted to Nmm). The hardness of cooked GBR indicates the hardness or softness of the rice. This was measured when the ball probe passed through 3 g of cooked rice and reached 1 mm above the bottom of the cylinder where the cooked rice grains were crushed to the greatest degree, while the texture meter was still in safe mode. The toughness of cooked GBR is the texture parameter which indicates the ability of the cooked rice to resist the stress applied to it by the ball probe. This stress deforms the cooked GBR from the beginning of the compression until the probe reaches the bottom of the cylinder, as in case of the hardness measurement. In the present study, toughness was indicated by the extent of the area under curve from the beginning of the compression until the probe was withdrawn from the cooked rice, i.e., the amount of energy required to deform 3 g of cooked rice. The stickiness of cooked GBR was measured by detecting the maximum negative force exerted when the ball probe was being withdrawn but the deformed sticky cooked rice remained attached to the probe, thereby exerting a pulling force (the opposite to a compressing force; therefore, a negative force) upon the probe. The adhesiveness of cooked GBR was measured by calculating the negative value of the area during withdrawal of the probe when deformed sticky cooked rice remained attached to the probe and the rice exerted a pulling force upon the probe until the rice was separated from the probe. Ninety-two samples were subjected to the back extrusion test, and an average value of 5 replications was recorded for each sample.

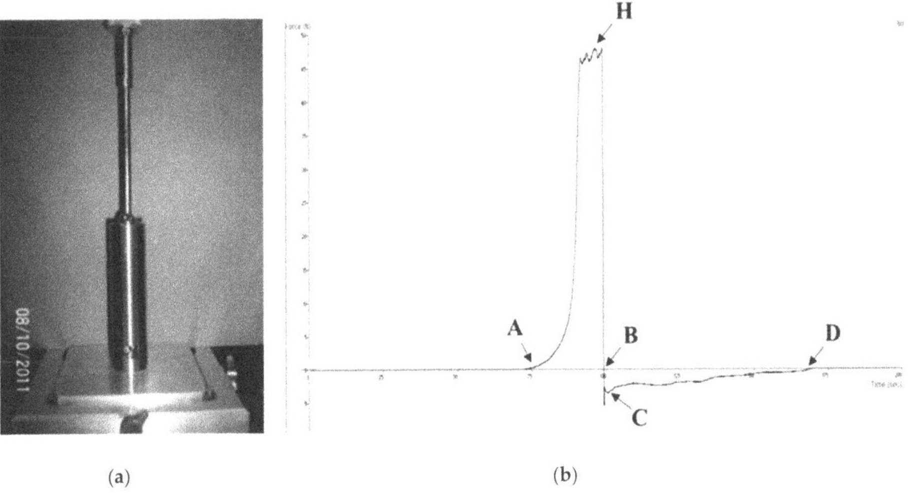

(a)　　　　　　　　(b)

Figure 1. (a) Back extrusion (BE) test rig. (b) Force–time curve from cooked germinated brown rice BE test. (A) The point at which the stainless-steel ball probe first touched and began to compress the cooked germinated brown rice; (B) the point at which the stainless-steel ball began to withdraw from the compressed cooked rice, at a distance of 99 mm from the top opening; (C) the negative force (N) applied by the cooked rice to pull the probe, due to the adhesiveness of the former; (D) the point at which the probe wholly separated from the cooked germinated brown rice; and (H) the maximum compression force (N).

2.4. The Repeatability and Reproducibility of the Measurements of Texture Properties

The repeatability and reproducibility of the measurement of texture properties were determined by measuring four duplicates (four pairs) that were randomly selected at different times during the experiment. Reproducibility was defined as the standard deviation of the differences observed between blind duplicate values. In addition, the repeatability of the reference test was determined as the standard deviation of the differences between the values obtained from four duplicates (four pairs) that were not blind samples. The repeatability indicates the precision of the analysis of the measurement methods and reproducibility indicates precision of the analyst practice.

2.5. Statistical Analysis

The means and standard deviations associated with the five replicates were calculated for hardness, toughness, stickiness and adhesiveness. The sensitivity of the test for each of the texture parameters was indicated by the coefficient of variation (%). One-way ANOVA was used to determine significant differences among the means for varieties of commercial GBR, and a mean comparison was obtained using Duncan's multiple range test with a confidence level of 95%.

3. Results and Discussion

3.1. The Repeatability and Reproducibility of the Measurements of Texture Properties

Tables 1–4 show the precision of the results of the reference testing of the textural properties of cooked GBR, i.e., hardness, toughness, stickiness and adhesiveness. It was obvious that the reproducibility was higher than repeatability because the blind samples were in the reproducibility test. The repeatability and reproducibility of the hardness, toughness and stickiness tests did not differ greatly compared with the corresponding results for the adhesiveness tests, indicating the lower precision of the method and analysis used for the adhesiveness test, which might have led to high fluctuation and scatter in the recorded values. However, the high precision of the BE testing with respect to hardness, toughness and stickiness might explain the high correlation with sensory properties found by Jindal and Limpisut [27] and Cheevitsopon et al. [15], and with the incubation times described in this study. The characteristics of brown rice that has germinated are influenced by the incubation time. Cho et al. [28] investigated the effect of germination on the physicochemical and textural properties of brown rice (BR) in different rice varieties (Samkwang, Misomi, Chindeul, and Hyeonpum). The results showed that hardness and toughness were decreased by germination, whereas stickiness and adhesiveness increased significantly. These results revealed that germination leads to improvements in the cooking and eating properties of BR.

Table 1. Repeatability and reproducibility of reference tests for the hardness of GBR samples (N).

Repeatability SD				Reproducibility SD			
Sample Numbers	Duplicate.a	Duplicate.b	Diff. a-b	Sample Numbers	Duplicate.a	Duplicate.b	Diff. a-b
5, 17	21.16	21.90	−0.74	11, 18	19.49	20.89	−1.40
28, 33	19.83	18.54	1.29	28, 34	18.81	18.54	0.27
38, 57	12.75	12.11	0.64	44, 58	25.02	23.46	1.56
35, 69	20.57	22.17	−1.60	47, 70	21.30	19.68	1.62
		Mean	−0.11			Mean	0.51
		SD	1.31			SD	1.42

a and b is a replication number.

Table 2. Repeatability and reproducibility of reference tests for toughness of GBR samples (Nmm).

Repeatability SD				Reproducibility SD			
Sample Numbers	Duplicate.a	Duplicate.b	Diff. a-b	Sample Number	Duplicate.a	Duplicate.b	Diff. a-b
5, 17	211.90	218.19	−6.28	11, 18	187.66	207.44	−19.78
28, 33	189.84	181.21	8.63	28, 34	181.45	181.21	0.25
38, 57	131.20	109.69	21.51	44, 58	246.03	233.45	12.58
35, 69	208.77	217.08	−8.31	47, 70	199.07	201.73	−2.66
		Mean	3.89			Mean	−2.40
		SD	13.97			SD	13.34

a and b is a replication number.

Table 3. Repeatability and reproducibility of reference tests for stickiness of GBR samples (N).

Repeatability SD				Reproducibility SD			
Sample Numbers	Duplicate.a	Duplicate.b	Diff. a-b	Sample Number	Duplicate.a	Duplicate.b	Diff. a-b
5, 17	−5.35	−6.55	1.20	11, 18	−5.74	−4.65	−1.09
28, 33	−5.76	−5.91	0.15	28, 34	−6.64	−5.91	−0.73
38, 57	−4.70	−4.11	−0.58	44, 58	−6.56	−5.79	−0.77
35, 69	−4.50	−3.97	−0.54	47, 70	−4.61	−4.44	−0.17
		Mean	0.06			Mean	−0.69
		SD	0.83			SD	0.38

a and b is a replication number.

Table 4. Repeatability and reproducibility of reference tests for adhesiveness of GBR samples (Nmm).

	Repeatability SD				Reproducibility SD		
Sample Numbers	Duplicate.a	Duplicate.b	Diff. a-b	Sample Numbers	Duplicate.a	Duplicate.b	Diff. a-b
5, 17	−75.53	−69.92	−5.61	11, 18	−60.35	−74.47	14.11
28, 33	−63.87	−65.19	1.32	28, 34	−65.40	−65.19	−0.22
38, 57	−51.85	−50.69	−1.16	44, 58	−72.51	−71.20	−1.30
35, 69	−61.33	−59.73	−1.60	47, 70	−77.68	−60.51	−17.17
		Mean	−1.76			Mean	−1.15
		SD	2.87			SD	12.79

a and b is a replication number.

3.2. Effects of Different Soaking and Incubation Durations on Cooked GBR Texture in the Production of Khao Dawk Mali 105 (KDML 105) GBR

With a soaking time of 24 h, there were no correlations between incubation times and the texture properties of hardness, toughness, stickiness and adhesiveness tested by BE; the coefficient of determination (R^2) values for these properties were 0.0357, 0.0087, 0.0064 and 0.1452, respectively. This indicated that the soaking time was too short to result in any linear variation in texture, even with a wide range of incubation times from 0 to 36 h. However, the texture of cooked GBR was softer, a result in line with the findings of Chao et al. [29], who found that the hardness of grains soaked for 12 h and germinated for 30 h was on average 24 N (39%) lower than that of brown rice. Paddy soaking in water at ambient temperature (20–30°C) requires 36 to 48 hours for a 30% moisture content level to be obtained. A soaking time less than this may not affect texture properties.

With a soaking time of 48 h, the R^2 relationships between incubation times and hardness, toughness, stickiness and adhesiveness were 0.8182, 0.8054, 0.6396 and 0.0312, respectively (Figure 2). These characteristics are all affected by the ratio of the starch constituents amylose and amylopectin. Jiamjariyatam et al. [30] reported that a higher amylose content and longer incubation time resulted in a greater hardness in puffed products made with rice starch. This might also apply to the cooked rice KDML 105 used in the present study.

We found no relationship between adhesiveness and incubation time. However, hardness and toughness both decreased with incubation times, and stickiness increased with longer incubation periods. In line with our findings, Jindal and Limpisut [27] reported reliable empirical models developed for estimating sensory hardness and stickiness with $R^2 = 0.96$, and an overall acceptability with $R^2 > 0.71$ from the BE force (hardness by BE) and water-to-rice ratio as independent variables. Our findings were also confirmed by the work of Cheevitsopon et al. [15], who reported a correlation coefficient (R) of −0.856 for the hardness–softness of cooked rice using a sensory test and hardness by BE.

Munarko et al. [31] reported that five varieties of Indonesia GBR experienced reductions in peak viscosity, trough viscosity, breakdown, setback, and final viscosity. The decrease in peak viscosity was attributed to the activity of endogenous hydrolytic enzymes such as amylase, which hydrolyzes starch to smaller molecules [32]. Both α-amylase and β-amylase increased as germination progressed, leading to a decrease in peak viscosity [31,33,34]. GBR peak viscosity might be related to the stickiness of cooked GBR which increases (becomes more viscous) as germination progresses, i.e., the incubation time increases (increased peak viscosity). Li et al. [35] reported that germination led to a decrease in amylose content, while the molecular weights of the germinated starches showed no significant changes; however, the relative crystallinity of grain starches decreased significantly during germination, and brown rice starches exhibited marginal increases in peak viscosities during germination. But Cho et al. [29] reported that the viscosity of germinated brown rice decreased and that germination percentages were linearly associated with reduced pasting characteristics (final, peak and setback viscosities); in addition, varieties with faster germination speed tended to have lower viscosities. These results suggest that the germination of certain varieties greatly reduced the final viscosity of the flour and the hardness of the cooked brown rice. Chao et al. [10] reported that early harvest and germination resulted in decreased pasting viscosities and cooked-grain hardness. A reduction in setback value indicates that GBR is more stable against retrogradation [31]. This explains why a decreased hardness in cooked GBR (i.e., a softer product) might be due to longer incubation duration causing retrogradation of the cooked GBR.

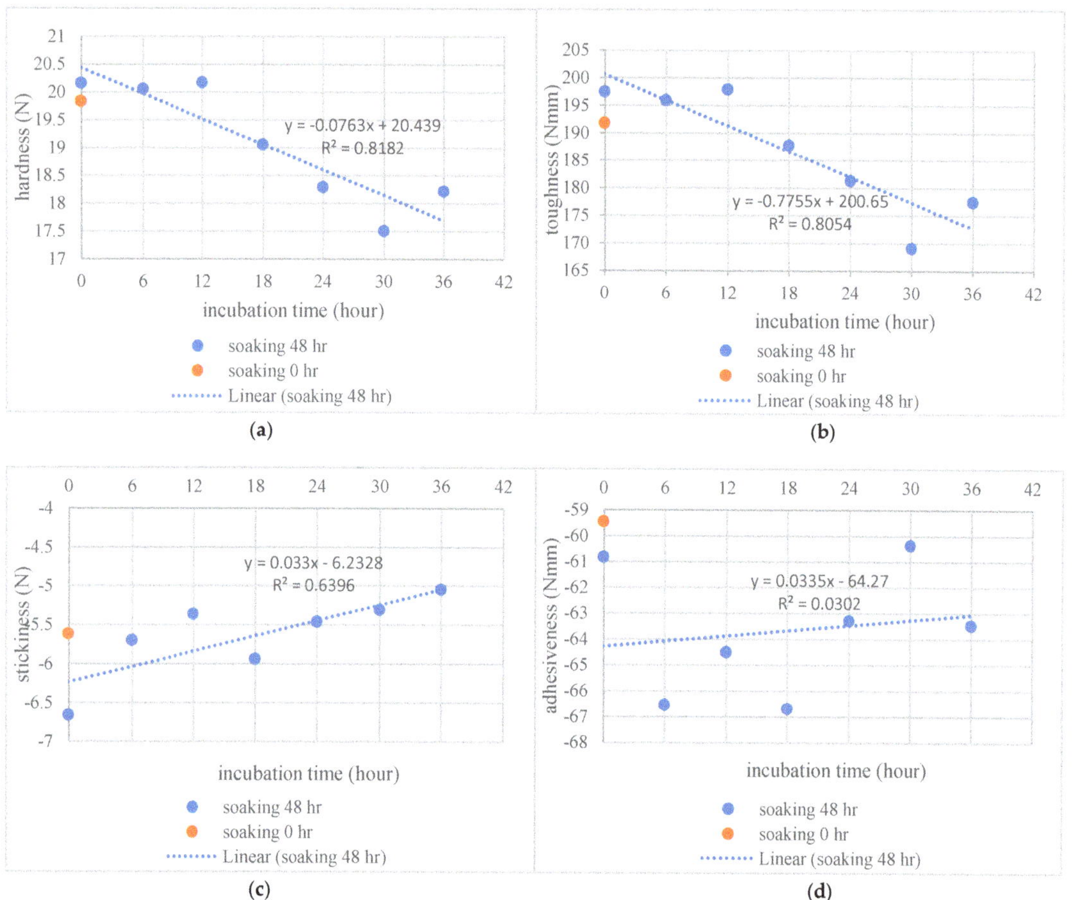

Figure 2. Effects of incubation time on the texture of cooked germinated brown rice: (**a**) hardness; (**b**) toughness; (**c**) stickiness; and (**d**) adhesiveness.

In the study of Oliveira et al. [36], the incubation process caused protein levels in GBR to decline due to hydrolysis of proteins to amino acids. Amylose was degraded and reduced in size, and starch granules had pitted surfaces due to the degradation of protein and starch. In another study, monounsaturated fatty acids (MUFAs) decreased while saturated fatty acids (SFAs) and polyunsaturated fatty acids (PUFAs) increased; however, total fat values did not change; they were transformed only in terms of their reactions [37]. Additionally, content levels of phytic acid, pyridoxine, niacin and thiamine contents have been found to decrease in brown rice after germination [38,39]. Dextrins and oligosaccharides are the primary breakdown products of both amylose and amylopectin when they are broken down by-amylase. The breakdown of starch also results in an increase in reducing sugars, a drop in total starch content, and a decrease in the pasting viscosity of germinated rice. The authors of [40] found that germination time (incubation time) was positively connected with the stickiness, sweetness and softness of cooked rice, but negatively correlated with hardness, cooking time, water uptake, and volume expansion [40]; these findings corresponded to the results of the present study.

According to Oliveira et al. [36], Rusydi Megat et al. [37], Jiamyangyuen and Ooraikul [40] and the study described in this paper, the reduction in protein, the degradation of amylose, and the decrease in MUFA, while SFAs and PUFAs increased, might be related to the decrease in hardness and toughness and the increase in stickiness. Jiamyangyuen and Ooraikul [40] found that a decrease in the pasting viscosity of germinated rice due to the breakdown of starch was related to the linear change in these texture properties.

To the best of our knowledge, no work has yet described the phenomenon of adhesiveness correlating with changes in the constituents of GBR during incubation. In the present study, adhesiveness fluctuated with different incubation times (R^2 was very small) (Figure 2d). In future studies, researchers may wish to consider how the adhesiveness of cooked rice or cooked GBR affects the attachment of rice to production equipment during mixing or stirring processes, with consequent impacts upon the operational time and cost during production.

3.3. The Sensitivity of BE Test on Texture Properties of Cooked GBR Rice

Figure 3 shows the textural properties—hardness, toughness, stickiness and adhesiveness—of cooked GBR of 32 different commercial brands in Thailand using some different varieties and some same varieties. Ranges of values for means, standard deviations and coefficients of variation are shown in Table 5. The coefficient of variation shows the degree of variation of the properties and can show the variation relativity of different sets of data even when means differ dramatically. The BE method for measuring the hardness, toughness, stickiness and adhesiveness of cooked GBR produced different degrees of variation within the same sample set, indicating different levels of sensitivity to different properties. A high coefficient of variation, where relatively wide variations in the absolute value of the property can be detected, indicates high sensitivity when small resolutions can be detected, and vice versa. The sensitivity of the BE tests for stickiness, toughness and hardness all ranked higher, in that order, than the sensitivity of the method for adhesiveness, which ranked lowest. This implies that BE testing is sensitive for measurements of stickiness, toughness and hardness. It can be seen in Figure 3 that, in the production of GRB, KDML 105 was used either on its own or mixed with other varieties. A total of 13 brands of KMDL 105 were used out of the 32 brands, Thai jasmine rice; fragrant rice (KDML 105 is one variety of it) was used in 12 brands and a further 7 brands used other fragrant rice, mountain rice and color rice varieties. Therefore, the texture properties of these cooked GBR products did not differ significantly overall. Brands using Thai jasmine rice mixed with Red Hawm rice and Homnil (m27), Red Hawm rice (m28) and Red Cargo rice (m31) had the highest values for hardness and toughness. However, brands including Red Hawm rice (m17, m18, m19 and m20) were characterized by a low hardness and low toughness, perhaps as a result of their different production methods. The softest and most easily deformed (least tough) cooked GRB was from KDML 105 (m2), the adhesiveness of which was also the lowest; it was also classified among the least-sticky group.

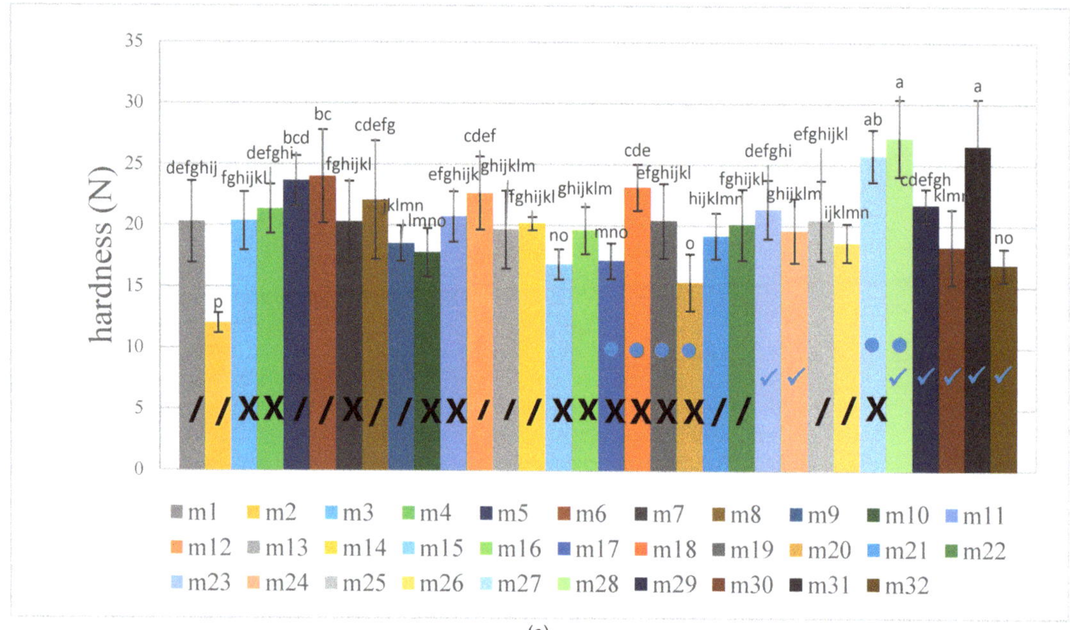

(**a**)

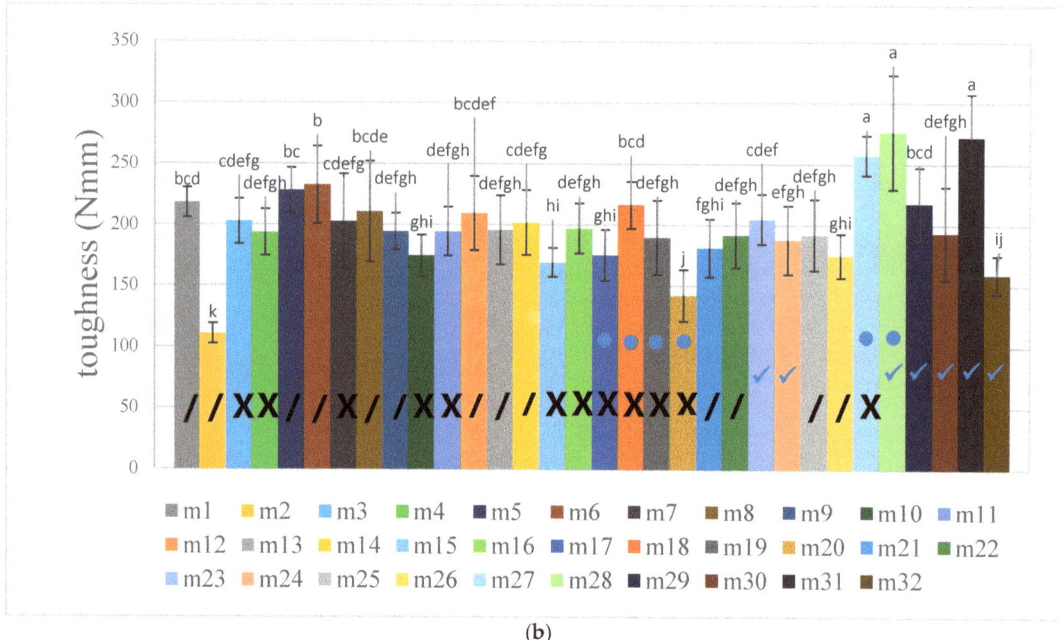

(**b**)

Figure 3. *Cont.*

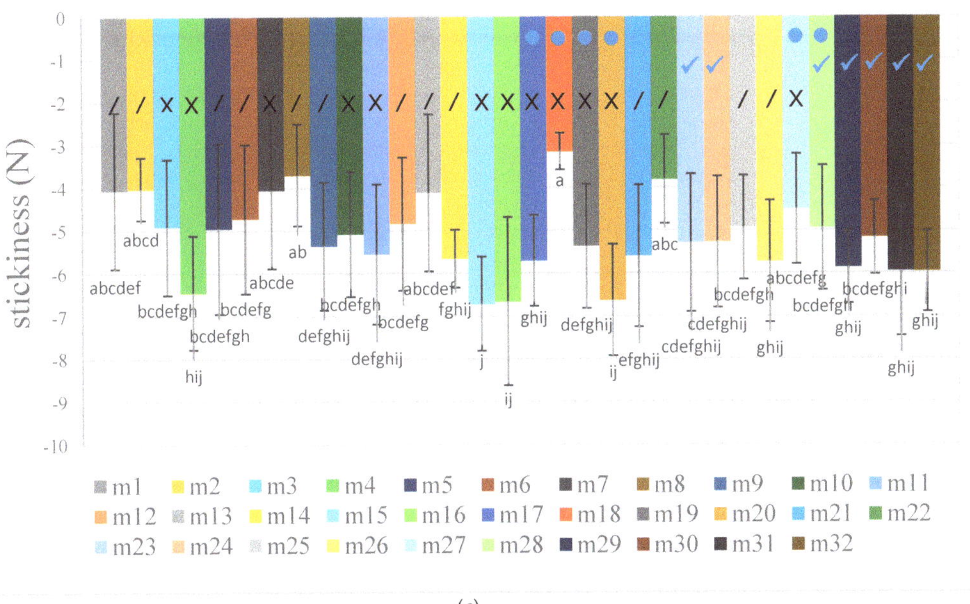

(c)

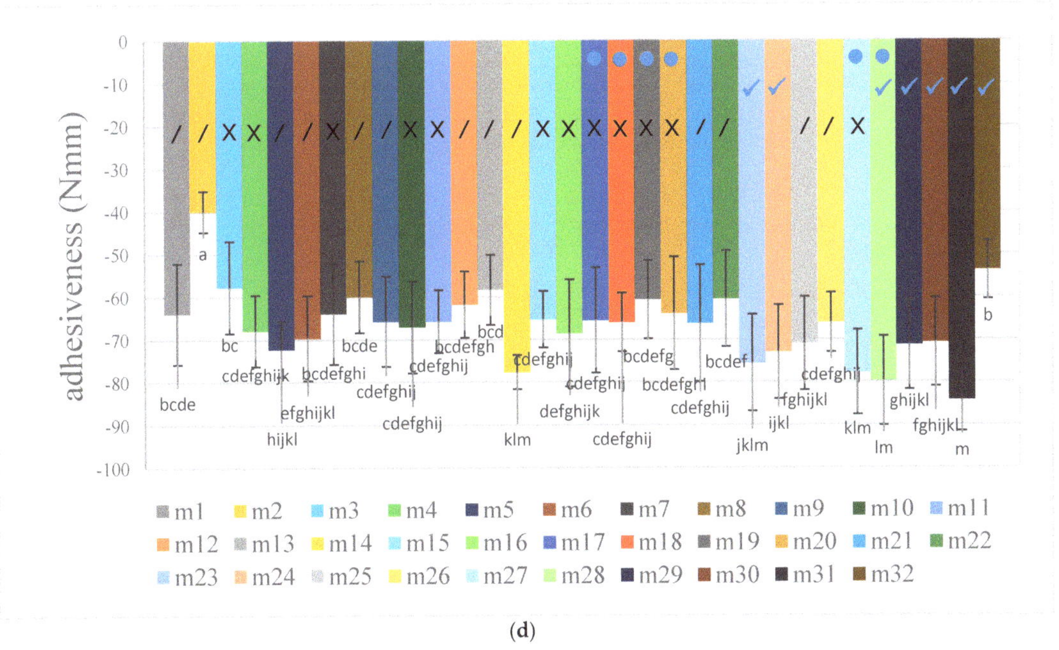

(d)

Figure 3. The texture properties of hardness (**a**), toughness (**b**) stickiness (**c**) and adhesiveness (**d**) of cooked germinated brown rice of different commercial brands in Thailand, with some varieties used alone or mixed with others. / indicates Khao Dawk Mali Rice (alone or mixed). ✗ indicates Thai Jasmine (alone or mixed). ● indicates Red Hawm (alone or mixed). ✓ indicates other varieties of fragrant, mountain and color rice. (m1) Khao Dawk Mali 105; (m2) Khao Dawk Mali 105; (m3) Thai jasmine rice; (m4) Thai jasmine rice; (m5) Khao Dawk Mali 105; (m6) Khao Dawk Mali 105; (m7) Thai jasmine rice; (m8) Khao Dawk Mali 105; (m9) Khao Dawk Mali 105; (m10) Thai jasmine rice; (m11) Thai jasmine rice; (m12) Khao Dawk Mali 105; (m13) Khao Dawk Mali 105; (m14) Khao Dawk Mali 105; (m15) Thai jasmine rice; (m16) Thai jasmine rice; (m17) Thai jasmine rice, Red Hawm Rice; (m18) Thai jasmine rice, Red Hawm Rice; (m19) Thai jasmine rice, Red Hawm Rice; (m20) Thai jasmine rice, Red Hawm Rice; (m21) Khao Dawk Mali 105, Red Hawm Rice, Hom Dang Sukhothai, Sinlek Rice, Homnil, Riceberry, Thai Pathumthani fragrant rice, Black sticky rice, RD6; (m22) Khao Dawk Mali 105, Homnil, Red Cargo rice; (m23) species not specified; (m24) Doi rice, Red Doi rice, Kum Doi Rice, Saren Rice, Ubon Ratchathani Hommali Rice; (m25) Khao Dawk Mali 105, Red Hawm Rice, RD6; (m26) Muser Purple Rice, Red Hawm Rice, RD6, Khao Dawk Mali 105; (m27) Thai jasmine rice, Red Hawm Rice, Homnil; (m28) Red Hawm Rice; (m29) Homnil; (m30) Homnil; (m31) Red Cargo rice; (m32) Muser Purple Rice.

Table 5. Statistics of texture parameters of commercial germinated brown rice in Thailand.

	Hardness (N)	Toughness (Nmm)	Stickiness (N)	Adhesiveness (Nmm)
Max	32.68	358.35	−1.86	−29.91
Min	10.83	100.90	−9.48	−97.85
Mean	20.39	198.95	−5.15	−66.45
SD	3.96	40.97	1.61	12.44
CV (%)	19.41	20.59	31.26	18.72

Table 6 shows the correlation between the texture parameters of commercial GBR in Thailand; these indicate the highest correlation coefficient was between hardness and toughness; other parameters correlated reasonably well with adhesiveness, but stickiness was not well correlated with other parameters. These results confirm that there is no relationship between the coefficient of variation (sensitivity) and the correlation between texture properties.

Table 6. Correlations between texture parameters of commercial germinated brown rice in Thailand.

	Hardness	Toughness	Stickiness	Adhesiveness
Hardness	1			
Toughness	0.966	1		
Stickiness	0.206	0.191	1	
Adhesiveness	−0.693	−0.732	0.277	1

4. Conclusions

In this study, we carried out an evaluation of the precision and sensitivity of the back extrusion (BE) test for measuring the texture of cooked germinated brown rice in a production process. Our first objective was to evaluate the effects of different soaking and incubation times in the production of Khao Dawk Mali 105 (KDML 105) GBR on the cooked GBR texture, as measured by BE testing. We found that a 24 h soaking time was too short to produce any linear variation in texture, even when incubation times ranged from 0 to 36 h. However, a soaking time of 48 h resulted in the hardness and toughness both decreasing with incubation times, while stickiness increased with incubation times. There was no relationship between adhesiveness and incubation time. Our second objective was to determine the precision and sensitivity of BE testing of the textural properties of cooked GBR rice. We found that BE testing gave highly precise measurements of hardness, toughness and stickiness, which were both repeatable and reproducible, and highly sensitive measurements of stickiness, toughness and hardness, which were confirmed by the coefficients of variation in the textural properties measured using the BE test. These findings might explain why the texture properties measured using BE correlated well with the incubation times during soaking in the GBR production process. This idea is supported by the good correlation with the texture properties obtained using sensory testing, and by the change in constituents in GBR during incubation time reported by other researchers and referred to in the Results and Discussion section of this paper. However, the correlation coefficient among the texture properties by BE was not related to the precision or sensitivity of the test. In this study, we proposed an original protocol for the determination of the precision and sensitivity of food-texture measurement; the results described above confirm the usability of this protocol. We invite the researchers in food texture studies to use our protocol developed to check the precision and sensitivity of texture measurement test rig using 10-15 samples pririor to real experiment to be conducted.

Author Contributions: Conceptualization, K.K., P.S. and M.T.; validation, K.K., P.P., W.K. and T.K.; formal analysis, K.K. and P.M.; investigation, K.K.; writing—original draft preparation, K.K. and P.M.; writing—review and editing, K.K., P.S., P.P., W.K. and M.T.; visualization, M.T. and T.K.; supervision, M.T. and T.K.; funding acquisition, P.S. All authors have read and agreed to the published version of the manuscript.

Funding: This research was funded by the NIRS research center for agricultural product and food, King Mongkut's Institute of Technology Ladkrabang, Bangkok, Thailand.

Institutional Review Board Statement: Not applicable for this study not involving humans or animals.

Informed Consent Statement: Not applicable for this study not involving humans.

Data Availability Statement: Data are contained within the article.

Acknowledgments: The authors thank the laboratory of physical and engineering properties of agricultural product and food in department of agricultural engineering, school of engineering, King Mongkut's Institute of Technology Ladkrabang, Bangkok, Thailand for the instrument support and to P.J. Brand germinated rough rice in Chonburi Province, Thailand, for the germinated brown rice production methodology.

Conflicts of Interest: The authors declare no conflict of interest. The funders had no role in the design of the study; in the collection, analysis, or interpretation of data; in the writing of the manuscript; or in the decision to publish the results.

References

1. Yodpitak, S.; Mahatheeranont, S.; Boonyawan, D.; Sookwong, P.; Roytrakul, S.; Norkaew, O. Cold plasma treatment to improve germination and enhance the bioactive phytochemical content of germinated brown rice. *Food Chem.* **2019**, *289*, 328–339. [CrossRef]
2. Nguyen, B.C.Q.; Shahinozzaman, M.; Tien, N.T.K.; Thach, T.N.; Tawata, S. Effect of sucrose on antioxidant activities and other health-related micronutrients in gamma-aminobutyric acid (GABA)-enriched sprouting Southern Vietnam brown rice. *J. Cereal Sci.* **2020**, *93*, 102985. [CrossRef]
3. Ren, C.; Hong, B.; Zheng, X.; Wang, L.; Zhang, Y.; Guan, L.; Yao, X.; Huang, W.; Zhou, Y.; Lu, S. Improvement of germinated brown rice quality with autoclaving treatment. *Food Sci. Nutr.* **2020**, *8*, 1709–1717. [CrossRef]
4. Munarko, H.; Sitanggang, A.B.; Kusnandar, F.; Budijanto, S. Germination of five Indonesian brown rice: Evaluation of antioxidant, bioactive compounds, fatty acids and pasting properties. *Food Sci. Technol.* **2021**, *42*, e19721. [CrossRef]
5. Cho, D.H.; Lim, S.T. Germinated brown rice and its bio-functional compounds. *Food Chem.* **2016**, *196*, 259–271. [CrossRef]
6. Cáceres, P.J.; Peñas, E.; Martinez-Villaluenga, C.; Amigo, L.; Frias, J. Enhancement of biologically active compounds in germinated brown rice and the effect of sun-drying. *J. Cereal Sci.* **2017**, *73*, 1–9. [CrossRef]
7. Techo, J.; Soponronnarit, S.; Devahastin, S.; Wattanasiritham, L.S.; Thuwapanichayanan, R.; Prachayawarakorn, S. Effects of heating method and temperature in combination with hypoxic treatment on γ-aminobutyric acid, phenolics content and antioxidant activity of germinated rice. *Int. J. Food Sci. Technol.* **2018**, *54*, 1330–1341. [CrossRef]
8. Neamsorn, N.; Inprasit, C.; Terdwongworakul, A. Water absorption of RD31 paddy during soaking in parboiling process. *TSAE J.* **2017**, *23*, 1–8.
9. Wang, Z.; Wang, J.; Chen, X.; Li, E.; Li, S.; Li, C. Mutual relations between texture and aroma of cooked rice-a pilot study. *Foods* **2022**, *11*, 3738. [CrossRef]
10. Chao, S.; Mitchell, J.; Prakash, S.; Bhandari, B.; Fukai, S. Effects of variety, early harvest, and germination on pasting properties and cooked grain texture of brown rice. *J. Texture Stud.* **2022**, *53*, 503–516. [CrossRef]

11. Kaosa-ard, T.; Songsermpong, S. Influence of germination time on the GABA content and physical properties of germinated brown rice. *As. J. Food Agro-Ind.* **2012**, *5*, 270–283.

12. Singh, K.; Simapisan, P.; Decharatanangkoon, S.; Utama-ang, N. Effect of soaking temperature and time on GABA and total phenolic content of germinated brown rice (Phitsanulok 2). *Curr. J. Appl. Sci. Technol.* **2017**, *17*, 224–232.

13. Zhu, C.; Yang, L.; Nie, P.; Zhong, L.; Wu, Y.; Sun, X.; Song, L. Effects of hydrogen-rich water on the nutritional properties, volatile profile and texture of germinated brown rice. *Int. J. Food Sci. Technol.* **2022**, *57*, 7666–7680. [CrossRef]

14. Guan, H.; Wu, Y.; Liu, X. Effect of microwave precooking and freeze-drying on the quality of the germinated brown rice. *J. Food Process. Preserv.* **2023**, *2023*, 3610644. [CrossRef]

15. Cheevitsopon, E.; Klakankhai, T.; Kladsuk, S.; Sonsanguan, N.; Phanomsophon, T.; Sirisomboon, P.; Pornchaloempong, P. Texture properties of cooked rice evaluated by sensory test interpreted by instrument tests. In Proceedings of the XX CIGR World Congress 2022, Sustainable Agricultural Production-Water, Land, Energy and Food, Kyoto International Conference Center, Kyoto, Japan, 5–9 December 2022.

16. Reyes, V.G.; Jindal, V.K. A small sample back extrusion test for measuring texture of cooked rice. *J. Food Qual.* **1990**, *13*, 109–118. [CrossRef]

17. Sirisoontaralak, P.; Noomhorm, A. Changes to physicochemical properties and aroma of irradiated rice. *J. Stored Prod. Res.* **2006**, *42*, 264–276. [CrossRef]

18. Srisawas, W.; Jindal, V. Sensory evaluation of cooked rice in relation to water-to-rice ratio and physicochemical properties. *J. Texture Stud.* **2007**, *38*, 21–41. [CrossRef]

19. Parnsakhorn, S.; Noomhorm, A. Changes in physicochemical properties of parboiled brown rice during heat treatment. *Agric. Eng. Int.* **2008**, *10*, 1–20.

20. Shin, S.H.; Choi, W.S. Analysis of the texture characteristics of cooked rice using extrusion tests under various measurement conditions. *J. Korean Soc. Food Sci. Nutr.* **2022**, *51*, 627–631. [CrossRef]

21. Parnsakhorn, S.; Langkapin, J. Effect of different rice varieties on changes in anthocyanin content and physical properties of Zongzi products. *J. Eng. Rmutt.* **2022**, *20*, 59–70. (In Thai)

22. Zhou, Z.; Robards, K.; Helliwell, S.; Blanchard, C. Ageing of stored rice: Changes in chemical and physical attributes. *J. Cereal Sci.* **2002**, *35*, 65–78. [CrossRef]

23. Keawpeng, I.; Venkatachalam, K. Effect of incubation on changes in rice physical qualities. *Int. Food Res. J.* **2015**, *22*, 2180–2187.

24. USDA Department of Agriculture. Available online: https://www.ers.usda.gov/topics/crops/rice/rice-sector-at-a-glance/ (accessed on 30 June 2023).

25. Kaewsorn, K.; Sirisomboon, P. Study on evaluation of gamma oryzanol of germinated brown rice by near infrared spectroscopy. *J. Innov. Opt. Health Sci.* **2014**, *7*, 1450002. [CrossRef]

26. Sirisomboon, P.; Kaewsorn, K.; Thanimkarn, S.; Phetpan, K. Non-linear viscoelastic behavior of cooked white, brown, and germinated brown Thai jasmine rice by large deformation relaxation test. *Int. J. Food Prop.* **2017**, *20*, 1547–1557. [CrossRef]

27. Jindal, V.K.; Limpisut, P. Back extrusion testing of cooked rice texture. In Proceedings of the 2002 ASAE Annual Meeting, Chicago, IL, USA, 28–31 July 2002. No. 026068.

28. Cho, D.H.; Park, H.Y.; Lee, S.K.; Park, J.; Choi, H.S.; Woo, K.S.; Kim, H.J.; Sim, E.Y.; Won, Y.J.; Lee, D.H.; et al. Differences in physiochemical and textural properties of germinated brown rice in various rice varieties. *Korean J. Crop Sci.* **2017**, *62*, 172–183. [CrossRef]

29. Chao, S.; Mitchell, J.; Prakash, S.; Bhandari, B.; Fukai, S. Effect of germination level on properties of flour paste and cooked brown rice texture of diverse varieties. *J. Cereal Sci.* **2021**, *102*, 103345. [CrossRef]

30. Jiamjariyatam, R.; Kongpensook, V.; Pradipasena, P. Effects of amylose content, cooling rate and aging time on properties and characteristics of rice starch gels and puffed products. *J. Cereal Sci.* **2015**, *61*, 16–25. [CrossRef]

31. Munarko, H.; Sitanggang, A.B.; Kusnandar, F.; Budijanto, S. Effect of different soaking and germination methods on bioactive compounds of germinated brown rice. *Int. J. Food Sci. Technol.* **2021**, *56*, 4540–4548. [CrossRef]

32. Wichamanee, Y.; Teerarat, I. Production of germinated red jasmine brown rice and its physicochemical properties. *Int. Food Res. J.* **2012**, *19*, 1649–1654.

33. Mohan, B.H.; Malleshi, N.G.; Koseki, T. Physico-chemical characteristics and non-starch polysaccharide contents of Indica and Japonica brown rice and their malts. *LWT Food Sci. Technol.* **2010**, *43*, 784–791. [CrossRef]

34. Pinkaew, H.; Thongngam, M.; Wang, Y.J.; Naivikul, O. Isolated rice starch fine structures and pasting properties changes during pre-germination of three Thai paddy (*Oryza sativa* L.) cultivars. *J. Cereal Sci.* **2016**, *70*, 116–122. [CrossRef]

35. Li, C.; Oh, S.G.; Lee, D.H.; Baik, H.W.; Chung, H.J. Effect of germination on the structures and physicochemical properties of starches from brown rice, oat, sorghum, and millet. *Int. J. Biol. Macromol.* **2017**, *105*, 931–939. [CrossRef] [PubMed]

36. Oliveira, M.E.A.S.; Coimbra, P.P.S.; Galdeano, M.C.; Carvalho, C.; Piler, W.; Takeiti, C.Y. How does germinated rice impact starch structure, products and nutrional evidences?—A review. *Trends Food Sci. Technol.* **2022**, *122*, 13–23. [CrossRef]

37. Rusydi Megat, M.R.; Noraliza, C.W.; Azrina, A.; Zulkhairi, A. Nutritional changes in germinated legumes and rice varieties. *Int. Food Res. J.* **2011**, *18*, 688–696.

38. Jabeen, R.; Hussain, S.Z.; Jan, N.; Fatima, T.; Naik, H.R.; Jabeen, A. Comparative study of brown rice and germinated brown rice for nutritional composition, in vitro starch digestibility, bioactive compounds, antioxidant activity and microstructural properties. *Cereal Chem.* **2023**, *100*, 434–444. [CrossRef]

39. Liang, J.; Han, B.Z.; Nout, M.; Hamer, R.J. Effects of soaking, germination and fermentation on phytic acid, total and in vitro soluble zinc in brown rice. *Food Chem.* **2008**, *110*, 821–828. [CrossRef]

40. Jiamyangyuen, S.; Ooraikul, B. The physico-chemical, eating and sensorial properties of germinated brown rice. *J. Food Agric. Environ.* **2008**, *6*, 119–124.

Article

Impact of Guar Gum and Locust Bean Gum Addition on the Pasting, Rheological Properties, and Freeze–Thaw Stability of Rice Starch Gel

Xuejiao Xu [1], Shuhui Ye [2], Xiaobo Zuo [3] and Sheng Fang [2,*]

1 College of Biology and Environmental Engineering, Zhejiang Shuren University, Hangzhou 310015, China
2 School of Food Science and Biotechnology, Zhejiang Gongshang University, Hangzhou 310018, China
3 Hangzhou Tea Research Institute, CHINA COOP/Zhejiang Key Laboratory of Transboundary Applied Technology for Tea Resources, Hangzhou 310016, China
* Correspondence: fangsheng@zjgsu.edu.cn; Tel.: +86-130-9375-2831

Abstract: Improving the gel texture and stability of rice starch (RS) by natural hydrocolloids is important for the development of gluten-free starch-based products. In this paper, the effects of guar gum and locust bean gum on the pasting, rheological properties, and freeze–thaw stability of rice starch were investigated by using a rapid visco analyzer, rheometer, and texture analyzer. Both gums can modify the pasting properties, revealed by an increment in the peak, trough, and final viscosities, and prevent the short-term retrogradation tendency of RS. Dynamic viscoelasticity measurements also indicated that the starch–gum system exhibits superior viscoelastic properties compared with starch alone, as revealed by its higher storage modulus (G'). Compared with the control, the hysteresis loop area of the guar gum-containing system and locust bean gum-containing system was reduced by 37.7% and 24.2%, respectively, indicating that the addition of gums could enhance shear resistance and structure recovery properties. The thermodynamic properties indicated that both gums retard short-term retrogradation as well as long-term retrogradation of the RS gels. Interestingly, the textural properties and freeze–thaw stability of the RS gel were significantly improved by the addition of galactomannans ($p < 0.05$), and guar gum was more effective than locust bean gum, which may be due to the different mannose to galactose ratio. The results provide alternatives for gluten-free recipes with improved texture properties and freeze–thaw stability.

Keywords: rice starch; guar gum; locust bean gum; rheological properties; RVA; freeze–thaw stability

Citation: Xu, X.; Ye, S.; Zuo, X.; Fang, S. Impact of Guar Gum and Locust Bean Gum Addition on the Pasting, Rheological Properties, and Freeze–Thaw Stability of Rice Starch Gel. *Foods* **2022**, *11*, 2508. https://doi.org/10.3390/foods11162508

Academic Editors: Sidonia Martínez and Javier Carballo

Received: 30 June 2022
Accepted: 16 August 2022
Published: 19 August 2022

1. Introduction

Rice starch (RS) products are common diets in China and Southeast Asian countries due to the consumer-friendly taste, texture, and ease-of-preparation feature [1,2]. RS products are also an important alternative diet for patients with celiac disease [3]. However, RS gels have deficiencies in ductility, elasticity, and extensibility in the processing and texture of products [4,5]. In addition, RS gel is prone to retrograde during storage, especially during freeze–thaw cycles [6]. Hydrocolloids are widely used in the food industry as functional additives [7–9]. Generally, hydrocolloids play critical roles in modifying the rheological and pasting properties of starch, with enhancement in production process efficiency and optimization of texture stability and sensory properties [8,9]. Mechanistic studies have revealed that the incorporation of hydrocolloids in native starches could improve their moisture content and water-holding capacity [10].

Galactomannans are linear hydrocolloids originally found in the endosperm of seeds of various legumes. Studies have revealed that most galactomannans form nonionic structures, which facilitate the formation of functional properties of pH resistance and stability in both ion-enriched solutions and heating processes [11]. In addition, galactomannans are also used as functional ingredients during digestion due to their ability to slow down the degradation rate of starch [12]. Of all the legume seed galactomannans, guar gum and locust bean gum are the most commonly used in the food industry.

Guar gum (*Cyamposis tetragonoloba*) and locust bean gum (*Ceretonia siliqua*) are classical galactomannans comprising a β-(1→4) linked mannopyranosyl linear backbone with α-(1→6) linked D-galactopyranosyl branched chains [13]. Guar gum and locust bean gum have similar molecular structures but differ in the ratio of D-mannosyl: D-galactosyl units, which are 2:1 and 4:1, respectively [14]. Guar gum and/or locust bean gum have been reported to modify the properties of yam starch [15], tapioca starch [16], acorn starch [17], and corn starch [18]. It has been found that guar gum generally exhibits pronounced elastic properties, whereas locust bean gum enhances the viscous properties of tapioca starch owing to the different chain extensions and hydrogen bond numbers [19,20]. Significant improvements in the freeze–thaw stability of corn starch with the incorporation of guar gum rather than that of locust bean gum have been reported, which have contributed to the more frequent interaction between guar gum and leached amylose [15]. Thus, the interactions between hydrocolloids and starch could be crucial for starch-based food products.

It has been reported that the different synergistic effects of starch and hydrocolloids could be related to the physicochemical differences of hydrocolloids such as solubility [21–23], intrinsic viscosity [24], extended conformation [25,26], antioxidant potential [27,28], hydrogen bonding capacity [29], flexibility [30], and temperature of gel formation [24,31,32]. The diversity of hydrocolloid molecular structures, especially for the differences in the side chain, usually contributes to the variations in physicochemical properties [33]. Therefore, comprehensive analyses of hydrocolloids and related starch would be beneficial for the further application of related products. However, to the best of our knowledge, the comprehensive understanding of guar gum and locust bean gum when added to rice starch remains unknown [34].

This study aimed to investigate the effect of two galactomannans (guar gum and locust bean gum) on the pasting, rheological properties, and freeze–thaw stability of RS and the possible interaction mechanism. The results may provide valuable information for the selection of gluten-free recipes with improved textural properties and freeze–thaw stability.

2. Materials and Methods

2.1. Materials

Guar gum (CAS No.: 9000-30-0) and locust bean gum (CAS No.: 9000-40-2) were purchased from Aladdin Industrial Corporation. Rice starch (food-grade) was supplied from Jiangxi Jinnong Co., Ltd. (Yichun, China). The main components of RS were analyzed by standard analytical methods [9]. The total starch content of RS was $90.5 \pm 1.35\%$, the amylose content was $24.67 \pm 0.27\%$, the fat content was $0.05 \pm 0.01\%$, the protein content was $0.90 \pm 0.06\%$, and the ash and moisture content were $0.22 \pm 0.02\%$ and $7.7 \pm 0.03\%$, respectively.

2.2. Pasting Properties

The pasting properties of RS gels were measured using a Rapid Visco Analyzer (RVA Tec Master, Perten instruments, Hägersten, Sweden) following previous methods [35]. RS alone slurries were obtained by dispersing 3 g RS powder in 25 g distilled water. In the case of RS/gum mixtures, weighed amounts of guar gum or locust bean gum powder were dispersed in distilled water firstly to prepare the hydrocolloid solutions (0.1 wt.%), with an hour of magnetic stirring at 80 °C. Then, 3 g RS powder was dispersed in the prepared solution (ca. 25 g) with mild stirring to obtain the mixture. The prepared slurries weighing 28 g were then transferred to aluminum RVA canisters to investigate the pasting properties. The starch slurry was at first held at 50 °C for 1 min, then heated to 95 °C within 3 min 45 s and maintained at 95 °C for 2 min 30 s. It was subsequently cooled to 50 °C within 3 min 45 s and maintained at 50 °C for 1 min 30 s. For the first 10 s before measurement, the speed of the plastic paddle was set as 960 rad/s for complete dispersion and then kept constant at 160 rad/s during measurement. RVA parameters such as peak viscosity (PV), trough viscosity (TV), final viscosity (FV), breakdown value (BV), pasting temperature (PT), and setback value (SBV) were obtained from the pasting curves. A pan containing the same mass of distilled water was used as a reference. RVA parameters were presented as the mean $\pm$ SD of triplicate experiments.

2.3. Rheological Properties

The rheological behaviors of RS gels were determined by using an AR-G2 rheometer (TA Instruments, New Castle, DE, USA) with 40 mm parallel plates according to previous methods [33]. The gelatinized slurries obtained from RVA analysis were then immediately transferred to the platform of the rheometer. Before testing, all samples were equilibrated at 25 °C for 120 s and then examined by both dynamic viscoelastic and steady flow measurements.

Dynamic viscoelastic tests were conducted in a frequency range from 0.1 to 10 Hz with a constant strain of 1%. The dynamic rheological data storage modulus (G') and loss modulus (G'') vs. angular frequency (ω) were obtained and analyzed by using the following equations [36]:

$$G' = K'(\omega)^{n'} \tag{1}$$

$$G'' = K''(\omega)^{n''} \tag{2}$$

where K' is constant, ω is the angular frequency, and n' is the frequency exponents.

Next, the steady flow tests were performed, and the shear rate as a function of shear stress was obtained by using the Data software. The shear rate ramps from 0.01 to 300 s^{-1} (upward flow curve) and then decreases from 300 s^{-1} to 0.01 s^{-1} after 1 min of equilibration (downward flow curve). Experiment results from the ascending and descending segments of the shear cycle were then fitted using the power-law model to characterize the flow properties [36]:

$$\sigma = K \times \gamma^n \tag{3}$$

where σ is the shear stress (Pa), K is the consistency coefficient (Pa·s^n), γ is the shear rate (s^{-1}), and n is the flow behavior index (dimensionless).

2.4. Thermodynamic Properties

Thermodynamic properties of the RS gels in the presence or absence of gums were determined by using a C80 differential scanning calorimeter (Setaram, Lyon, France) following previous methods with some modification [37]. Samples were prepared by dispersing 5 g RS powder in 15 g distilled water or hydrocolloid solutions (0.15 wt.%) and stirred for 2 h. Thereafter, the well-stirred suspensions (ca. 4 g) were transferred to an aluminum crucible and tested. The heating temperature ramped from 30 to 100 °C with an acceleration of 0.5 °C/min, and then the ramp was reversed to 30 °C at a rate of −2 °C/min. The instrument was calibrated by using indium and an empty pan as reference. After gelatinization, samples were cooled down and stored at 4 °C for 0, 3, 5, and 12 days, then heated again to investigate the effect of guar gum and locust bean gum on the retrogradation of RS gels. The retrogradation degree was calculated by the ratio of retrogradation enthalpy in the second run heating (ΔH_2) to the gelatinization enthalpy in the first run test (ΔH_1) [38].

2.5. Texture Properties

The textural properties of the RS in the presence or absence of gums were determined by using a TA-XTplus Texture Analyzer (Stable Micro System Ltd., Godalming, UK). The gel samples were prepared as described in Section 2.1 and stored at 4 °C for 0, 3, and 5 d. After being conditioned at room temperature, the samples were analyzed by using texture profile analysis (TPA). The TPA tests were performed on samples 25 mm in diameter and 20 mm in height. The sample was placed on the text platform and squeezed twice to 15 mm with a 25.0 mm diameter cylinder probe P/25. The speed of the probe was 3 mm/s, and the trigger force was 5 g. The interval between two compressions was 3 s, and the data acquisition rate was 200 pps.

2.6. Freeze–Thaw Stability

The freeze–thaw stability of RS in the presence or absence of gums was determined following the method of Zhai et al. [39] with some modifications. Samples were prepared as described in Section 2.2. The slurry was cooled to 30 °C and then transferred to a preweighed centrifuge tube (10 mL) to record its total weight. The samples were then frozen at −21 °C for 24 h, thawed at 30 °C for 2 h, and centrifuged at 8000 rpm for 20 min. The supernatant was discarded and then weighed. These steps were repeated five times to determine the freeze–thaw stability of samples. The syneresis rate was calculated from the following Equation (4):

$$\text{Syneresis } (\%) = \frac{(M_2 - M_3) \times 100}{M_2 - M_1} \tag{4}$$

where M_1 (g) is the weight of the centrifugal tube, M_2 (g) is the weight of starch paste and centrifuge tube, and M_3 (g) is the weight of starch paste and centrifuge tube after pouring out the supernatant.

2.7. Statistical Analysis

The experimental data were analyzed by variance analysis for a completely random design using the SPSS package (19.0, SPSS Inc., Armonk, NY, USA). Duncan's multiple range tests were conducted to analyze the difference between means with a statistical significance of $p < 0.05$.

3. Results and Discussion

3.1. Pasting Properties

The pasting curves of RS in the presence or absence of guar gum or locust bean gum are shown in Figure 1. According to the results of pasting behaviors (Table 1), a significant increment was determined in peak and trough viscosity in the presence of galactomannans ($p < 0.05$). This observation was interpreted by the thickening properties of galactomannans. Guar gum and locust bean gum would enhance the shearing forces exerted on the starch granules [40] and defer the hydrolysis rate of starch [41], which is related to the higher viscosity. Moreover, the effective starch concentration was increased by the immobilization of the water molecules [11], enhancing a strong entanglement with amylose and hydrocolloids [24]. Particularly, the viscosity increment of the starch system with guar gum was more pronounced than that with locust bean gum. The primary and secondary OH (hydroxyl groups) located at the exterior branch of guar gum preferred to form more hydrogen bonds, which exhibited the highest PV and BV in the starch-related system.

The breakdown viscosity (BV) of the RS mixture ranged between 878.33 and 1137.33 cP. Higher BV was observed in the guar gum-related system, which indicated its relatively lower thermostability and compatibility compared with the locust bean gum. Similar observations have been obtained in that the BV of wheat flour pastes was increased along with an increment in guar gum levels [41].

SBV could be used as an indicator to measure the syneresis level of starch during the cooling process [42]. Starch systems containing both guar gum and locust bean gum exhibited a lower tendency to retrograde, as evidenced by the lower SBV value. The extension of the network structure in the paste system was inhibited due to the interaction between the colloidal molecules and the leached amylose [43]. Meanwhile, the addition of strong hydrophilic colloids could reduce the free water content, thereby hindering the rearrangement of starch.

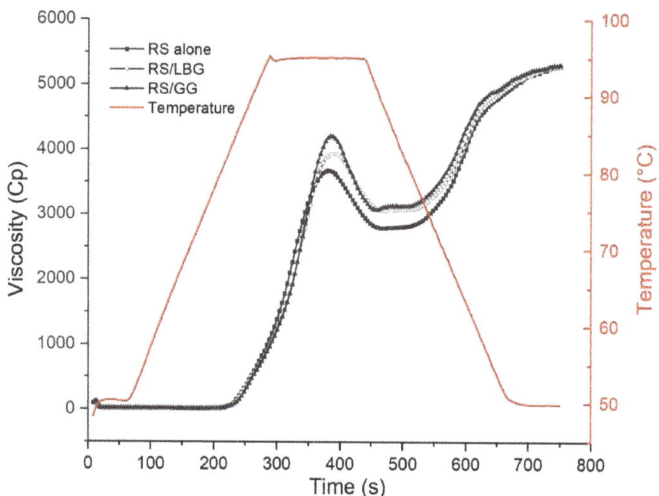

Figure 1. RVA pasting curves of RS alone, RS containing locust bean gum (RS/LBG), and RS containing guar gum (RS/GG).

Table 1. Pasting parameters of RS alone, RS containing locust bean gum (RS/LBG), and RS containing guar gum (RS/GG) *.

Sample	PV (cP)	TV (cP)	BV (cP)	FV (cP)	SBV (cP)	PT (°C)
RS alone	3679.67 ± 17.47 [c]	2790.67 ± 8.33 [b]	889.00 ± 19.29 [b]	5287.33 ± 25.89 [a]	2496.67 ± 26.50 [a]	82.43 ± 0.42 [b]
RS/LBG	3940.33 ± 4.04 [b]	3062.00 ± 23.90 [a]	878.33 ± 21.03 [b]	5291.67 ± 7.23 [a]	2229.67 ± 24.17 [b]	82.42 ± 0.49 [b]
RS/GG	4210.00 ± 36.29 [a]	3072.67 ± 43.68 [a]	1137.33 ± 22.81 [a]	5308.67 ± 10.69 [a]	2236.00 ± 39.00 [b]	83.25 ± 0.43 [a]

* Each value represents the mean ± SD of triplicate experiments. Different superscripts with the same columns are significantly different ($p < 0.05$ by Duncan's multiple range test).

3.2. Dynamic Viscoelastic Properties

The storage modulus (G') and loss modulus (G'') of RS gels in the presence or absence of gums are shown in Figure 2. It shows that the values of G' and G'' increased with an increasing angular frequency for all samples. Similar results have also been reported with other starch-related systems [44]. In addition, a significant increment in G' and G'' was determined with the addition of galactomannans. For example, the G' values of the guar gum-containing system increased from 356.9 to 459.9 Pa with respect to the control. The viscoelastic properties of starch-related systems were enhanced owing to the thickening properties of galactomannans [45].

Interestingly, it seemed different hydrocolloids modified the elastic and viscous properties of starch systems variously. For the guar gum-containing system, the increasing rate of the G' value was much greater than that of the G'' value, indicating remarkable elastic properties that could be considered an enhancement of the weak gel network due to more OH groups on the galactose chain. It has been reported that more galactose substitutions of guar gum would hinder intramolecular hydrogen bond formation, thereby exhibiting a more extended conformation than locust bean gum [20]. It is known that noncovalent intermolecular interactions such as hydrogen bonding are the most common and important interaction types between biopolymers [46,47]. It is predicted that the extended chain form promoted the interaction between amylose and guar gum via a noncovalent bond, thereby enhancing the elasticity and pseudo-plasticity of the system [39,48]. On the contrary, the locust bean gum-containing system exhibited pronounced viscous properties, as evidenced by the higher tan(δ) and G'' values (Table 2). A similar trend has also been reported with a corn starch–galactomannan system [17].

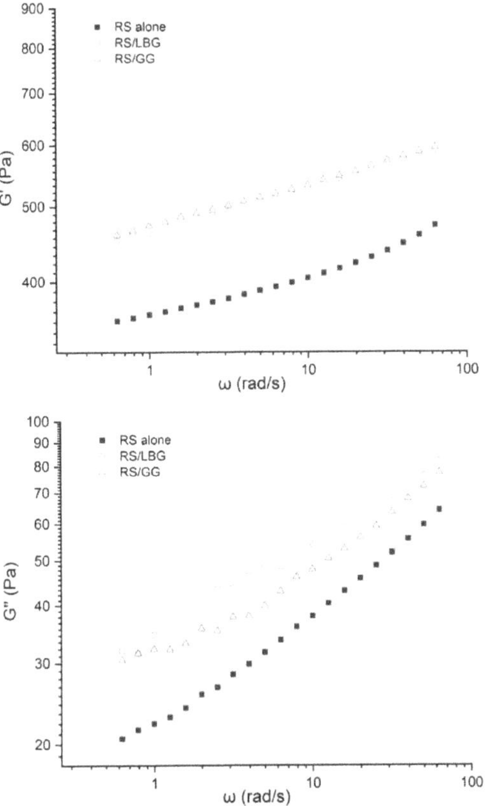

Figure 2. Storage modulus and loss modulus of RS alone, RS containing locust bean gum (RS/LBG), and RS containing guar gum (RS/GG).

Table 2. Storage (G'), loss moduli (G''), and loss tangent (tan δ) at 6.28 rad s^{-1} for RS alone, RS containing locust bean gum (RS/LBG), and RS containing guar gum (RS/GG) *.

Sample	G' (Pa)	G'' (Pa)	tan(δ)
RS alone	394.47 ± 21.27 [b]	33.65 ± 1.35 [c]	0.085 ± 0.003 [b]
RS/LBG	512.77 ± 27.98 [a]	48.51 ± 0.41 [a]	0.095 ± 0.006 [a]
RS/GG	519.60 ± 30.47 [a]	42.95 ± 1.63 [b]	0.083 ± 0.005 [b]

* Mean ± SD. Different superscripts with the same columns are significantly different ($p < 0.05$ by Duncan's multiple range test).

For all RS-related systems, $\ln(G', G'')$ as a function of ln ω were conducted using linear regression and are summarized in Table 3. It was observed that each starch-related system exhibited weak gel-like behavior with positive slopes (n' and n'') [48]. The value of K' and K'' increased significantly after the addition of galactomannans, indicating enhanced viscoelasticity of the continuous phase caused by the thickening properties of gums. Such observation is in agreement with maize starch–guar gum mixtures [45] and rice starch–xanthan gum mixtures [34]. The presence of galactomannans facilitated the associations of ordered chain segments, thereby enhancing the weak three-dimensional network of RS–gum mixtures [49].

Table 3. Parameters of RS alone, RS containing locust bean gum (RS/LBG), and RS containing guar gum (RS/GG) at 25 °C as determined from Equations (1) and (2) *.

Sample	Storage Modulus G'			Loss Modulus G''		
	k' (pa·s^n)	n''	R^2	k'' (pa·s^n)	n''	R^2
RS alone	358.76 ± 20.01 [b]	0.06 ± 0.00 [a]	0.970	21.25 ± 0.66 [c]	0.26 ± 0.01 [a]	0.996
RS/LBG	461.77 ± 28.63 [a]	0.06 ± 0.00 [a]	0.997	34.29 ± 0.42 [a]	0.20 ± 0.00 [b]	0.988
RS/GG	470.98 ± 33.50 [a]	0.06 ± 0.01 [a]	0.998	29.69 ± 1.40 [b]	0.22 ± 0.01 [c]	0.974

* Mean ± SD. Different superscripts with the same columns are significantly different ($p < 0.05$ by Duncan's multiple range test).

3.3. Steady Shear Properties

The effects of gums on the steady shear properties of RS gels are shown in Figure 3. Thixotropic behaviors of all samples were observed within the range of shear rates (0.01–300 s^{-1}). The first peak that appeared in the upward flow curves was related to the stress required to break the gel structure and caused the solution to recover. Similar results have been reported in the RS–glucans system [38]. The collected data showed a good fit to the power-law model with R^2 between 0.976 and 0.997. For all samples, the consistency coefficient (K), the yield stresses (σ), flow behavior indices (n), and the apparent viscosity at 300 s^{-1} ($\eta_{a,\,300}$), as well as hysteresis loop areas between the upward and downward curves are summarized in Table 4. Obvious hysteresis loop areas were observed in all starch-related systems, which could be explained by the structural breakdown of the shear field to change the original structure or build up a new one, which then maintained the shear-thinning properties in subsequent shear sweeps [50]. The decomposition of the original structures was observed during gel shearing, as evidenced by $n < 1$. After shearing, the gel structure could only be partially recovered, which is related to the lower $\eta_{a,\,300}$ values of the downward curves compared with the upward ones [51]. For the downward curve, it showed that the addition of both gums markedly increased the K values, which reflects that the gums mainly enhanced viscoelastic properties, especially for guar gum, owing to the thickening effect.

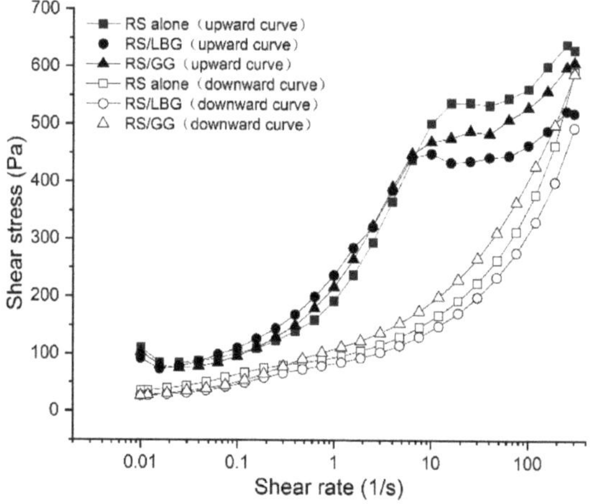

Figure 3. Steady flow curves of RS alone, RS containing locust bean gum (RS/LBG), and RS containing guar gum (RS/GG).

Compared with the control, the gum-containing systems exhibited more pseudoplastic properties, as evidenced by the higher K values (Table 4). Specifically, this effect was more pronounced for the RS–guar gum systems than the RS–locust bean gum systems. This observation is in agreement with the creep recovery results, showing that the guar gum-containing system exhibited higher shear resistance (please see Supplementary Materials Figure S1).

Table 4. The steady flow fitting parameters of RS alone, RS containing locust bean gum (RS/LBG), and RS containing guar gum (RS/GG) *.

Sample	Hla (Pa·s)	Upward Curve				Downward Curve			
		K (Pa·s^n)	n	$\eta_{a,300}$ (Pa·s)	R^2	K (Pa·s^n)	n	$\eta_{a,300}$ (Pa·s)	R^2
RS alone	53,245	236.57 ± 2.41 [a]	0.187 ± 0.004 [ab]	2.15 ± 0.05 [a]	0.909	81.84 ± 1.6 [c]	0.334 ± 0.002 [a]	2.02 ± 0.03 [a]	0.976
RS/LBG	40,385	236.51 ± 32.32 [a]	0.152 ± 0.056 [b]	2.23 ± 0.03 [a]	0.915	86.64 ± 6.6 [b]	0.328 ± 0.004 [b]	2.15 ± 0.02 [a]	0.987
RS/GG	33,150	237.62 ± 16.60 [a]	0.180 ± 0.033 [ab]	2.02 ± 0.56 [a]	0.951	101.9 ± 12.6 [a]	0.312 ± 0.002 [c]	1.96 ± 0.54 [a]	0.997

* Hla: hysteresis loop area (Pa·s); power-law parameters: K, consistency coefficient; n, flow behavior index; $\eta_{a,300}$, the apparent viscosity at 300 s^{-1}. Different superscript letters with the same columns are significantly different ($p < 0.05$ by Duncan's multiple range test).

The hysteresis loop area is an indicator to value the structural breakdown during shearing [52]. Compared with the control, the hysteresis loop area of RS gels containing guar gum and locust bean gum were reduced by 37.7% and 24.2%, respectively, indicating remarkable shear resistance, which has been attributed to the enhancement of the three-dimensional structure of the starch [53]. A similar trend was observed in the dynamic viscoelastic results (Figure 2), which, in turn, reflected the retrogradation extent of RS gels (Table 6).

3.4. Thermodynamic Properties

Various information such as the starch–hydrocolloid interaction, the effects of water, and related properties could be provided by using DSC analysis. The thermodynamic properties of RS gels in the presence or absence of gums and their corresponding retrograded gels during refrigerated storage (4 °C for 3, 5, and 12 days) are listed in Table 5.

Table 5. The retrogradation thermodynamic parameters of RS alone, RS containing locust bean gum (RS/LBG), and RS containing guar gum (RS/GG) *.

Samples	First Run	Second Run (3 d at 4 °C)		Third Run (5 d at 4 °C)		Fourth Run (12 d at 4 °C)	
	ΔH_1	ΔH_2	$\Delta H_2/\Delta H_1$	ΔH_3	$\Delta H_3/\Delta H_1$	ΔH_4	$\Delta H_4/\Delta H_1$
RS alone	2.66	0.30	0.11	1.48	0.56	2.32	0.87
RS/LBG	2.76	0.40	0.14	1.38	0.50	1.43	0.52
RS/GG	2.74	0.33	0.12	1.11	0.41	1.47	0.54

* ΔH_1, gelatinization enthalpy; ΔH_2, retrogradation enthalpy; $\Delta H_2/\Delta H_1$, retrogradation ratio.

Starch–galactomannan interactions affected the retrogradation of samples. The gelatinization enthalpies (ΔH_1) of the starch system, when added with gums, were slightly increased at first, which may be related to the higher viscosity. In the third and last run, the retrogradation enthalpies (ΔH_2) and retrogradation ratios ($\Delta H_2/\Delta H_1$) of retrograded RS gels decreased significantly in the presence of both guar gum and locust bean gum, indicating that gums slowed the retrogradation rate of RS gels during refrigerated storage. Starch retrogradation is generally considered a liquid state event, which requires orientational mobility of the polymer chains in the amylopectin molecule [54]. The presence of guar gum and locust bean gum decreased the orientational mobility of the starch gels and, thus, decreased the retrogradation rate. Moreover, the addition of hydrocolloids could hinder the formation of spongy structures in the starch system, where the gum combined with the starch separated from the starch granules.

The reduction in free water content would inhibit the rearrangement of starch chains, thereby effectively suppressing retrogradation [43]. Interestingly, it seems guar gum suppressed, more than locust bean gum, the short-term retrogradation of RS samples stored for the first 5 days, which was attributed to lower water availability due to the stronger hydration capacity of guar gum [55]. Instead, locust bean gum exhibited the potential to suppress long-term retrogradation, which may be related to its viscous properties, as mentioned above.

The melting enthalpy of recrystallized starch was lower than that of gelatinization, which is consistent with the easier melting properties of recrystallized starch than that of native starch granules [56]. It also found that guar gum and locust bean gum increased or decreased the melting temperatures of recrystallized starch differently during storage time (please see Supplementary Materials Table S1). Similar research [57] has been reported in gum–tapioca starch systems. The results indicated that the addition of guar gum and locust bean gum could modify the thermal properties of RS dependent on the gum type and storage time.

3.5. Determination of Texture Properties

The texture properties of RS gels in the presence or absence of gums during refrigerated storage are presented in Table 6. The hardness of starch gel is always an indicator of the degree of starch retrogradation [58]. The higher hardness values of the starch gel during the initial 3 days indicated its rapid retrogradation. This observation is similar to the retrogradation of amaranth starch in that retrogradation is accelerated at refrigerated temperatures [59]. As a result, a more compact and ordered structure is formed by the amylose and amylopectin in the gelatinized starch, thereby increasing the gel hardness of the RS [60].

Table 6. TPA texture parameters of RS alone, RS containing locust bean gum (RS/LBG), and RS containing guar gum (RS/GG) during cold storage *.

Sample	Days	Hardness	Springiness	Adhesiveness	Cohesiveness	Gumminess	Chewiness
RS alone	3	1134.16 ± 58.91 [a]	0.37 ± 0.02 [a]	77.64 ± 4.16 [b]	0.24 ± 0.02 [a]	269.40 ± 38.87 [a]	99.06 ± 20.13 [a]
	5	907.52 ± 6.93 [a]	0.26 ± 0.01 [a]	17.63 ± 3.21 [a]	0.1 ± 0.01 [a]	94.42 ± 5.81 [a]	24.80 ± 2.97 [a]
RS/LBG	3	931.82 ± 14.62 [b]	0.36 ± 0.05 [a]	123.20 ± 14.03 [a]	0.25 ± 0.02 [a]	233.56 ± 24.78 [a]	84.93 ± 20.51 [a]
	5	861.36 ± 71.83 [a]	0.32 ± 0.05 [a]	11.90 ± 2.06 [a]	0.10 ± 0.01 [a]	84.02 ± 1.88 [a]	26.73 ± 3.90 [a]
RS/GG	3	861.02 ± 66.74 [b]	0.32 ± 0.01 [a]	54.32 ± 5.98 [c]	0.24 ± 0.04 [a]	202.81 ± 19.53 [b]	64.49 ± 4.68 [a]
	5	762.69 ± 0.54 [b]	0.28 ± 0.04 [a]	18.66 ± 4.97 [a]	0.10± 0.02 [a]	79.81 ± 13.04 [a]	21.81 ± 0.23 [a]

* Values represent the mean ± SD of triplicate tests. Columns with different superscripts are significantly different during different samples ($p < 0.05$ by Duncan's multiple range test).

Both guar gum and locust bean gum exhibited excellent potential in the texture modification of RS products during cold storage time. The numerical reduction in hardness was observed with the addition of gums after 3 days as compared with the control. This observation could be interpreted as the possible interaction between gums and the RS granules. The presence of galactomannans tends to combine with RS molecules, thereby inhibiting the leaching of amylose and hindering the retrogradation of starch-related systems [61]. The arrangement of leached amylose is also retarded by the combination, which further suppresses the recrystallization of starch molecules. Generally, the addition of hydrocolloids would improve water retention and inhibit the rearrangement of amylose of starch, which helps to maintain better texture characteristics during refrigerated storage [62].

To specify, the guar gum-containing system exhibited better textural properties during cold storage time, as shown in Table 6. It has also been reported that hydrogen bonding plays a critical role in the gelatinization and retrogradation of starch [11]. A large amount of hydroxyl groups in guar gum, rather than locust bean gum, tends to transform the nucleus of starch recrystallization from amylose to an amylose–gum mixture. As a result, the arrangement of starch molecules is retarded, leading to a lower retrogradation rate of starch [63]. In general, the hardness of RS alone gel would be increased during cold storage owing to starch retrogradation. Meanwhile, the addition of the gums to RS gels helped to slow down the changes in textural characteristics during refrigerated storage in the order of guar gum > locust bean gum.

3.6. Freeze–Thaw Stability

The syneresis value of freeze–thaw RS is often determined as an indicator to evaluate its ability to maintain desirable physical properties during the freezing and thawing process [64,65]. The freeze–thaw stability of RS in the presence or absence of locust bean gum or guar gum is shown in Figure 4. For the RS gels alone, a high syneresis value (37.2%) was observed after the first freeze–thaw cycle. With an increase in freeze–thaw cycles, the syneresis values of gels increased consequently. During repeated freeze–thaw cycles, the starch molecules will recombine, coagulate, and even form a spongy structure, and water will precipitate from the starch body, resulting in dehydration and condensation [24,66].

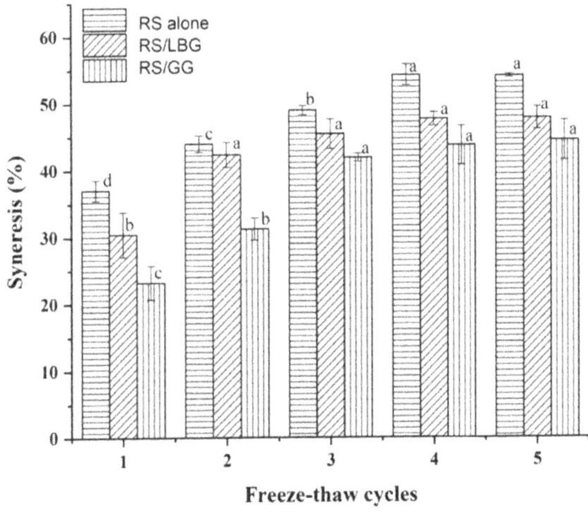

Figure 4. Freeze–thaw stability of RS alone, RS containing locust bean gum (RS/LBG), and RS containing guar gum (RS/GG). Different letters indicate significant differences among group ($p < 0.05$ by Duncan's multiple range test).

The evidence is demonstrated by the lower syneresis value of the RS/gums gels than of the control, showing better freeze–thaw stability in the gels containing gums during freeze–thaw cycles. It could be interpreted by the thickening properties of guar gum and locust bean gum, which would reduce the ice crystal size [43]. The presence of hydrocolloids in starch gel could also bind to water molecules, which reduces the syneresis degree during the freeze–thaw cycles [67].

Compared with locust bean gum, guar gum exhibit better stability during all the freeze–thaw cycles. This is consistent with previous research in which guar gum showed pronounced freeze–thaw stability and a more remarkable synergistic combination with sweat potato starch [43] and corn starch [15]. Although locust bean gum showed similar gelatinization viscosity (Table 1), it was not as effective as guar gum. This is because of their different physical structures [68], especially the different ratios of galactose branching to the mannose backbone [69]. For the locust bean gum-containing system, more bulky phase water would be generated during the repeated FT cycles owing to the ease in the chain association [43].

4. Conclusions

In the present study, the effects of guar gum and locust bean gum on the gelatinization properties, rheological properties, and freeze–thaw stability of RS gels were investigated. RVA results showed that both guar gum and locust bean gum had inhibitory effects on the retrogradation of RS, with a lower SBV value and higher PV. Higher viscoelastic behavior and less thixotropic behavior were reviewed in the presence of galactomannans in the rheological measurements, which related to the enhanced weak gel structure and higher shear resistance. The thermal properties of RS gels could be modified with the addition of galactomannans, which were dependent on the gum type and storage time. Moreover, the textural properties and freeze–thaw stability of the RS gel were significantly improved by the addition of galactomannans. Particularly, the guar gum-containing system exhibits a more significant effect rather than that of locust bean gum, which could be attributed to the different mannose to galactose ratios. From these findings, guar gum could be a better alternative for gluten-free products, as it confers a higher retrogradation inhibition effect, freeze–thaw stability, and fewer texture changes in gels. Future attempts with more detailed parameters optimization are strongly required to unveil the possible interaction mechanism between hydrocolloids and starch and, finally, pinpoint their effects on human digestibility and the sensory properties of food products.

Supplementary Materials: The following supporting information can be downloaded at: https://www.mdpi.com/article/10.3390/foods11162508/s1, Figure S1: The creep-recovery curves of RS alone, RS containing locust bean gum (RS/LBG) and RS containing guar gum (RS/GG); Table S1: Gelatinization temperature and enthalpy and retrogradation ratio for RS alone, RS containing locust bean gum (RS/LBG) and RS containing guar gum (RS/GG) measured by the differential scanning calorimeter (DSC).

Author Contributions: Conceptualization, X.X.; data curation, S.Y.; formal analysis, X.Z. and X.X.; funding acquisition, S.F.; methodology, X.Z. and S.Y.; project administration, S.F.; resources, X.X.; software, X.X.; supervision, S.F. and X.X.; validation, S.F.; writing-original draft, S.Y. and S.F. All authors have read and agreed to the published version of the manuscript.

Funding: This study was funded by the Natural Science Foundation of Zhejiang Province (LQ22C200010) and Zhejiang Shuren University Basic Scientific Research Special Funds (2022XZ008).

Data Availability Statement: The data presented in this study are available on request from the corresponding author.

Conflicts of Interest: The authors declare that they have no conflict of interest.

References

1. Tarique, J.; Sapuan, S.M.; Khalina, A.; Ilyas, R.A.; Zainudin, E.S. Thermal, flammability, and antimicrobial properties of arrowroot (*Maranta arundinacea*) fiber reinforced arrowroot starch biopolymer composites for food packaging applications. *Int. J. Biol. Macromol.* **2022**, *213*, 1–10. [CrossRef] [PubMed]
2. Gatade, A.A.; Sahoo, A.K. Effect of additives and steaming on quality of air dried noodles. *J. Food Sci.Technol.* **2015**, *52*, 8395–8402. [CrossRef] [PubMed]
3. Benkadri, S.; Salvador, A.; Sanz, T.; Nasreddine Zidoune, M. Optimization of xanthan and locust bean gum in a gluten-free infant biscuit based on rice-chickpea flour using response surface methodology. *Foods* **2020**, *10*, 12. [CrossRef]
4. Anil, M.; Durmus, Y.; Tarakci, Z. Effects of different concentrations of guar, xanthan and locust bean gums on physicochemical quality and rheological properties of corn flour tarhana. *Nutr. Food Sci.* **2020**, *51*, 137–150. [CrossRef]
5. Pongjaruvat, W.; Methacanon, P.; Seetapan, N.; Fuongfuchat, A.; Gamonpilas, C. Influence of pregelatinised tapioca starch and transglutaminase on dough rheology and quality of gluten-free jasmine rice breads. *Food Hydrocoll.* **2014**, *36*, 143–150. [CrossRef]
6. Kirchmajer, D.M.; Steinhoff, B.; Warren, H.; Clark, R.; In het Panhuis, M. Enhanced gelation properties of purified gellan gum. *Carbohydr. Res.* **2014**, *388*, 125–129. [CrossRef] [PubMed]
7. Fang, S.; Zhao, X.; Liu, Y.; Liang, X.; Yang, Y. Fabricating multilayer emulsions by using OSA starch and chitosan suitable for spray drying: Application in the encapsulation of β-carotene. *Food Hydrocoll.* **2019**, *93*, 102–110. [CrossRef]
8. Hu, Q.H.; Wu, Y.L.; Zhong, L.; Ma, N.; Zhao, L.Y.; Ma, G.X.; Cheng, N.H.; Nakata, P.A.; Xu, J. In vitro digestion and cellular antioxidant activity of β-carotene-loaded emulsion stabilized by soy protein isolate-Pleurotus eryngii polysaccharide conjugates. *Food Hydrocoll.* **2021**, *112*, 106340. [CrossRef]
9. Sun, J.; Zuo, X.B.; Fang, S.; Xu, H.N.; Chen, J.; Meng, Y.C.; Chen, T. Effects of cellulose derivative hydrocolloids on pasting, viscoelastic, and morphological characteristics of rice starch gel. *J. Texture Stud.* **2017**, *48*, 241–248. [CrossRef]
10. Clemens, R.A.; Pressman, P. Food gums: An overview. *Nutr. Today* **2017**, *52*, 41–43. [CrossRef]
11. Li, J.X.; Yadav, M.P.; Li, J.L. Effect of different hydrocolloids on gluten proteins, starch and dough microstructure. *J. Cereal Sci.* **2019**, *87*, 85–90. [CrossRef]

12. Charoenrein, S.; Tatirat, O.; Rengsutthi, K.; Thongngam, M. Effect of konjac glucomannan on syneresis, textural properties and the microstructure of frozen rice starch gels. *Carbohydr. Polym.* **2011**, *83*, 291–296. [CrossRef]

13. Busch, V.; Rozycki, V.; Buera, M.P. Galactomannans from different prosopis species: Extraction, characterization, and applications. In *Prosopis as a Heat Tolerant Nitrogen Fixing Desert Food Legume*; Academic Press: Cambridge, MA, USA, 2022; pp. 305–317.

14. Das, A.; Kundu, S.; Gupta, M.; Mukherjee, A. Guar gum propionate-kojic acid films for *Escherichia coli* biofilm disruption and simultaneous inhibition of planktonic growth. *Int. J. Biol. Macromol.* **2022**, *211*, 57–73. [CrossRef] [PubMed]

15. Sudhakar, V.; Singhal, R.S.; Kulkarni, P.R. Starch-galactomannan interactions: Functionality and rheological aspects. *Food Chem.* **1996**, *55*, 259–264. [CrossRef]

16. Huang, C.C. Physicochemical, pasting and thermal properties of tuber starches as modified by guar gum and locust bean gum. *Int. J. Food Sci. Technol.* **2009**, *44*, 50–57. [CrossRef]

17. Kim, W.W.; Yoo, B. Rheological and thermal effects of galactomannan addition to acorn starch paste. *LWT Food Sci. Technol.* **2011**, *44*, 759–764. [CrossRef]

18. Bahnassey, Y.A.; Breene, W.M. Rapid visco-analyzer (RVA) pasting profiles of wheat, corn, waxy corn, tapioca and amaranth starches *(A. hypochondriacus and A cruentus)* in the presence of konjac flour, gellan, guar, xanthan and locust bean gums. *Starch-Stärke* **1994**, *46*, 134–141. [CrossRef]

19. Prajapati, V.D.; Jani, G.K.; Moradiya, N.G.; Randeria, N.P.; Nagar, B.J.; Naikwadi, N.N.; Variya, B.C. Galactomannan: A versatile biodegradable seed polysaccharide. *Int. J. Biol. Macromol.* **2013**, *60*, 83–92. [CrossRef]

20. Borries-Medrano, E.V.; Jaime-Fonseca, M.R.; Aguilar-Méndez, M. Tapioca starch-galactomannan systems: Comparative studies of rheological and textural properties. *Int. J. Biol. Macromol.* **2018**, *122*, 1173–1183. [CrossRef]

21. Richardson, P.H.; Willmer, J.; Foster, T.J. Dilute solution properties of guar and locust bean gum in sucrose solutions. *Food Hydrocoll.* **1998**, *12*, 339–348. [CrossRef]

22. Salehi, F. Effect of common and new gums on the quality, physical, and textural properties of bakery products: A review. *J. Texture Stud.* **2020**, *51*, 361–370. [CrossRef] [PubMed]

23. Tarique, J.; Zainudin, E.S.; Sapuan, S.M.; Ilyas, R.A.; Khalina, A. Physical, mechanical, and morphological performances of arrowroot (*Maranta arundinacea*) fiber reinforced arrowroot starch biopolymer composites. *Polymers* **2022**, *14*, 388. [CrossRef] [PubMed]

24. Qiu, S.; Yadav, M.P.; Liu, Y.; Chen, H.; Tatsumi, E.; Yin, L. Effects of corn fiber gum with different molecular weights on the gelatinization behaviors of corn and wheat starch. *Food Hydrocoll.* **2016**, *53*, 180–186. [CrossRef]

25. Nazrin, A.; Sapuan, S.M.; Zuhri, M.Y.M.; Tawakkal, I.S.M.A.; Ilyas, R.A. Flammability and physical stability of sugar palm crystalline nanocellulose reinforced thermoplastic sugar palm starch/poly(lactic acid) blend bionanocomposites. *Nanotechnol. Rev.* **2021**, *11*, 86–95. [CrossRef]

26. Culetu, A.; Duta, D.E.; Papageorgiou, M.; Varzakas, T. The role of hydrocolloids in gluten-free bread and pasta; rheology, characteristics, staling and glycemic index. *Foods* **2021**, *10*, 3121. [CrossRef]

27. Syafiq, R.M.O.; Sapuan, S.M.; Zuhri, M.Y.M.; Othman, S.H.; Ilyas, R.A. Effect of plasticizers on the properties of sugar palm nanocellulose/cinnamon essential oil reinforced starch bionanocomposite films. *Nanotechnol. Rev.* **2022**, *11*, 423–437. [CrossRef]

28. Hamdani, A.M.; Wani, I.A. Guar and Locust bean gum: Composition, total phenolic content, antioxidant and antinutritional characterisation. *Bioact. Carbohydr. Diet. Fibre* **2017**, *11*, 53–59. [CrossRef]

29. Schorsch, C.; Garnier, C.; Doublier, J.L. Viscoelastic properties of xanthangalactomannan mixtures: Comparison of guar gum with locust bean gum. *Carbohydr. Polym* **1997**, *34*, 165–175. [CrossRef]

30. Maier, H.; Anderson, M.; Karl, C.; Magnuson, K.; Whistler, R.L. Guar, locust bean, Tara, and fenugreek gums: Industrial gums. In *Industrial Gums*; Academic Press: Cambridge, MA, USA, 1993; pp. 181–226.

31. Ge, H.; Wu, Y.; Woshnak, L.L.; Mitmesser, S.H. Effects of hydrocolloids, acids and nutrients on gelatin network in gummies. *Food Hydrocoll.* **2021**, *113*, 106549. [CrossRef]

32. Xu, X.; Fang, S.; Li, Y.; Zhang, F.; Shao, Z.; Zeng, Y.; Chen, J.; Meng, Y. Effects of low acyl and high acyl gellan gum on the thermal stability of purple sweet potato anthocyanins in the presence of ascorbic acid. *Food Hydrocol.* **2018**, *86*, 116–123. [CrossRef]

33. Fang, S.; Wang, J.; Xu, X.J.; Zuo, X.B. Influence of Low Acyl and High Acyl Gellan Gums on Pasting and Rheological Properties of Rice Starch Gel. *Food Biophys.* **2018**, *13*, 116–123. [CrossRef]

34. Kim, C.; Yoo, B. Rheological properties of rice starch–xanthan gum mixtures. *J. Food Eng.* **2006**, *75*, 120–128. [CrossRef]

35. Meng, Y.C.; Sun, M.H.; Fang, S.; Chen, J.; Li, Y.H. Effect of sucrose fatty acid esters on pasting, rheological properties and freeze–thaw stability of rice flour. *Food Hydrocoll.* **2014**, *40*, 64–70. [CrossRef]

36. Wang, B.; Wang, L.J.; Li, D.; Özkan, N.; Li, S.J.; Mao, Z.H. Rheological properties of waxy maize starch and xanthan gum mixtures in the presence of sucrose. *Carbohydr. Polym.* **2009**, *77*, 472–481. [CrossRef]

37. Kamaruddin, Z.H.; Jumaidin, R.; Ilyas, R.A.; Selamat, M.Z.; Alamjuri, R.H.; Yusof, F.A.M. Biocomposite of cassava starch-cymbopogon citratus fibre: Mechanical, thermal and biodegradation properties. *Polymers* **2022**, *14*, 514. [CrossRef] [PubMed]

38. Zhang, D.; Lin, Z.; Lei, W.; Zhong, G. Synergistic effects of acetylated distarch adipate and sesbania gum on gelatinization and retrogradation of wheat starch. *Int. J. Biol. Macromol.* **2020**, *156*, 171–179. [CrossRef]

39. Zhai, Y.H.; Xing, J.L.; Luo, X.H.; Zhang, H.; Yang, K.; Shao, X.F.; Li, Y.N. Effects of pectin on the physicochemical properties and freeze–thaw stability of waxy rice starch. *Foods* **2021**, *10*, 2419. [CrossRef]

40. Sasaki, T.; Yasui, T.; Matsuki, J. Influence of non-starch polysaccharides isolated from wheat flour on the gelatinization and gelation of wheat starches. *Food Hydrocoll.* **2000**, *14*, 295–303. [CrossRef]

41. Brennan, C.S.; Suter, M.; Luethi, T.; Matia-Merino, L.; Qvortrup, J. The relationship between wheat flour and starch pasting properties and starch hydrolysis: Effect of non-starch polysaccharides in a starch gel system. *Starch Strke* **2008**, *60*, 23–33. [CrossRef]

42. Torres, M.D.; Moreira, R.; Chenlo, F.; Morel, M.H. Effect of water and guar gum content on thermal properties of chestnut flour and its starch. *Food Hydrocoll.* **2013**, *33*, 192–198. [CrossRef]

43. Lee, M.H.; Baek, M.H.; Cha, D.S.; Park, H.J.; Lim, S.T. Freeze–thaw stabilization of sweet potato starch gel by polysaccharide gums. *Food Hydrocoll.* **2002**, *16*, 345–352. [CrossRef]

44. Ptaszek, P.; Grzesik, M. Viscoelastic properties of maize starch and guar gum gels. *J. Food Eng.* **2007**, *82*, 227–237. [CrossRef]

45. Alloncle, M.; Doublier, J.L. Viscoelastic properties of maize starch/hydrocolloid pastes and gels. *Food Hydrocoll.* **1991**, *5*, 455–467. [CrossRef]

46. Gao, J.; Mao, Y.; Xiang, C.; Cao, M.; Ren, G.; Wang, K.; Ma, X.; Wu, D.; Xie, H. Preparation of β-lactoglobulin/gum arabic complex nanoparticles for encapsulation and controlled release of EGCG in simulated gastrointestinal digestion model. *Food Chem.* **2021**, *354*, 129516. [CrossRef] [PubMed]

47. Wu, X.; Hu, Q.; Liang, X.; Fang, S. Fabrication of colloidal stable gliadin-casein nanoparticles for the encapsulation of natamycin: Molecular interactions and antifungal application on cherry tomato. *Food Chem.* **2022**, *391*, 133288. [CrossRef] [PubMed]

48. Gao, J.; Liu, C.; Shi, J.; Ni, F.; Shen, Q.; Xie, H.; Wang, K.; Lei, Q.; Fang, W.; Ren, G. The regulation of sodium alginate on the stability of ovalbumin-pectin complexes for vd3 encapsulation and in vitro simulated gastrointestinal digestion study. *Food Res. Int.* **2021**, *140*, 110011. [CrossRef] [PubMed]

49. Ptaszek, A.; Berski, W.; Ptaszek, P.; Witczak, T.; Repelewicz, U.; Grzesik, M. Viscoelastic properties of waxy maize starch and selected non-starch hydrocolloids gels. *Carbohydr. Polym.* **2009**, *76*, 567–577. [CrossRef]

50. Zhang, F.; Fu, S.L.; Jin, T. Rheological properties of maize starch pastes. *Transactions of the CSAE* **2008**, *24*, 294–297.

51. Hussain, S.; Mohamed, A.A.; Alamri, M.S.; Ibraheem, M.A.; Qasem, A.A.A.; Shahzad, S.A.; Ababtain, I.A. Use of gum cordia (*cordia myxa*) as a natural starch modifier; effect on pasting, thermal, textural, and rheological properties of corn starch. *Foods* **2020**, *9*, 909. [CrossRef]

52. Banchathanakij, R.; Suphantharika, M. Effect of different β-glucans on the gelatinisation and retrogradation of rice starch. *Food Chem.* **2009**, *114*, 5–14. [CrossRef]

53. Achayuthakan, P.; Suphantharika, M. Pasting and rheological properties of waxy corn starch as affected by guar gum and xanthan gum. *Carbohydr. Polym.* **2008**, *71*, 9–17. [CrossRef]

54. Slade, L.; Levine, H. Recent advances in starch retrogradation. In *Idustrial polysaccharides-The impact of biotechnology and advanced methodologies*; Stivala, S.S., Crescenzi, V., Dea, I.C.M., Eds.; Gordon and Breach Science Publishers: New York, NY, USA, 1987; pp. 387–430.

55. Lee, S.; Yoo, B. Effect of sucrose addition on rheological and thermal properties of rice starch-gum mixtures. *Int. J. Food Eng.* **2014**, *10*, 849–856. [CrossRef]

56. Babić, J.; Šubarić, D.; Ackar, D.; Kovačević, D.; Piližota, V.; Kopjar, M. Preparation and characterization of acetylated tapioca starches. *Dtsch. Lebensm. -Rundsch.* **2007**, *103*, 580–585.

57. Šubarić, D.; Babić, J.; Ačkar, Đ.; Piližota, V.; Kopjar, M.; Ljubas, I.; Ivanovska, S. Effect of galactomannan hydrocolloids on gelatinization and retrogradation of tapioca and corn starch. *Croatian J. Food Sci. Technol.* **2011**, *3*, 26–31.

58. Luo, Y.M.; Niu, L.Y.; Li, D.M.; Xiao, J.H. Synergistic effects of plant protein hydrolysates and xanthan gum on the short- and long-term retrogradation of rice starch. *Int. J. Biol. Macromol.* **2019**, *144*, 967–977. [CrossRef]

59. Baker, L.A.; Rayas-Duarte, P. Retrogradation of amaranth starch at different storage temperatures and the effects of salt and sugars. *Cereal Chem.* **1998**, *75*, 308–314. [CrossRef]

60. Wu, Y.; Chen, Z.X.; Li, X.X.; Mei, L. Effect of tea polyphenols on the retrogradation of rice starch. *Food Res. Int.* **2009**, *42*, 221–225. [CrossRef]

61. Chen, L.; Ren, F.; Zhang, Z.P.; Tong, Q.Y.; Rashed, M.M.A. Effect of pullulan on the short-term and long-term retrogradation of rice starch. *Carbohydr. Polym.* **2015**, *115*, 415–421. [CrossRef]

62. Chen, L.; Tian, Y.Q.; Tong, Q.Y.; Zhang, Z.P.; Jin, Z.Y. Effect of pullulan on the water distribution, microstructure and textural properties of rice starch gels during cold storage. *Food Chem.* **2016**, *214*, 702–709. [CrossRef]

63. Tian, Y.Q.; Xu, X.M.; Li, Y.; Jin, Z.Y.; Chen, H.Q.; Wang, H. Effect of β-cyclodextrin on the long-term retrogradation of rice starch. *Eur. Food Res. Technol.* **2009**, *228*, 743–748. [CrossRef]

64. Yuan, R.C.; Thompson, D.B. Freeze–thaw stability of three waxy maize starch pastes measured by centrifugation and calorimetry. *Cereal Chem.* **1998**, *75*, 571–573. [CrossRef]

65. Wang, Y.J.; Jane, J. Correlation between glass transition temperature and starch retrogradation in the presence of sugars and maltodextrins. *Cereal Chem.* **1994**, *71*, 527–531.

66. Teng, L.Y.; Chin, N.L.; Yusof, Y.A. Rheological and textural studies of fresh and freeze–thawed native sago starch-sugar gels. II. Comparisons with other starch sources and reheating effects. *Food Hydrocoll.* **2013**, *31*, 156–165. [CrossRef]

67. Arunyanart, T.; Charoenrein, S. (Eds.) Effects of sucrose, xylitol and konjac glucomannan on stability of frozen rice starch gels. In Proceedings of the 34th Congress on Science and Technology of Thailand, Bangkok, Thailand, 31 October–2 November 2008.

68. Sprenger, M. New stabilising systems using galactomannans. *Dairy Ind. Int.* **1990**, *55*, 19–20.
69. Choi, H.M.; Yoo, B. Rheology of Mixed Systems of Sweet Potato Starch and Galactomannans. *Starch Stärke.* **2008**, *60*, 263–269. [CrossRef]

157

Article

Study of the Technological Properties of *Pedrosillano* Chickpea Aquafaba and Its Application in the Production of Egg-Free Baked Meringues

Paula Fuentes Choya [†], Patricia Combarros-Fuertes [†], Daniel Abarquero Camino, Erica Renes Bañuelos, Bernardo Prieto Gutiérrez, María Eugenia Tornadijo Rodríguez and José María Fresno Baro *

Lactic Acid Bacteria and Technological Application Group (BALAT), Food Hygiene and Technology Department, Faculty of Veterinary, University of León, 24071 León, Spain
* Correspondence: jmfreb@unileon.es
† These authors contributed equally to this work.

Abstract: Aquafaba is a by-product derived from legume processing. The aim of this study was to assess the compositional differences and the culinary properties of *Pedrosillano* chickpea aquafaba prepared with different cooking liquids (water, vegetable broth, meat broth and the covering liquid of canned chickpeas) and to evaluate the sensory characteristics of French-baked meringues made with the different aquafaba samples, using egg white as a control. The content of total solids, protein, fat, ash and carbohydrates of the aquafaba samples were quantified. Foaming and emulsifying capacities, as well as the foam and emulsions stabilities were determined. Instrumental and panel-tester analyses were accomplished to evaluate the sensory characteristics of French-baked meringues. The ingredients added to the cooking liquid and the intensity of the heat treatment affected the aquafaba composition and culinary properties. All types of aquafaba showed good foaming properties and intermediate emulsifying capacities; however, the commercial canned chickpea's aquafaba was the most similar to egg white. The aquafaba meringues showed less alveoli, greater hardness and fracturability and minimal color changes after baking compared with egg white meringues; the meat and vegetable broth's aquafaba meringues were the lowest rated by the panel-tester and those prepared with canned aquafaba were the highest scored in the sensory analysis.

Keywords: *Pedrosillano* chickpea; aquafaba; chemical composition; functional properties; egg-free meringues; sensory characteristics

Citation: Fuentes Choya, P.; Combarros-Fuertes, P.; Abarquero Camino, D.; Renes Bañuelos, E.; Prieto Gutiérrez, B.; Tornadijo Rodríguez, M.E.; Fresno Baro, J.M. Study of the Technological Properties of *Pedrosillano* Chickpea Aquafaba and Its Application in the Production of Egg-Free Baked Meringues. *Foods* **2023**, *12*, 902. https://doi.org/10.3390/foods12040902

Academic Editor: Cornelia Witthöft

Received: 22 December 2022
Revised: 10 February 2023
Accepted: 15 February 2023
Published: 20 February 2023

1. Introduction

Legumes are a group of edible seeds from the *Leguminoseae* family, which include, among others, beans, chickpeas, peas, lentils or soybeans. According to the FAO [1], the production of legumes has increased 25% between 2010 and 2020 in the world; however, their consumption continues to be scarce. This low intake is associated with several factors: social changes and current lifestyle (people eating more outside of the home) [2]; very long preparation times (long soaking and cooking times) [3]; or difficulty for some people to digest them, causing belly swelling due to the production of large amounts of gases [4]. However, in recent years, consumers are demanding more and more plant origin foods in their diet instead of animal origin foods, mainly for environmental sustainability, animal health and animal welfare reasons [5,6], which may contribute to increasing their intake, either in traditional recipes or in the form of flour, processed purees, creams and prepared and/or canned legumes.

One of the typical legumes of Spanish Gastronomy are chickpeas (*Cicer arietinum*) and Castilla y León; Castilla-La Mancha and Andalucia are the regions with the largest area of cultivation [7]. In Spain, one of the most appreciated varieties, both in traditional and in avant-garde cuisine (stews, salads, broths or stir-fries), is the *Pedrosillano* chickpea (a variety of small size and almost spherical-shaped chickpeas, which are characterized by their small beak and a yellow-orange coloration and a firm structure that acquires a buttery texture after cooking).

The use of products or by-products derived from the processing of legumes would be a further step in the improvement of sustainable food production. One of the most important by-products generated during the legume processing is the aquafaba. Aquafaba is defined as the liquid obtained after cooking legumes, which includes both that obtained in traditional cooking (in water or broths) and the covering liquid of canned legumes [8]. The presence of some proteins, complex carbohydrates and other flavour components turns aquafaba into an ingredient of exceptional culinary quality due to its functional properties (foaming, emulsifying, binding, gelling or thickening) [9]. In fact, the use of aquafaba in the formulation of new vegan products, both sweet and savoury, such as foams, emulsions or dairy substitutes, has increased nowadays [10].

In most of the vegan products, aquafaba is used as an egg substitute, not only by sharing several technological properties but also because the production of legumes causes a lower environmental impact than the production of eggs or another animal protein sources [11]. Studies carried out by Nette et al. [12] reported that products made with legumes generate 35% less greenhouse gases than those made with eggs. However, it is necessary to consider that a more recent study [13] questioned the environmental benefits of aquafaba as an egg substitute in food processing, highlighting that it could have a negative impact on the environment footprint as a result of higher electricity requirements. In addition, the use of aquafaba would also be compatible with the concept of upcycling, which refers to the process of finding a new use for by-products [14] that could become innovative culinary options.

The use of aquafaba as an alternative to eggs and dairy products not only satisfies the needs of vegan people, but also of a growing number of consumers who prefer dietary ingredients rich in fibre and/or are allergen-free [5,10]. Egg is one of the 14 food allergens with mandatory declaration in the list of ingredients [15]. The egg white, particularly the ovomucoid, is more allergenic than the yolk and is problematic for allergic people in the preparation of those recipes and elaborations in which it is the main ingredient [16]. In this sense, the aquafaba has garnered attention in the food industry for a new generation of natural substitutes and egg- and dairy-free labels, without the need for protein isolates, edible gums or modified starches [10,17–19].

The foaming and stabilizing capacity of the classical meringues are defined by the egg proteins due to its amphiphilic nature [20]. In the aquafaba, the foam production is also possible due to the presence of the chickpeas' proteins. Those proteins require a partial denaturation so that the hydrophobic amino acids were oriented towards the air bubbles and the hydrophilic ones towards the aqueous phase favouring the maintenance of the structure [21,22]. In the preparation of aquafaba, it is necessary that enough chickpea proteins are dissolved in the cooking broth so that when the air was mechanically introduced, the foams were generated and stable [23]. Other factors, such as the content of intrinsic carbohydrates (especially polysaccharides), the pH of the medium or the presence of fat in the aquafaba, could play a very important role in the foam formation and stability [9].

In recent years, some works on the foaming and emulsifying properties of aquafaba [23–29], the application of ultrasound technology in the extraction [30], the optimization of cooking conditions and pH [31] and the impact of concentration methods on the characteristics of spray-dried powders of soybean aquafaba [32] were published. Most of these studies were focused on the use of aquafaba in the preparation of vegan mayonnaise with limited studies on its chemical composition or its applications in other culinary elaborations, such as meringues or sponge cakes [23,33]. In general, the studies on aquafaba have been based on the use of aquafaba derived from commercial canned legumes; to the best of the authors' knowledge, the studies of the chemical characteristics and functional properties of the aquafaba obtained during the preparation of culinary dishes and their impact on the functional and sensory properties are very limited or have not been addressed [34].

From the scientific point of view, the original hypothesis of this research work was: "could the cooking liquid of *Pedrosillano* chickpeas under different culinary recipes, widely established in gastronomy, be a good substitute for eggs as an ingredient in the manufacture of sweet or salty baked meringues?".

In order to test this hypothesis, the aim of this study was to assess the compositional differences, as well as the foaming and the emulsifying properties of *Pedrosillano* chickpea aquafaba, prepared with different cooking liquids (water, vegetable broth, meat broth and the covering liquid of canned chickpeas) and to evaluate the sensory characteristics of French-type baked meringues made with the different aquafaba samples using egg white as a control.

2. Materials and Methods

2.1. Preparation the Aqueous Broths for Cooking Chickpeas

The different aqueous broths used to cook the chickpeas were prepared as follows

(1) Water: 4 g salt were dissolved in 3 L distilled water.

(2) Vegetable broth: prepared with 210 g cabbage, 105 g turnip, 175 g onion, 160 g leek, 260 g carrot, 20 g garlic and 4 g salt. All the ingredients were washed, drained, chopped, weighed and, finally, cooked with 3 L of distilled water for 40 min in a pressure cooker at 100 kPa (Excellent BRA, Valls, Spain). Next, the broth was filtered through a cotton cloth and stored refrigerated (4 °C) for a maximum of 24 h until it was used to cook the chickpeas.

(3) Meat broth: was prepared using the same ingredients and amounts used for the vegetable broth and 97 g fresh chorizo, 204 g marinated and cured pork ribs, 222 g beef shank, 219 g chicken, 153 g fresh salted pork ear and 210 g ham bone (these last two products were previously desalted in water for 12 h). All the ingredients were introduced into a pressure cooker (Excellent BRA, Spain) with 3 L of distilled water where they were cooked for 40 min at 100 kPa. Then, the broth was filtered, kept refrigerated (4 °C) for 24 h and defatted before being used to cook the chickpeas.

2.2. Cooking Conditions of Chickpea to Obtain Aquafaba

Pedrosillano chickpeas (Polifer S.L., La Bañeza, León, Spain) were used in all the experiments. Hydration was carried out at room temperature by introducing 300 g chickpeas into a container with 2 L distilled water and 6 g salt for 12 h. Then, the chickpeas were drained, washed and weighed.

Each broth described in the Section 2.1 (except for the canned chickpea aquafaba) was used to cook the chickpeas for 35 min in a pressure cooker (Excellent BRA, Spain) at 100 kPa using a ratio of chickpea–broth of 1:2 (*w/v*). Two replicates were carried out for each elaboration. The different types of aquafaba were obtained by draining the chickpeas.

The canned chickpea aquafaba was obtained by draining the covering liquid from 9 jars of 570 g commercial canned *Pedrosillano* chickpeas (Legumbres Penelas S.L., Villarejo del Órbigo, León, Spain).

All types of aquafaba were frozen at −30 °C in jars of approximately 100 g until analysis.

2.3. Analysis of the Chemical Composition of Aquafaba

The chemical composition of the different types of aquafaba was determined in duplicate following the methods described by Stantiall et al. [25]. Total solids content was analysed by drying the samples in an oven at 100 °C until constant weight. Proteins were quantified following the Kjeldahl method using an automatic distiller (Digestor Kjeldahl Tecator 1002, FOSS IBERIA, S.A, Barcelona, Spain); a value of 6.25 was used as the conversion factor [35]. The fat content was determined following the AOAC standard 920.39 method using a Soxhlet extraction system [36]. The ash content was quantified by gravimetry [37] using a dry ashing method with a muffle at 500 °C for 6 h. Finally, the quantification of carbohydrates was carried out by subtracting the content of protein, fat and ash from total solids content.

2.4. Analysis of Foaming and Emulsifying Properties of Aquafaba and Egg White

The foaming properties were determined in duplicate following the method described by Stantiall et al. [25]. A total of 80 mL of aquafaba or egg white (as control) was taken in a glass and mixed with a hand blender (Miniquick 5 MQ5000, Braun, Neu-Isenburg, Germany) for 4 min. Next, the content was transferred to a 250 mL cylinder and the volume of the foam generated at zero time and after 40 min was measured. The foaming capacity (FC) and the foam stability (FS) were calculated using the following equations and expressed as the percentage of overrun, which is defined as the amount of air incorporated into the liquid phase.

$$FC\ (\%) = ((V_{Final} - V_{Initial})/V_{Initial}) \times 100 \tag{1}$$

where V_{Final} is the volume of the foam generated after homogenization and $V_{Initial}$ is the volume of aquafaba or egg white before homogenization.

$$FS\ (\%) = ((V_{40} - V_{Initial})/V_{Initial}) \times 100 \tag{2}$$

where V_{40} is the volume of the foam after 40 min and $V_{Initial}$ is the initial volume of the foam generated after homogenization.

The emulsifying properties were determined in duplicate by spectrophotometry following the protocol described by Karaca et al. [38]. To compare the emulsifying capacity of aquafaba and egg white, 10 mL of each sample was mixed with 10 mL of olive oil using an Ultraturrax homogenizer (Unidrive 1000D Ingenieurbüro CAT, M.Zipperer GmbH, Ballrechten-Dottingen, Germany) for 1 min at 3000 rpm. A total of 5 mL of a 1% sodium dodecyl sulfate (SDS) solution was placed in a beaker, then 50 μL of the emulsion was added. Then the mixture was homogenized, and the absorbance was measured at 500 nm for 40 min taking the absorbance values at 10 min intervals. The emulsifying capacity (EC) and the emulsion stability (ES) were calculated using the following equations:

$$EC\ (m^2/g) = ((2 \times 2.303 \times A_0)/\%\ Protein) \tag{3}$$

where A_0 is the absorbance of the diluted emulsion immediately after homogenization and (% Protein) is the weight of protein per volume (g/mL).

$$ES\ (min) = A_0/\Delta A \times t \tag{4}$$

where A_0 is the absorbance of the diluted emulsion immediately after homogenization, ΔA is the change in absorbance between 0 and 10 min (A_0–A_{10}) and t is the time interval (10 min).

2.5. Baked French Meringue Preparation

The culinary applicability of the different types of aquafaba compared with egg white was assessed by preparing, in duplicate, different batches of French meringues using the following recipe: 100 g aquafaba or egg white, 100 g icing sugar and 1 mL natural lemon juice. All the ingredients were mixed with a hand blender (Miniquick 5 MQ5000, Braun, Neu-Isenburg, Germany) for 1 min at medium speed and, later, at maximum power until they were stiff. The meringue mixture was deposited on a *silpat* with the help of metal moulds to obtain meringues of 5 cm in diameter and 1 cm of height. Next, the meringues were baked in an oven (Conterm, Selecta, S.A., Barcelona, Spain) at 120 °C for 1.5 h, cooled to room temperature and stored in hermetically closed containers to protect them from the moisture until their analysis.

2.6. Sensory Analysis of Meringues

The colour analysis of the meringues was accomplished using a Konica CM reflectance colourimeter (Minolta, Osaka, Japan). The MAV (measurement/illumination area) and MAV mask pattern were read with 8 mm diameter glass. Data was processed using the program Color Data Software CM-S100w SpectraMagic TM NX ve. 1.9, Pro USB (Konica, Minolta, Osaka, Japan). Nine measurements on the surface of each meringue were carried out for each type of meringue and each batch elaborated. In all colour determinations, the equipment was previously calibrated for zero and white using standard plates (illuminate D65 and the $10°$ SCI observer). The colour parameters of the CIE*Lab* scale were studied: lightness (L^*), red-green component (a^*) and yellow-blue component (b^*).

The instrumental analysis of the meringue's texture was performed following the method proposed by Stantiall et al. [25] with some modifications. A TA.XT2i texturometer and the Texture Expert program, v1.20 (Stable Micro Systems, Godalming, Surrey, UK) were used. The assays were carried out with a p 40 probe at a constant speed of 0.5 mm/s and using a degree of compression of 50%. Fracturability, brittleness, hardness and elasticity were quantified. In each type of meringue and each batch, 8 determinations of the texture profile were carried out.

The proportion of gas cells in the meringues (alveoli), expressed as the percentage of porosity, was determined by taking cross section photographs of the meringues, which were subsequently divided into portions of 1 cm length. The images were processed with the program Gimp 2.10.10 (GNOME, 2019) using the histogram tool. The alveoli percentage was obtained through the percentage of certain colour strip pixels [39]. For each meringue batch, the alveoli analyses were carried out in four meringues and three areas of each meringue.

Finally, the analysis of four sensory attributes (appearance, smell, taste and texture) was carried out using a seven-point hedonic scale for each attribute (1 = dislike extremely; 2 = dislike very much; 3 = dislike somewhat; 4 = neither like nor dislike; 5 = like somewhat; 6 = like a lot; and 7 = like very much) [40]. Likewise, a global impression score was accomplished using a scale from 1 (very bad) to 10 (very good). These analyses were performed by a panel of 114 untrained tasters.

2.7. Statistical Analysis

The statistical analysis was performed using the software SPSS statistics v.26.0 [41]. All variables were tested for the assumptions of normality and homoscedasticity. An analysis of variance (ANOVA) was applied to study the differences on the chemical composition of the different types of aquafaba and the differences on the sensory parameters of the meringues (data sets with normal distribution); subsequently, in those cases in which significant differences were observed, the Tukey test was applied to compare differences among groups. The Kruskal–Wallis test was used to study the differences on the foaming and emulsifying properties of the samples (data sets without normal distribution); in the cases in which significant differences were observed the U of the Mann–Whitney test was performed to define the differences among groups. A $p < 0.05$ was considered to be significant.

3. Results and Discussion

3.1. Chemical Composition of Aquafaba

The results of the chemical composition analysis of the different types of aquafaba were collected in Table 1.

The water aquafaba as well as the vegetable broth aquafaba showed a very similar total solids, protein and carbohydrate contents, with values ranging 5.63–5.84, 1.19–1.21 and 3.36–3.79 g/100 g aquafaba, respectively. On the contrary, the ash contents showed significant differences ($p < 0.05$) between them. The vegetable broth aquafaba had an ash content approximately 33% higher than water aquafaba due to the presence of minerals from the vegetables. Neither of the aquafaba samples contained fat in its composition.

Table 1. Mean values (expressed as g/100 g aquafaba) and standard deviation of the chemical composition of different types of chickpeas aquafaba.

	Total Solids (%)	Protein (%)	Carbohydrates (%)	Ash (%)	Fat (%)
Water Aquafaba	5.84 ± 0.08 [a]	1.21 ± 0.01 [a]	3.79 ± 0.08 [c]	0.77 ± 0.02 [a]	nd
Vegetable broth Aquafaba	5.63 ± 0.15 [a]	1.19 ± 0.04 [a]	3.36 ± 0.13 [c]	1.03 ± 0.08 [b]	nd
Meat broth Aquafaba	7.84 ± 0.11 [c]	2.36 ± 0.05 [b]	2.20 ± 0.11 [b]	1.12 ± 0.01 [b]	2.14 ± 0.23 [b]
Canned Aquafaba	6.24 ± 0.05 [b]	2.48 ± 0.03 [b]	1.46 ± 0.09 [a]	1.15 ± 0.03 [b]	1.15 ± 0.72 [a]

[a–c] Means in the same column with different superscripts showed significant differences ($p < 0.05$); nd: not detected.

The results of total solids, protein, ash and fat contents in the water aquafaba were very similar to those described by Bird et al. [42], Stantiall et al. [25] and He et al. [26]. The minor variations observed among the studies would be associated with differences in the soaking and cooking conditions of the chickpeas (time or water–chickpea ratio) and also related to the chickpea variety [9].

The meat broth aquafaba contained the highest total solids value due to the elevated protein and fat contents, which together represented almost 60% of the total solids. An important part of the proteins come from the meat sarcoplasm and connective tissue that were solubilized during the cooking. Furthermore, although the meat broth was defatted before being used to cook the chickpeas, the resulting aquafaba showed the highest fat content. The difficulty to eliminate the fat of the meat broth is probably related to its incorporation as an emulsion, stabilized by the solubilized proteins after the cooking process, in the aqueous phase. The ash content was similar to that reported for the vegetable broth aquafaba, but its carbohydrate concentration was a 35% lower with a value of 2.20 g/100 g aquafaba.

The canned and the meat broth aquafaba showed similar ash and protein contents; however, significant differences ($p < 0.05$) on their carbohydrates and fat contents were observed. The proteins represented approximately 40% of total solids in the canned chickpeas aquafaba and 30% in the meat broth aquafaba. The total solids content of the commercial aquafaba used in this study was very similar to that described by Shim et al. [34]; however, it showed between a 45 and 90% higher protein content, about 2.6 times more ash content and 50% less carbohydrates, than those described by other authors [27,29,34]. These differences might be associated to the chickpea variety; He et al. [26] reported important variations in the chemical composition and the properties of aquafaba samples obtained from different types of chickpeas. Likewise, since there is a relationship between the proportion of broth and the concentration of the extracted compounds [43], these differences may be also associated with the chickpea–water ratio used in the preparation. In the same way, another important factor that may affect the chemical composition of the aquafaba would be the variations in the cooking conditions. High temperatures combined with prolonged cooking times would lead to significant changes in the seed coat of the chickpeas, which allows the passage of a greater or lesser amount of solute into the cooking liquid [44]. Some authors reported that a reduced gelatinization of the starch would avoid chickpea breakage and, consequently, the migration of complex carbohydrates but not minerals and hydrophilic proteins [9,27]. Most of the proteins present in aquafaba have a molecular weight of less than 23 kDa, hence they can easily pass through the pores of the chickpea covering [27].

3.2. Foaming and Emulsifying Properties of the Different Types of Aquafaba

The foaming capacity and the foam stability of the different types of aquafaba and the egg white are depicted in Figure 1.

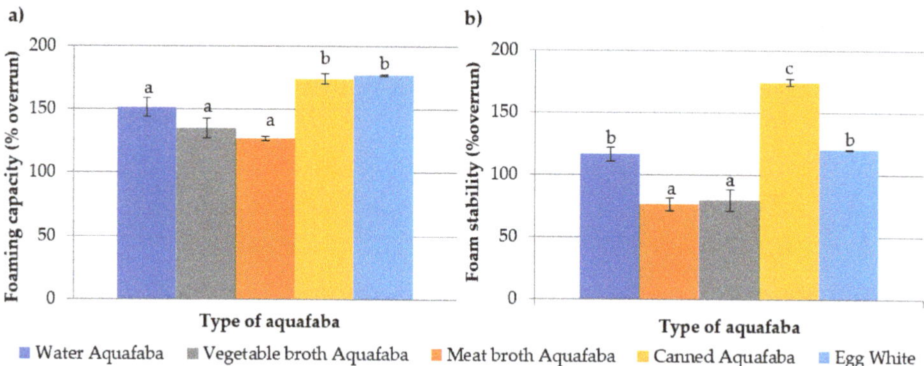

Figure 1. Foaming capacity (**a**) and foam stability (**b**) expressed in % of overrun of *Pedrosillano* chickpea aquafaba obtained with different broths and egg white. [a–c] Means with different superscripts showed significant differences ($p < 0.05$).

The highest foaming capacity was observed in the egg white sample with overrun values of around 180%. This foaming capacity is associated with the quantity (approximately 10–11 g/100 g) and the type of proteins present in its composition. These proteins allow the integration of numerous air bubbles during the shake and gave rise to foams with a large volume [25].

Regarding to the aquafaba samples, all of them presented a good foaming capacity with overrun values greater than 100%, although slight variations between them were observed. The results obtained in this study were similar to those described by Lafarga et al. [31], Aslan et al. [45] and Alsalman et al. [46] who reported foaming capacities between 120 and 324%.

The foaming capacity of aquafaba is related to the content of chickpeas proteins, which reduce the surface tension between the air droplets and the aqueous medium through their amphiphilic nature [47]. Stantiall et al. [25] found a very high positive correlation between the foaming capacity of aquafaba and its protein concentration; however, some discrepancies in the results obtained in our study were observed. The aquafaba samples with the lowest protein concentration (water and vegetable broth aquafaba with 1.21 and 1.19 g/100 g, respectively) were those with intermediated foaming capacity ($p > 0.05$). The canned aquafaba showed the highest foaming capacity with values practically equal ($p > 0.05$) to those of the egg white; this sample showed the highest values of protein content (2.48 g/100 g). The meat broth aquafaba was the next highest sample for its protein content (2.36 g/100 g); nevertheless, this type of aquafaba had the lowest foaming capacity. This inconsistency could be largely explained by the fat content of the sample (2.14 g/100 g). Immonen et al. [48] reported that fat in an aqueous medium hinders the formation of the foam by competing with proteins in the formation of stable films around the air droplets, interfers with the surfactant role of the proteins and reduces their foaming capacity [49]. The fat content in the meat broth aquafaba represented around the 30% of the total solids.

Relative to the foam stability, significant differences ($p < 0.05$) among the aquafaba samples and the egg white were observed. The most stability was found in the foam prepared with canned aquafaba (174% overrun), followed by the egg white (120% overrun) and the water aquafaba (116% overrun); the lowest values corresponded to the foams made with the meat and the vegetable broth aquafaba (80% and 76% overrun). Our results were similar to those described by other authors [23,31,45,50].

The vegetable and meat broth aquafaba samples showed a lower stabilizing capacity, with volume reductions (% overrun) which ranged from 37 to 44%. The low stability of the meat broth aquafaba foam was probably also related to the fat present in the dispersion. The fat competes with proteins for the adsorption at the air–water interface, resulting in the formation of interfacial films with poor viscoelastic properties and steric stability [51]. The low foam stability of vegetable broth aquafaba would be related to the lower carbohydrates content, which leads a lower foam viscosity and a higher salt concentration (vegetable broth aquafaba contains a 33% more ash content than water aquafaba), which might suppress the viscosity and the foam stability of the aquafaba [34].

The canned aquafaba showed the highest stabilizing properties and the foam maintained the same overrun percentage values as freshly prepared. This fact might be explained with a greater presence of complex polysaccharides among the total carbohydrates. The application of more intense heat treatments in the canned chickpeas than those used in the conventional pressure cooker, together with the use of less water for cooking, could accelerate the starch gelatinization as well as the solubilization and diffusion of pectins into the aquafaba, increasing the viscosity of the aqueous medium and the three-dimensional structure of the foam [52]. In addition, the higher protein concentration in the canned aquafaba, compared to the water aquafaba (2.48% versus 1.19% respectively), would facilitate the formation of more resistant films to the coalescence of the air bubbles of the foam.

The stability of the egg white foam was lower than the stability of the water and canned aquafaba foams. The foam volume remained stable or decreased by approximately 21% for canned and water aquafaba foams vs. 33% for egg white foams. This lower foaming stability could be explained by differences in the physicochemical properties between the egg and the aquafaba proteins. Soto-Madrid et al. [22] described a greater flexibility and hydrophobicity in chickpea proteins compared to egg proteins, which would contribute to create more stable films, avoiding the coalescence and separation of air bubbles. Furthermore, Mustafa et al. [23] also attributed the greater stability of the canned aquafaba foams, compared to egg white foams, to the lower content of sulfur-containing amino acids in the proteins, which better resist overwhipping by limiting the formation of disulfide bonds and the aggregation of aquafaba proteins. Finally, the pH can also play an important role in foam stability. Tufaro and Cappa observed a higher degree of syneresis in the egg white foam compared to aquafaba foam (47% vs. 27%) due to differences in their pH [33]. In our study, the pH values of egg white were approximately 8 and those of aquafaba was approximately 6, which is closer to the isoelectric point of their proteins. The approximation of the pH of aquafaba to the isoelectric point of the proteins contribute to reduce their surface charge, generating more resistant films around the air bubbles and thereby contribute to the formation of more stable foams as described by other authors [19,33]. This could be due to the higher molecular weight of globulins in plant proteins, which helps to form adsorption films with good rheological properties [19].

The emulsifying capacity and the emulsion stability of the different types of aquafaba and egg white are depicted in Figure 2.

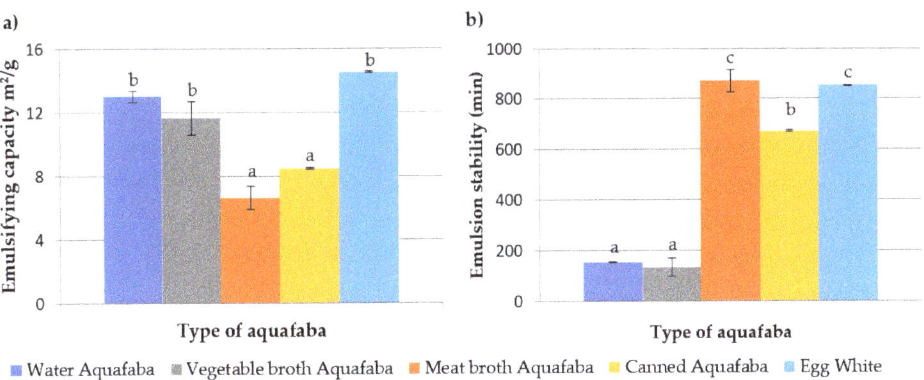

Figure 2. Emulsifying capacity, expressed as m²/g (**a**), and emulsion stability, expressed in min (**b**), of *Pedrosillano* chickpea quafaba obtained with different broths and egg white. [a-c] Means with different superscripts showed significant differences ($p < 0.05$).

In general, the emulsifying capacity of the different types of aquafaba were similar to those described by other authors, while the emulsion stability was much higher [24,27,53]. The egg white samples showed the highest emulsifying properties, followed by the water and vegetable broth aquafaba ($p > 0.05$); the emulsifying capacities of meat broth and canned aquafaba were significantly ($p < 0.05$) lower (around 40–50% lower than those described for egg white) [24,27,53]. The lower concentration of protein and complex polysaccharides in the aqueous phase of the water and the vegetable broth aquafaba could justify the higher emulsifying properties; in this way, the proteins could faster access the oil–water interface generated during the application of mechanical energy [54]. In addition, flexible proteins adsorbed at the interface at low concentrations are more easily denatured, favouring polymer–interface contact versus polymer–aqueous phase [54]. Moreover, the fibrous proteins present in the meat broth aquafaba and derived from the cooking of the meat are more rigid in their structure than the globular ones, so their access and adhesion to the interface would be limited and, consequently, a smaller number of emulsified fat droplets were formed. The lower emulsifying capacity of canned aquafaba could be related to the presence of complex polysaccharides in its composition. Complex polysaccharides could act, not only by hindering the adhesion of denatured proteins to the interface due to molecular entanglement, but also by competing with the proteins through the interface, preventing the emulsification of a greater proportion of oil due to the lower interfacial activity of the polysaccharides compared with proteins [54].

The emulsions prepared with water and vegetable broth aquafaba showed the lowest stability values, and the separation of the phases was observed after 120–150 min. In contrast, the emulsions prepared with meat broth and canned chickpeas aquafaba took approximately 700 to 800 min to separate; this time range was very similar to that obtained for the egg white emulsions. The gelatine present on the meat broth aquafaba and the complex polysaccharides in the case of canned aquafaba were able to retain large amounts of water, even at low concentrations, and act by increasing the viscosity or viscoelasticity of the continuous aqueous phase stabilizing the emulsion [9,55]. In addition, the higher protein content in both types of aquafaba would act to reinforce the interfaces by establishing interactions between the protein chains. Some authors, such as Bergenstahl and Claesson [56], reported a sequential mechanism for strengthening the film that covers the oil droplets derived from the interaction of the proteins attached to the interface with polysaccharides, which could help to stabilize the emulsion for longer.

3.3. Structure and Sensory Properties of Meringues

The alveoli percentage in cross sections of the different types of meringues made with aquafaba samples and egg white are depicted in Figure 3.

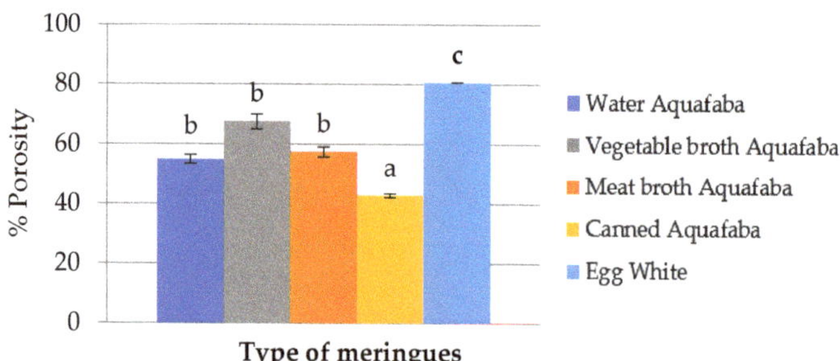

Figure 3. Mean values of alveoli degree of meringues, expressed as % of porosity, made with four types of *Pedrosillano* chickpea aquafaba and egg white. [a–c] Means with different superscripts showed significant differences ($p < 0.05$).

The highest alveoli degree was observed in the meringues made with egg white, followed by those prepared with vegetable broth, meat broth and water aquafaba. The lowest alveoli degree was observed in canned aquafaba meringues, which were 50% lower than those made with egg white. Our results were in accordance with those described by Mustafa et al. [23] who observed that sponge cakes made with aquafaba showed lower height and volume.

The high protein content of egg white allowed us to obtain a more homogeneous and stable foam, which was reinforced with the added sugar and could better support the baking conditions. During baking, the three-dimensional structure compacts when dehydrated and then stabilizes [39]. On the contrary, the smaller alveoli observed in the meringues made with aquafaba could be explained by the different behaviour of the chickpea proteins. Shim et al. [34] reported that the aquafaba contains more heat-stable proteins, which not be fully polymerized during the cooking of the meringues and can cause the foam to collapse.

Figure 4 shows the mean values of the three colour parameters determined in meringues: L^* (lightness), a^* (red-green index) and b^* (yellow-blue index). The results of the L^* parameter were similar in all the meringues (around 80); the lowest values were obtained in the meringues made with vegetable broth aquafaba, while the highest was with canned aquafaba. The parameters a^* and b^* showed a very similar behaviour, corresponding to the highest values for the vegetable and meat broth aquafaba meringues and the lowest for the canned aquafaba. The positive values of a^* and b^* were indicative of the predominance in the meringues of reddish and yellowish colours, respectively. These results were associated with, more or less, the development of non-enzymatic browning reactions mainly by the Maillard reaction but also by caramelization [23]. The aquafaba proteins together with large amounts of sugar and the application of temperatures of 90 °C for a prolonged time in baking cause a reduction in the water activity and a certain degree of hydrolysis of the sucrose glycosidic bonds, which increased the sugar reducing power and, consequently, the development of the Maillard reaction. The Maillard reaction was very similar in all types of meringues; the yellowish colorations stood out (b^* values of approximately +20) above the reddish ones, which were only more accentuated in the meringue samples made with vegetable broth aquafaba, which also showed a lower value in their luminosity.

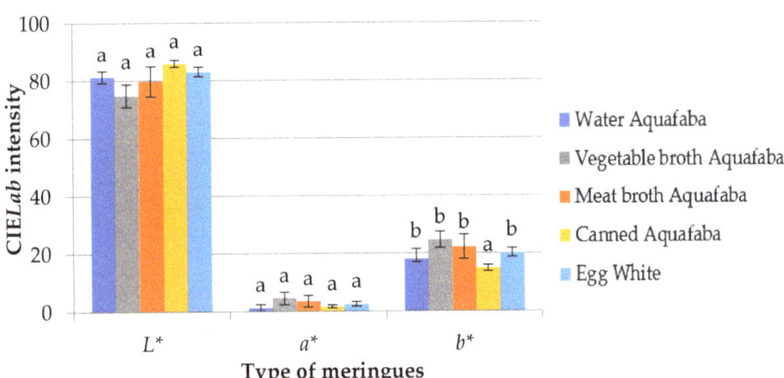

Figure 4. Mean values of CIE*Lab* parameters in meringues made with four types of *Pedrosillano* chickpea aquafaba and egg white. Luminosity (L^*), red-green (a^*), yellow-blue (b^*). [a,b] Means for each CIE*Lab* parameter with different superscripts showed significant differences ($p < 0.05$).

Lafarga et al. [31] also described no differences in the L^*, a^* and b^* values and the chroma between fresh meringues made with aquafaba and egg white. In our study, the small differences observed in the colour parameters might be associated with the baking process, probably due to irregular heat distribution inside the oven as well as the differences in the reducing sugar content of the aquafaba. Our results were similar to those described by Mustafa et al. [23] and Nguyet et al. [28] for cupcakes, although different from those described by Stantiall et al. [25] and Tufaro and Cappa [33] for meringues made with aquafaba under similar conditions. Higher luminosity values and lower red ($a*$) and yellow ($b*$) indices were observed; these differences could be influenced by the type of chickpea variety and also by the cooking conditions (temperature, time, air speed or the type of oven).

The results obtained in the texture profile analysis of baked meringues prepared with the different samples of aquafaba as well as with egg white are shown in Table 2.

Table 2. Mean values and standard deviation obtained in the texture profile analysis of meringues made with four types of *Pedrosillano* chickpea aquafaba and egg white.

	Water Aquafaba	Vegetable Broth Aquafaba	Meat Broth Aquafaba	Canned Aquafaba	Egg White
Fracturability (N)	3.76 ± 0.41 [a]	9.02 ± 2.71 [c]	12.73 ± 2.44 [d]	9.26 ± 2.79 [c]	5.23 ± 1.09 [b]
Hardness (N)	39.33 ± 4.89 [b]	49.18 ± 2.72 [c]	56.17 ± 4.83 [d]	45.28 ± 2.28 [c]	14.34 ± 1.51 [a]
Springiness	0.05 ± 0.01 [a]	0.08 ± 0.02 [ab]	0.17 ± 0.04 [b]	0.07 ± 0.01 [ab]	0.05 ± 0.02 [a]

[a–d] Means in the same row with different superscripts showed significant differences ($p < 0.05$). N = Newtons.

In all texture parameters, significant differences ($p < 0.05$) were observed between the meringues made with aquafaba and with egg white.

The fracturability values (the force necessary for the meringues collapse) were higher in the meringues made with meat broth aquafaba and showed significant differences ($p < 0.05$) with vegetable broth and canned aquafaba meringues; the water aquafaba and egg white meringues had similar values in the order of 2–3 times lower than those of the rest of the meringues. A very similar behaviour was observed in the hardness values, although in this case, all the meringues made with aquafaba showed similar results but were still significantly different ($p < 0.05$); the egg white meringues had the lowest hardness values.

The higher values of hardness and fracturability of the meringues made with aquafaba could be associated with a greater collapse of the bubbles during baking, which resulted in a greater compaction of the structure. The lower dehydration of the aquafaba proteins would contribute to this behaviour, allowing a greater association between the different protein chains, which could be reinforced by the formation of salt bonds with divalent ions released from the chickpeas during cooking. Likewise, complex polysaccharides, such as starch or pectin, and oligosaccharides solubilized during the chickpeas cooking would contribute to stabilizing the protein network of the meringue by crystallizing or forming glassy amorphous states during the cooling and causing more compact meringues, which require the application of more force to break them [47]. Finally, it should be noted that the elasticity values were practically the same and very low probably because the samples were destroyed in the first compression cycle.

Our results were very similar to those described by Stantiall et al. [25] for meringues and Aslan et al. [45] for cupcakes but differed from those described by Meurer et al. [30] and Nguyet et al. [28] who described that meringues and cupcakes made with aquafaba were less hard than those prepared with egg white.

Figure 5 shows the appearance, odour, taste and texture data of the meringues elaborated with different types of aquafaba and with egg white.

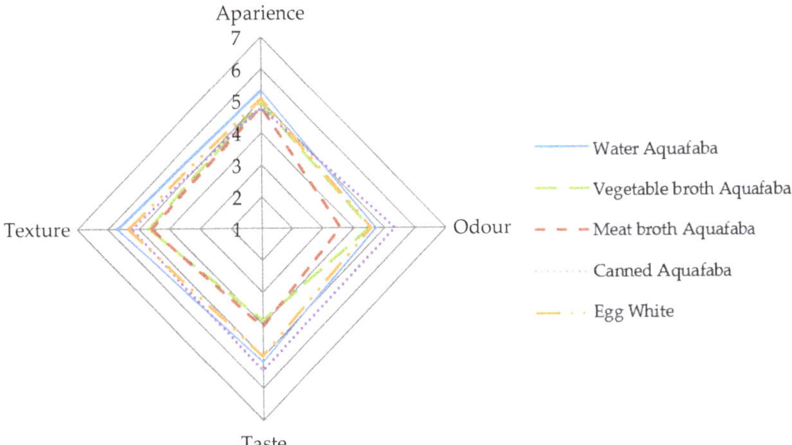

Figure 5. Average scores obtained during the sensory analysis of the different types of meringues made with *Pedrosillano* chickpea aquafaba and egg white using a 7-points hedonic scale (1 = dislike extremely; 7 = like very much).

The appearance values were practically the same in all types of meringues with average scores of approximately 5. The texture values for the egg white and canned and water aquafaba meringues (scores between 5.2 and 5.7) showed significant differences ($p < 0.05$) relative to the meat and vegetable broth aquafaba meringues (scores of approximately 4.6–4.7). The greatest differences among samples were observed in the odour and taste attributes. The meringue samples prepared with meat broth and vegetable broth aquafaba had the lowest scores (approximately 4) and were significantly different ($p < 0.05$) from the other meringues. The canned aquafaba meringues were the highest scored.

These ratings were confirmed in the global impression analysis whose results are shown in Figure 6.

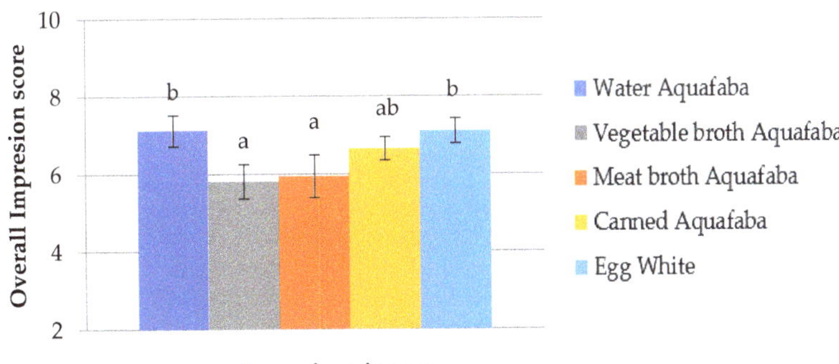

Figure 6. Overall impression scores of meringues made with *Pedrosillano* chickpea aquafaba and egg white using a rating from 1 (very bad) to 10 (very good). [ab] Means with different superscripts showed significant differences ($p < 0.05$).

The meringues prepared with egg white, water and canned aquafaba obtained the highest scores (approximately 7 points out of 10), while those made with meat broth and vegetable broth aquafaba ranged between 5.8 and 6. The sapid and aromatic components from vegetables and meat contributed to the meringues aromatic scents (enhanced during the cooking) that were qualified as "strange" by the tasters, as they were not associated with what is expected for a sweet dish; therefore, these meringues received the worst scores in smell and flavour. Despite the results obtained, we consider that this aspect could be a starting point for the use of vegetable and meat broth aquafaba in the development of savoury dishes. Finally, it should be noted that the highest flavour score of the meringues made with canned aquafaba might be related with their greater compaction because of the low alveoli degree of the dough that would allow a greater contact of the meringue with the papillae during chewing.

All the meringues made with *Pedrosillano* chickpea aquafaba showed similar sensory properties to those made with egg white. Analogous results were described by Stantiall et al. [25] and Lafarga et al. [31] who found no differences in flavour, texture and the overall impression between meringues prepared with egg white and canned chickpea aquafaba.

4. Conclusions

The ingredients added to the cooking liquid of the chickpeas, as well as the intensity of the heat treatment during the cooking, influenced the aquafaba composition, especially the proteins, carbohydrates and fat content. Broth meat aquafaba and canned chickpeas aquafaba showed the highest concentrations for these two latter parameters.

All types of aquafaba showed good foaming properties and intermediate emulsifying capacities. The presence of proteins, complex polysaccharides and fat in the aquafaba affected the foaming and emulsifying capacities, as well as the foam and emulsion stability.

All meringues made with aquafaba displayed lower alveoli degree, greater hardness and fracturability and minimal color changes after baking than the meringues prepared with egg white. Similarly, regarding the sensory analysis, all aquafaba meringues got high scores and were similar to those obtained with egg white, except for meat and vegetable broth aquafaba, which were the lowest rated.

The results obtained in this study proved that the *Pedrosillano* chickpea aquafaba could be an egg substitute in the development of different cooking recipes (sweet or savory), which involved foams or emulsions and could be an alternative for consumers with an egg allergy. The aquafaba derived from commercial canned chickpeas showed, in general, more similarities to the properties of egg white and had the highest sensory scores from the meringues prepared with it.

Author Contributions: Conceptualization: P.F.C., M.E.T.R. and J.M.F.B.; methodology: P.F.C., P.C.-F., E.R.B., B.P.G., M.E.T.R. and J.M.F.B.; formal analysis: P.F.C., P.C.-F., D.A.C. and E.R.B.; investigation: P.C.-F., B.P.G., M.E.T.R. and J.M.F.B.; data curation: P.C.-F., B.P.G., M.E.T.R. and J.M.F.B.; writing—original draft preparation: P.F.C., P.C.-F. and J.M.F.B.; writing—review and editing: P.C.-F., D.A.C., B.P.G., M.E.T.R. and J.M.F.B.; visualization: B.P.G., M.E.T.R. and J.M.F.B.; supervision. J.M.F.B. All authors have read and agreed to the published version of the manuscript.

Funding: This research received no external funding.

Data Availability Statement: Data is contained within the article.

Acknowledgments: Authors would like to thank Yolanda León and Juanjo Peréz, from the restaurant COCINANDOS (León, Spain) with a Michelin Star, for providing the seed for the development of this work and to Julio Claro for his help in carrying out some analyses. They would also like to thank Marina Abajo Delgado, production manager of Legumbres Penelas S.L., (Villarejo del Órbigo, León, Spain) for supplying the canned chickpeas for this study.

Conflicts of Interest: The authors declare no conflict of interest.

References

1. FAOSTATS. Índice de Producción de Legumbres. Available online: https://www.fao.org/faostat/es/#data (accessed on 10 December 2022).
2. Cussó, X.; Ramón, S.; Segura, G. La Transición Nutricional en la España Contemporánea: Las Variaciones en el Consumo de Pan, Patatas y Legumbres (1850–2000). *Investig. Hist. Econ.* **2007**, *3*, 69–100. [CrossRef]
3. FAO. *Legumbres un Viaje Por Todas las Regiones del Planeta Brasil, China, España, India, Marruecos, México, Pakistán, Tanzania, Turquía, USA y Las Recetas de Algunos de Los Chefs Más Prestigiosos del Mundo*; FAO: Rome, Italy, 2016.

4. Peralta, R.B.; Veas, R.E.A. Garbanzo: Usos Alternativos para Generar Valor Agregado al Descarte. 2014. Available online: https: //rdu.unc.edu.ar/bitstream/handle/11086/1808/Peralta%20_%20Veas%20-%20Garbanzo.pdf?sequence=1&isAllowed=y (accessed on 10 December 2022).
5. Boukid, F.; Gagaoua, M. Vegan Egg: A Future-Proof Food Ingredient? *Foods* **2022**, *11*, 161. [CrossRef] [PubMed]
6. Saget, S.; Costa, M.; Santos, C.S.; Vasconcelos, M.W.; Gibbons, J.; Styles, D.; Williams, M. Substitution of Beef with Pea Protein Reduces the Environmental Footprint of Meat Balls Whilst Supporting Health and Climate Stabilisation Goals. *J. Clean. Prod.* **2021**, *297*, 126447. [CrossRef]
7. Mercasa. *Alimentación en España 2021. Producción, Industria, Distribución y Consumo*, 24th ed.; Mercasa–Distribución y Consumo: Madrid, Spain, 2022.
8. The Official Aquafaba Site. Available online: http://aquafaba.com (accessed on 10 December 2022).
9. He, Y.; Meda, V.; Reaney, M.J.T.; Mustafa, R. Aquafaba, a New Plant-Based Rheological Additive for Food Applications. *Trends Food Sci. Technol.* **2021**, *111*, 27–42. [CrossRef]
10. Mustafa, R.; Reaney, M.J.T. Aquafaba, from Food Waste to a Value-Added Product. In *Food Wastes and By-Products*; Campos-Vega, R.B., Oomah, D., Vergara-Castañeda, H.A., Eds.; John Wiley & Sons, Inc.: Hoboken, NJ, USA, 2020; pp. 93–126. [CrossRef]
11. Poore, J.; Nemecek, T. Reducing Food's Environmental Impacts through Producers and Consumers. *Science* **2018**, *360*, 987–992. [CrossRef]
12. Nette, A.; Wolf, P.; Schlüter, O.; Meyer-Aurich, A. A Comparison of Carbon Footprint and Production Cost of Different Pasta Products Based on Whole Egg and Pea Flour. *Foods* **2016**, *5*, 17. [CrossRef]
13. Saget, S.; Costa, M.; Styles, D.; Williams, M. Does Circular Reuse of Chickpea Cooking Water to Produce Vegan Mayonnaise Reduce Environmental Impact Compared with Egg Mayonnaise? *Sustainability* **2021**, *13*, 4726. [CrossRef]
14. Serventi, L. *Upcycling Legume Water: From Wastewater to Food Ingredients*; Springer Nature: Cham, Switzerland, 2020; pp. 1–174. [CrossRef]
15. Regulation (EU) No 1169/2011 of the European Parlament and of the Council. *Off. J. Eur. Union* **2011**, *304*, 18–63.
16. Duan, X.; Li, M.; Ji, B.; Liu, X.; Xu, X. Effect of Fertilization on Structural and Molecular Characteristics of Hen Egg Ovalbumin. *Food Chem.* **2017**, *221*, 1340–1345. [CrossRef]
17. Cabrita, M.; Simões, S.; Álvarez-Castillo, E.; Castelo-Branco, D.; Tasso, A.; Figueira, D.; Guerrero, A.; Raymundo, A. Development of Innovative Clean Label Emulsions Stabilized by Vegetable Proteins. *Int. J. Food Sci. Technol.* **2023**, *58*, 406–422. [CrossRef]
18. Yazıcı, G.; Taspinar, T.; Ozer, M.S.; Yazıcı, G.; Taspinar, T.; Ozer, M.S. Aquafaba: A Multifunctional Ingredient in Food Production. *Foods* **2022**, *18*, 24. [CrossRef]
19. Tang, Q.; Roos, Y.H.; Miao, S. Plant Protein versus Dairy Proteins: A PH-Dependency Investigation on Their Structure and Functional Properties. *Foods* **2023**, *12*, 368. [CrossRef]
20. Montejano, J.G. Estado de Dispersión. In *Química de los Alimentos*; Pearson Educación: Atlacomulco, Mexico, 2006; pp. 547–563. ISBN 9788477384519.
21. Huang, Z.; Wang, X.; Zhang, J.; Liu, Y.; Zhou, T.; Chi, S.; Bi, C. High-Pressure Homogenization Modified Chickpea Protein: Rheological Properties, Thermal Properties and Microstructure. *J. Food Eng.* **2022**, *335*, 111196. [CrossRef]
22. Soto-Madrid, D.; Pérez, N.; Gutiérrez-Cutiño, M.; Matiacevich, S.; Zúñiga, R.N. Structural and Physicochemical Characterization of Extracted Proteins Fractions from Chickpea (*Cicer arietinum* L.) as a Potential Food Ingredient to Replace Ovalbumin in Foams and Emulsions. *Polymers* **2023**, *15*, 110. [CrossRef]
23. Mustafa, R.; He, Y.; Shim, Y.Y.; Reaney, M.J.T. Aquafaba, Wastewater from Chickpea Canning, Functions as an Egg Replacer in Sponge Cake. *Int. J. Food Sci. Technol.* **2018**, *53*, 2247–2255. [CrossRef]
24. Damian, J.J.; Huo, S.; Serventi, L. Phytochemical Content and Emulsifying Ability of Pulses Cooking Water. *Eur. Food Res. Technol.* **2018**, *244*, 1647–1655. [CrossRef]
25. Stantiall, S.E.; Dale, K.J.; Calizo, F.S.; Serventi, L. Application of Pulses Cooking Water as Functional Ingredients: The Foaming and Gelling Abilities. *Eur. Food Res. Technol.* **2017**, *244*, 97–104. [CrossRef]
26. He, Y.; Shim, Y.Y.; Mustafa, R.; Meda, V.; Reaney, M.J.T. Chickpea Cultivar Selection to Produce Aquafaba with Superior Emulsion Properties. *Foods* **2019**, *8*, 685. [CrossRef]
27. Buhl, T.F.; Christensen, C.H.; Hammershøj, M. Aquafaba as an Egg White Substitute in Food Foams and Emulsions: Protein Composition and Functional Behavior. *Food Hydrocoll.* **2019**, *96*, 354–364. [CrossRef]
28. Nguyen, T.M.N.; Nguyen, T.P.; Tran, G.B.; Le, P.T.Q. Effect of Processing Methods on Foam Properties and Application of Lima Bean (*Phaseolus lunatus* L.) Aquafaba in Eggless Cupcakes. *J. Food Process. Preserv.* **2020**, *44*, e14886. [CrossRef]
29. Raikos, V.; Hayes, H.; Ni, H. Aquafaba from Commercially Canned Chickpeas as Potential Egg Replacer for the Development of Vegan Mayonnaise: Recipe Optimisation and Storage Stability. *Int. J. Food Sci. Technol.* **2020**, *55*, 1935–1942. [CrossRef]
30. Meurer, M.C.; de Souza, D.; Ferreira Marczak, L.D. Effects of Ultrasound on Technological Properties of Chickpea Cooking Water (Aquafaba). *J. Food Eng.* **2020**, *265*, 109688. [CrossRef]
31. Lafarga, T.; Villaró, S.; Bobo, G.; Aguiló-Aguayo, I. Optimisation of the PpH and Boiling Conditions Needed to Obtain Improved Foaming and Emulsifying Properties of Chickpea Aquafaba Using a Response Surface Methodology. *Int. J. Gastron. Food Sci.* **2019**, *18*, 100177. [CrossRef]
32. Echeverria-Jaramillo, E.; Shin, W.S. Effects of Concentration Methods on the Characteristics of Spray-Dried Black Soybean Cooking Water. *Int. J. Food Sci. Technol.* **2022**, *57*, 7330–7339. [CrossRef]

33. Tufaro, D.; Cappa, C. Chickpea Cooking Water (Aquafaba): Technological Properties and Application in a Model Confectionery Product. *Food Hydrocoll.* **2023**, *136*, 108231. [CrossRef]
34. Shim, Y.Y.; Mustafa, R.; Shen, J.; Ratanapariyanuch, K.; Reaney, M.J.T. Composition and Properties of Aquafaba: Water Recovered from Commercially Canned Chickpeas. *J. Vis. Exp.* **2018**, *2018*, e56305. [CrossRef]
35. Wang, N.; Hatcher, D.W.; Tyler, R.T.; Toews, R.; Gawalko, E.J. Effect of Cooking on the Composition of Beans (*Phaseolus vulgaris* L.) and Chickpeas (*Cicer arietinum* L.). *Food Res. Int.* **2010**, *43*, 589–594. [CrossRef]
36. AOAC 920.39-1920, Fat (Crude) or Ether Extract in Animal Feed—$14.15: AOAC Official Method. Available online: http://www.aoacofficialmethod.org/index.php?main_page=product_info&products_id=1088 (accessed on 25 January 2023).
37. Association of Official Analytical Chemists (AOAC). *Official Methods of Analysis of the Association of Official's Analytical Chemists*, 17th ed.; AOAC: Arlington, VA, USA, 2003.
38. Karaca, A.C.; Low, N.; Nickerson, M. Emulsifying Properties of Chickpea, Faba Bean, Lentil and Pea Proteins Produced by Isoelectric Precipitation and Salt Extraction. *Food Res. Int.* **2011**, *44*, 2742–2750. [CrossRef]
39. Juliá Medina, M. Características Reológicas, Estructurales y Sensoriales de Panes Elaborados a Base de Harina de Chufa. Bachelor's Thesis, Universitat Politècnica de València, Valencia, Spain, 2016.
40. Molina García, M.E. Evaluación de La Preferencia del Merengue con Canela Como Muestra Patrón, Frente a Cuatro Productos Similares Producidos Artesanalmente en Portoviejo. Master's Thesis, Universidad Laica Eloy Alfaro de Manabí, Manta, Ecuador, 2008.
41. IBM Corp. *IBM SPSS Statistics for Windows, Version 26.0*; IBM Corp.: Armonk, NY, USA, 2019.
42. Bird, L.G.; Pilkington, C.L.; Saputra, A.; Serventi, L. Products of Chickpea Processing as Texture Improvers in Gluten-Free Bread. *Food Sci. Technol. Int.* **2017**, *23*, 690–698. [CrossRef]
43. Mokni Ghribi, A.; Sila, A.; Maklouf Gafsi, I.; Blecker, C.; Danthine, S.; Attia, H.; Bougatef, A.; Besbes, S. Structural, Functional, and ACE Inhibitory Properties of Water-Soluble Polysaccharides from Chickpea Flours. *Int. J. Biol. Macromol.* **2015**, *75*, 276–282. [CrossRef]
44. Gowen, A. Development of Innovative, Quick-Cook Legume Products: An Investigation of the Soaking, Cooking and Dehydration Characteristics of Chickpeas (*Cicer arietinum* L.) and Soybeans (*Glycine max* L. Merr.). Ph.D. Thesis, Technological University Dublin, Dublin, Ireland, 2006.
45. Aslan, M.; Ertaş, N. Possibility of Using "chickpea Aquafaba" as Egg Replacer in Traditional Cake Formulation. *Harran Tarım Gıda Bilim. Derg.* **2020**, *24*, 1–8. [CrossRef]
46. Alsalman, F.B.; Tulbek, M.; Nickerson, M.; Ramaswamy, H.S. Evaluation and Optimization of Functional and Antinutritional Properties of Aquafaba. *Legume Sci.* **2020**, *2*, e30. [CrossRef]
47. Ngouémazong, E.D.; Christiaens, S.; Shpigelman, A.; van Loey, A.; Hendrickx, M. The Emulsifying and Emulsion-Stabilizing Properties of Pectin: A Review. *Compr. Rev. Food Sci. Food Saf.* **2015**, *14*, 705–718. [CrossRef]
48. Immonen, M.; Chandrakusuma, A.; Hokkanen, S.; Partanen, R.; Mäkelä-Salmi, N.; Myllärinen, P. The Effect of Deamidation and Lipids on the Interfacial and Foaming Properties of Ultrafiltered Oat Protein Concentrates. *LWT* **2022**, *169*, 114016. [CrossRef]
49. Ghanimah, M.A. Functional and Technological Aspects of Whey Powder and Whey Protein Products. *Int. J. Dairy Technol.* **2018**, *71*, 454–459. [CrossRef]
50. Thi Minh Nguyet, N.; Pham Tan Quoc, L.; Gia Buu, T. Evaluation of Textural and Microstructural Properties of Vegan Aquafaba Whipped Cream from Chickpeas. *Chem. Eng. Trans.* **2021**, *83*, 421–426. [CrossRef]
51. Damodaran, S. Protein Stabilization of Emulsions and Foams. *J. Food Sci.* **2005**, *70*, R54–R66. [CrossRef]
52. Fidantsi, A.; Doxastakis, G. Emulsifying and Foaming Properties of Amaranth Seed Protein Isolates. *Colloids Surf. B Biointerfaces* **2001**, *21*, 119–124. [CrossRef]
53. Włodarczyk, K.; Zienkiewicz, A.; Szydłowska-Czerniak, A. Radical Scavenging Activity and Physicochemical Properties of Aquafaba-Based Mayonnaises and Their Functional Ingredients. *Foods* **2022**, *11*, 1129. [CrossRef]
54. Muñoz García, J.; Alfaro Rodríguez, M.C.; Zapata, I. Avances en la Formulación de Emulsiones. *Grasas Aceites* **2007**, *58*, 64–73.
55. Friberg, S.; Larsson, K.; Sjöblom, J. *Food Emulsions*, 4th ed.; Revised and Expanded; CRC Press: New York, NY, USA, 2003; ISBN 9780824746964.
56. Bergenstahl, B.A.; Claesson, P.M. Surfaces forces in emulsions. In *Food Emulsions*, 3rd ed.; Friberg, S.E., Larsson, K., Eds.; Marcel Dekker: New York, NY, USA, 1997; pp. 141–188. ISBN 0-8247-9983-6.

 *foods*

Article

Development of Healthy Vegan Bonbons Enriched with Lyophilized Peach Powder

Dasha Mihaylova [1], **Aneta Popova** [2,*], **Zhivka Goranova** [3] **and Pavlina Doykina** [2]

1   Department of Biotechnology, Technological Faculty, University of Food Technologies, 4002 Plovdiv, Bulgaria; dashamihaylova@yahoo.com
2   Department of Catering and Nutrition, Economics Faculty, University of Food Technologies, 4002 Plovdiv, Bulgaria; pavlina_doikina@abv.bg
3   Department of Food Technologies, Institute of Food Preservation and Quality-Plovdiv, Agricultural Academy, 4002 Plovdiv, Bulgaria; jivka_goranova@abv.bg
*   Correspondence: popova_aneta@yahoo.com

Abstract: Changing nutritional demands, in combination with the global trend for snacking, sets a goal for preparing food products for direct consumption with certain beneficial properties. This study was designed to investigate the quality characteristics of raw vegan bonbons enriched with lyophilized peach powder. Three types of formulations were prepared in which 10%, 20%, and 30% of lyophilized peach powder were, respectively, added. The newly developed vegan products were characterized in terms of their physical (moisture, ash, color, water activity), microbiological, and nutritional characteristics. Their antioxidant activity, flavonoid, and phenolic content were also evaluated. Considering the content of the bonbons, the reported health claims indicate that they are sources of fiber, with no added sugar, and contain naturally occurring sugars. The color measurements demonstrated similarity in the values. This study showed that there is significant potential in the production of healthy snacks for direct consumption, with beneficial properties.

Keywords: health enhancing; raw snack; health claims; healthy ingredients

Citation: Mihaylova, D.; Popova, A.; Goranova, Z.; Doykina, P. Development of Healthy Vegan Bonbons Enriched with Lyophilized Peach Powder. *Foods* **2022**, *11*, 1580. https://doi.org/10.3390/foods11111580

Academic Editors: Manuel Viuda-Martos and Lisa M. Duizer

Received: 12 April 2022
Accepted: 26 May 2022
Published: 27 May 2022

1. Introduction

Food choice may be influenced by many factors such as calories in meals, personal preferences, flavor, diet practice, etc. [1]. The eating behavior is directly linked to the food choices a person makes throughout his life cycle. The COVID-19 outbreak has reportedly led to unhealthy changes in eating patterns [2], which emphasizes the need to promote healthy habits and foods.

The effect of plant-based diets, with their diverse range of dietary patterns, has been extensively studied throughout the years [3]. They are widely associated with a reduced risk of noncommunicable diseases and early mortality [4]. Some attribute their rapid spread to the sustainability of the food production system, and the welfare of animals [5]. Whatever the primary reason, plant-based diets continue to grow in popularity.

Fruit and nuts are undoubtedly beneficial to one's dietary intake. Fruits contain a palette of vitamins, minerals, and antioxidants [6], which provide obtainable advantages from their consumption, i.e., better cardiovascular health, improved response to some diseases, weight control, protective properties, etc. [7]. Nuts are usually integrated into daily diets because of their nutritional composition and health-promoting properties [8,9]. Their increased consumption, however, may lead to weight gain because of their fat content [10].

The "Evmolpiya" peach is a native Bulgarian late-season variety with health-promoting properties. It has a rich total phenolic content and total flavonoid content [11]. Moreover, its extracts showed potency to inhibit certain enzymes (α-glucosidase, lipase, α-amylase, and acetylcholinesterase) [12]. To date, it has not been extensively studied or used as an ingredient in culinary products, which makes the "Evmolpiya" peach an interesting object of research in the field.

All population groups eat between meals, which makes healthy snacking extremely important to maintaining good health [13]. Healthy snacks comprise the recommended nutrients and are associated with positive effects on the human body [14]. Scientists have presented and characterized several varieties of raw bars—with ingredients from the Amazon [15], papaya and tomato [16], sunroot, potato, and oat [17], or with added protein [18], which proves that this area of research is trending and presents endless opportunities for healthy nutrition [19].

Thus, the aim of the current research was to characterize and present raw vegan bonbons as healthy snack alternatives in the human diet. Three types of formulations were prepared in which 10%, 20%, and 30% of lyophilized peach powder were, respectively, added. The newly developed vegan products were characterized in terms of their physical (moisture, ash, color, water activity), microbiological, and nutritional characteristics. Their antioxidant activity, flavonoid, and phenolic content were also evaluated.

2. Materials and Methods

2.1. Materials

Fresh peach samples of the "Evmolpiya" variety were provided from the Fruit Growing Institute, Plovdiv, Bulgaria. The samples were lyophilized and powdered with a Tefal GT110838 grinder. Raw nuts and dried fruit were purchased from a local "Lidl" store (Plovdiv, Bulgaria). Both nuts and fruit are produced and packaged (200 g) by Lidl Stiftung & Co. KG., Neckarsulm, Germany. The raw cocoa butter was produced and packaged by "Dragon superfoods" and purchased at a local "dm drogerie" store in Plovdiv, Bulgaria.

2.2. Preparation of Bonbons

The bonbons were prepared in laboratory conditions at the University of Food Technologies. Table 1 provides information about the percentage distribution of the ingredients used to prepare the formulations.

Table 1. Bonbon formulations: LPP—lyophilized peach powder.

Type of Bonbon	Walnut, %	Almond, %	Raisin, %	Cranberry, %	Cocoa Butter, %	LPP, %
Control	18	18	18	18	28	-
LPP10	15	15	15	15	30	10
LPP20	12	12	12	12	32	20
LPP30	9	9	9	9	34	30

The nuts and fruit were finely chopped with the use of Silver Crest chopper SMZ 260 J4 (260 W) at the turbo boost button speed for approximately 30 s. The cocoa butter was heated in a water bath in order to be poured over at the quantity needed. The ingredients were then hand-mixed until resulting in a soft plastic mass. In order to produce bonbons of similar size and weight, the soft mass was placed in a mold and after that was hand-rounded. The bonbons (Figure 1) were stored in a refrigerator upon their further usage.

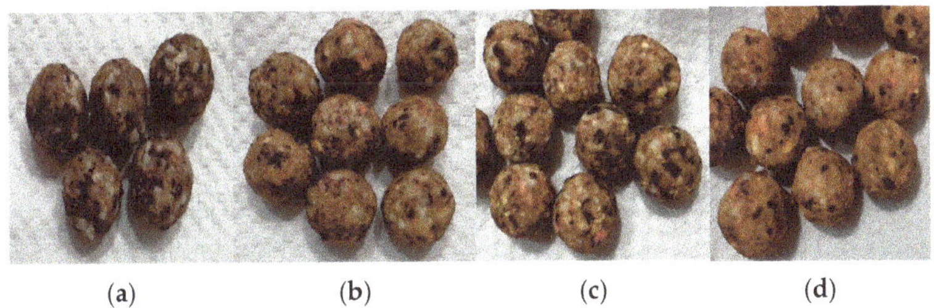

Figure 1. Bonbon formulations: (**a**) control sample; (**b**) LPP10; (**c**) LPP20; (**d**) LPP30.

2.3. Size and Weight Measurements

Weight was measured on a digital scale (KERN, EMB 1000-2). The diameter was measured with the use of a digital caliper SD-150.

2.4. Ash Content

Ash content was determined by burning in a muffle furnace according to AOAC 945.46 [20].

2.5. Moisture Content

The total moisture content of the samples was determined according to the procedure described in AACC method 44-15A [21].

2.6. Nutritional Data

The calculation method was used to determine the nutritional data. Supplier specifications for each of the ingredients (nuts, fruit, oil) were used to calculate the nutritional value of the finished products per 100 g. Information about the "Evmolpiya" peach was retrieved from previous research [11].

2.7. Color

A PCE-CSM 2 (PCE-CSM instruments, Deutschland) with a measuring aperture of 8 mm was used to analyze the color parameters. The L* (lightness; ranging from 0 to 100), a (representing red-green opponent colors), b (representing blue-yellow opponent colors), chroma (color saturation), and hue angle (color tone) were estimated.

2.8. Texture Profile Analysis (TPA)

TPA was performed by Texture Analyzer (Stable Microsystems, TAXT-2i Texture Analyzer, Godalming, UK), as described by Mazumder et al. [22]. The texture parameters (hardness, fracturability, maximum compressive force, and adhesiveness) were determined in a texture profile analysis mode (TPA) with speed before test 1.0 mm/s, trigger 5 g, speed after test 10 mm/s, voltage 60%, probe with a diameter 5 mm, distance 5 mm and specialized software "Texture Exponent".

2.9. Determination of Total Polyphenolic Content (TPC)

An extraction procedure was performed to evaluate the total polyphenolic content, total flavonoid content, and antioxidant activity of the bonbon formulations. Specifically, 5 g of each formulation was subjected to extraction with 25 mL 96% ethanol at 25 °C and 200 rpm for 2 h. The mixtures were then centrifuged at $4000\times g$ for 10 min, and the supernatant from each extraction was collected and used for analyses. The TPC was analyzed following a modified method of Kujala et al. [23], with some modifications [11]. The absorbance was measured at 765 nm and the TPC was expressed as mg gallic acid equivalents (GAE) per g dw.

2.10. Determination of Total Flavonoid Content (TFC)

The total flavonoid content was evaluated according to the method described by Kivrak et al. [24]. Results are expressed as µg QE/g dw, and quercetin (QE) was used as a standard.

2.11. Determination of Antioxidant Activity (AOA)
2.11.1. DPPH• Radical Scavenging Assay

The ability of the extracts to donate an electron and scavenge 2,2-diphenyl-1-picrylhydrazyl (DPPH) radical was determined by the slightly modified method of Brand-Williams et al. [25] as described by Mihaylova et al. [26]. The DPPH radical scavenging activity was presented as a function of the concentration of Trolox—Trolox equivalent antioxidant capacity (TEAC)—and was defined as the concentration of Trolox having equivalent antioxidant activity expressed as µM TE/g dw.

2.11.2. ABTS•+ Radical Scavenging Assay

The radical scavenging activity of the extracts against 2,2'-azino-bis(3-ethylbenzothiazoline-6-sulfonic acid) (ABTS•+) was estimated according to Re et al. [27]. The results are expressed as TEAC values (µM TE/g dw).

2.11.3. Ferric-Reducing Antioxidant Power (FRAP) Assay

The FRAP assay was carried out according to the procedure of Benzie and Strain [28]. The results are expressed as TEAC values (µM TE/g dw).

2.11.4. Cupric-Ion-Reducing Antioxidant Capacity (CUPRAC) Assay

The CUPRAC assay was carried out according to the procedure of Apak et al. [29]. Trolox was used as a standard, and the results are expressed as TEAC values (µM TE/g dw).

2.12. Water Activity

The water activity (a_w) was assessed using a Rotronic HP23-AW-A Lachen, Bassersdorf, Switzerland.

2.13. Microbial Count—Product Shelf Life

Bonbons were tested using the spread-plate method on days 1, 3, and 5 of storage to determine yeasts and molds (YM) using potato dextrose agar. Potato dextrose agar plates were incubated at 30 °C and counted after 72 h. The aerobic mesophilic microorganisms (AMM) count was evaluated according to ISO 4833-1:2013 [30] using plate count agar as a culture medium. The results are expressed as colony-forming units (CFUs)/mL.

2.14. Microscopic Imaging

The photographs of the surface of the bonbon were produced via a USB digital pocket microscope MX200-B, with a 1000× LED magnification endoscope camera and a focus range of 1–9 mm.

2.15. Statistical Analysis

MS Excel software was used for data analysis. All assays were performed in at least triplicates. Results are presented as mean ± SD (standard deviation). Relevant statistical analyses of the data were presented using one-way ANOVA and a Tukey–Kramer post hoc test ($\alpha = 0.05$), as described by Assaad et al. [31].

3. Results and Discussion

In order to characterize the bonbons, they were first evaluated in terms of their moisture and ash content, as well as their size (diameter, mm) and weight (Table 2). After that, the proximate nutritional data were gathered, which are presented in Table 3.

Table 2. Weight (g), size (mm), ash (%), and moisture (%) content of bonbons.

Bonbon Formulations	Weight, g	Diameter, mm	Ash Content, %	Moisture Content, %
Control	8.87 ± 0.60 [a]	25.15 ± 0.39 [a]	1.20 ± 0.31 [a]	7.51 ± 0.03 [a]
LPP10	8.31 ± 0.54 [a]	25.12 ± 0.67 [a]	1.44 ± 0.34 [a]	5.05 ± 0.05 [d]
LPP20	7.92 ± 0.69 [a]	24.61 ± 0.88 [a]	1.41 ± 0.08 [a]	7.07 ± 0.09 [b]
LPP30	7.86 ± 0.79 [a]	24.70 ± 0.80 [a]	1.47 ± 0.00 [a]	6.45 ± 0.06 [c]

Different letters in the same column indicate statistically significant differences ($p < 0.05$), according to ANOVA (one-way) and the Tukey test.

Table 3. Nutritional data of bonbon formulations.

Bonbon Formulations, 100 g	Proteins, g	Carbohydrates, g	Sugars, g	Fiber, g	Fat, g	Monosaturated Fats, g	ω 3, g	Energy, kcal
Control sample	7.77	29.16	25.70	4.96	50.54	18.96	1.73	612.36
LPP10	6.48	25.50	21.89	4.38	48.76	19.80	1.44	574.99
LPP20	5.18	21.87	18.09	3.80	46.98	20.64	1.15	537.63
LPP30	3.89	18.22	14.27	3.22	45.21	21.48	0.86	500.27

The different formulations had relatively the same moisture content, but LPP20 was the most similar to the control sample. When the ash content is considered, again no significant differences were observed, and all three enriched formulations had relatively analogous values. These values correspond well to the ones reported in the literature concerning the ash and moisture content of nuts [32], a major component in the formulations.

In terms of weight and size, no notable differences exist to the naked eye. Hand-made products are usually very difficult to prepare in order to make them visibly the same. Even though the products were thoroughly mixed prior to shaping, it has to be noted that different ingredients can exist in different quantities. A 160× magnification of the bonbon's surface proves this point (Figure 2). The micrographs very clearly show parts of the dried fruit, cocoa butter, and microscopic peach particles. Parts of the nuts can also be recognized.

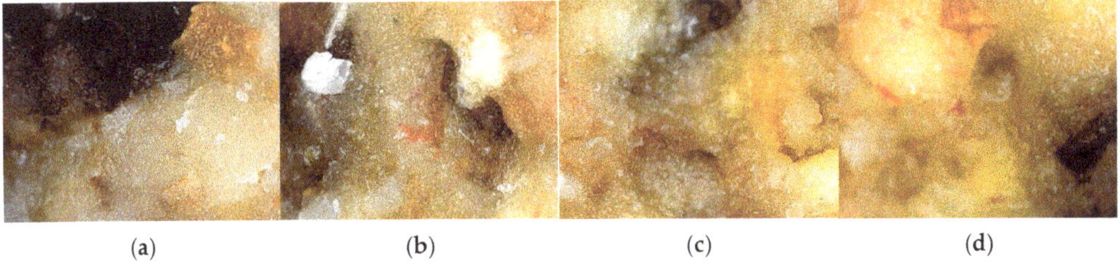

(a) (b) (c) (d)

Figure 2. Electronic microscopic photographs of bonbon formulations' surface (160×): (**a**) control sample; (**b**) LPP10; (**c**) LPP20; (**d**) LPP30.

Raw bars are versatile, ready-to-eat products that are highly appreciated for their convenience and healthy nutrients [33]. The currently prepared formulations can be related to raw bars, with the exception of their shape.

Considering the content of the bonbons, health claims reported in regulation (EC)1924/2006 [34] of the European Parliament and the Council of 20 December 2006 state that they are sources of fiber (at least 3 g fiber per 100 g), with no added sugar, and contain naturally occurring sugars.

Concerning the protein content, it ranged from 3.89 g/100 g (LPP30) to 7.77 g/100 g (control sample). The main contributors to the protein content are the nuts present in the formulations. The products cannot be considered sources of protein, as they do not provide enough to cover the daily needs of healthy individuals. For example, in order to provide enough protein in snack bars, other authors have used whey protein isolate or whey protein concentrate as an ingredient [18,35].

Regarding carbohydrates, the inclusion of more lyophilized peach powder (such as in LLP30) led to the lowering of the sugar content by 1.8 times. Carbohydrates in the formulations are mostly due to the fruit content of the formulations—dried cranberries and raisins. Compared to some of the commercially available raw bars in which the average content is 40 g/100 g, the currently presented formulations are more favorable in terms of sugar intake and energy provided from carbohydrates.

Regarding the lipid content, it varied from 45.21 to 50.54 g/100 g. This content is approximately 30% higher than the commercially available bars. The current lipid content is also higher than the one reported for high-protein bars [18]. However, compared to the control sample, it can be seen that the incorporation of lyophilized peach powder contributed to the lowering of the lipid content by 5%. It has to be noted, however, that there was a 13% increase in the monosaturated fat content in the LPP30 formulation, compared to the control.

The amount of energy obtained by consuming 100 g of the bonbon formulations will contribute significantly to the daily energy intake of healthy individuals. The energy value of formulation LPP30 was the most similar to commercially available raw bars. All of the formulations have a higher energy count than raw bars presented by other researchers [18]. It has to be mentioned that the lipid content is often excluded from the data presented in some papers, so a comparison is not always possible. This can be seen as a limitation for providing a broader discussion.

Color is an important determinant when it comes to deciding if a food is appealing and desirable to eat. Thus, the CIELAB color spectra of the studied formulations were determined, which are presented in Table 4.

Table 4. CIELAB color spectra of bonbon formulations.

Bonbon Formulations	L	a	b	c	h
Control sample	47.74 ± 4.26 [a]	9.46 ± 1.97 [a]	15.66 ± 2.33 [b]	18.44 ± 1.63 [b]	58.63 ± 8.13 [a]
LPP10	50.88 ± 2.13 [a]	11.38 ± 1.31 [a]	21.75 ± 1.19 [ab]	24.56 ± 1.57 [ab]	62.44 ± 1.89 [a]
LPP20	54.96 ± 2.65 [a]	13.12 ± 1.62 [a]	28.04 ± 2.01 [a]	30.96 ± 2.44 [a]	64.98 ± 1.55 [a]
LPP30	53.85 ± 2.50 [a]	12.42 ± 1.67 [a]	26.57 ± 1.83 [a]	29.39 ± 1.21 [a]	64.86 ± 4.22 [a]

Different letters in the same column indicate statistically significant differences ($p < 0.05$), according to ANOVA (one-way) and the Tukey test.

The color of the formulations was formed by the ingredients present in them, and no artificial colorants were used. The highest values for brightness belonged to the LPP20 formulation. The "h" value ranging from 62.44 ± 1.89 (LPP10) to 64.98 ± 1.55 (LPP20) suggested the presence of an orangey shade. This is seen very well in the micrographs in Figure 2. Natural colorants are commonly observed by lower "c" values and higher L values [36], which is supported by the current results, suggesting that the formulations are interpreted as natural in color. Some authors propose that the lightness can increase with the addition of more fruits and dry ingredients in general [37]. This proposition is supported in the current study, as the L value of the control was lower than the L value of the formulations with added peach powder.

All formulations, including the control sample, did not have significant differences in the measured parameters, which can indicate that certain consumers will perceive the formulations as similar or the same in color. However, the calculated ΔE value for formulations LPP10 (8.49), LPP20 (14.89), and LPP30 (14.99) suggests that the human eye should perceive a difference between the control sample and the newly developed ones, although LPP20 and LPP30 were indeed very similar. Limited or no data about the color spectra of other raw bars exist; thus, a comparison was not applicable. Moreover, in order to make a parallel comparison, similar ingredients should be used.

The TPC content of the formulations is presented in Table 5. The TPC showed the highest value in the control sample and the lowest value in LPP30 in which 30% of lyophilized peach powder was present. This hints at the fact that nuts and other fruit (raisins and cranberries) might contribute more to the TPC content. All of the ingredients used were proven to contain health-promoting phenolic compounds [38,39].

Table 5. Total flavonoid content (TFC) and total phenolic content (TPC) of bonbon formulations.

Bonbon Formulations	Total Flavonoid Content, µgQE/g fw	Total Phenolic Content, mgGAE/g dw
Control sample	84.64 ± 1.69 [c]	1.89 ± 0.03 [a]
LPP10	78.13 ± 1.36 [d]	1.33 ± 0.00 [c]
LPP20	117.63 ± 1.37 [a]	1.40 ± 0.04 [b]
LPP30	100.29 ± 2.55 [b]	1.21 ± 0.01 [d]

Different letters in the same column indicate statistically significant differences ($p < 0.05$), according to ANOVA (one-way) and the Tukey test.

When the TFC is considered (Table 5), LPP20 was the formulation with the highest values. Here, again, the heterogenic distribution of the ingredients found in the bonbon might lead to the established results, since all nuts and fruit parts of the formulation were found to contain certain flavonoids. Dried fruits, in general, are valuable sources of bioactive compounds, i.e., flavonoids [40].

A strong association between polyphenols and antioxidant properties exists [41–44]. The formulations were subjected to antioxidant analysis using ABTS, DPPH, FRAP, and CUPRAC methods in order to gain a better understanding of their antioxidant activity (Figure 3).

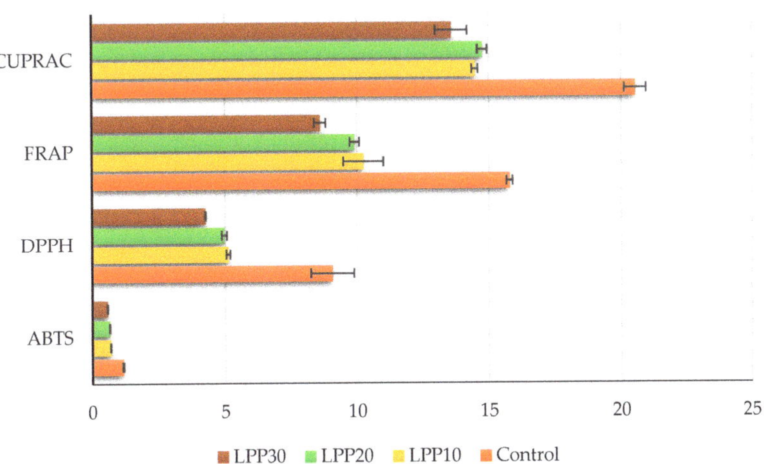

Figure 3. Antioxidant properties of bonbon formulations: ABTS—mMTE/g dw; DPPH, FRAP, CUPRAC—µMTE/g dw.

As seen in Figure 3, the control sample showed the most promising results in all assays. The highest values were reported in the ABTS assay (0.57–1.15 mMTE/g dw). All formulations containing lyophilized peach powder showed similar results, which may lead to the conclusion that the percentage incorporated does not influence the overall antioxidant activity.

Texture is an important multi-parameter property that is commonly used for food quality control [45]. Table 6 provides a visual presentation of the formulations' texture profile analysis on days 1 and 5 of their production, in terms of their hardness, fracturability, maximum compressive force (MCF), and adhesiveness. Variations in the shape and size of the ingredients used for the formulations, as well as the interaction they initiate in their preparation, can justify the relatively high observed standard deviations for some of the parameters assessed by the texture analyzer [46].

Table 6. Texture profile analysis of bonbon formulations.

Bonbon Formulations	Hardness/MCF, N		Fracturability, N		Adhesiveness, J	
	Day 1	**Day 5**	**Day 1**	**Day 5**	**Day 1**	**Day 5**
Control	31.14 ± 1.96 [a]	38.03 ± 3.11 [a]	16.10 ± 2.26 [a]	18.02 ± 0.51 [a]	0.18 ± 0.05 [a]	0.36 ± 0.09 [a]
LPP10	33.89 ± 4.79 [b]	42.80 ± 2.80 [b]	19.88 ± 1.92 [a]	19.46 ± 1.28 [a]	0.32 ± 0.03 [b]	0.73 ± 0.07 [b]
LPP20	41.90 ± 2.93 [c]	46.81 ± 0.62 [c]	22.25 ± 1.22 [ab]	28.42 ± 6.47 [b]	0.35 ± 0.04 [b]	1.14 ± 0.63 [c]
LPP30	48.94 ± 1.79 [d]	54.02 ± 2.94 [d]	24.91 ± 2.27 [c]	30.50 ± 1.39 [b]	0.38 ± 0.03 [c]	0.98 ± 0.15 [d]

Different letters in the same column indicate statistically significant differences ($p < 0.05$), according to ANOVA (one-way) and the Tukey test.

The hardness of the bonbon formulations increased progressively for 5 days of storage. The presence of more sugars usually causes increased hardness [47]. This is not supported by the current results according to which LPP30 had the least amount of sugar, and the control sample had the most. The highest value of hardness was recorded for LPP30, which contained higher amounts of lyophilized peach powder and cocoa butter. The lowest level of hardness was recorded in the control sample, in which no peach powder was present. Moisture migration may be responsible for the increased hardness, because it is induced by the formation of bonds between sugars and proteins, and moisture acts as a plasticizer and reduces the formation of this bond [48]. Thus the less the moisture present in the sample the more the hardness values. This is seen very well in formulation LPP10 in which there is the least amount of moisture (%), compared to the other formulations, and the greatest change in the hardness for 5 days of storage—1.3 times.

Hardness, fracturability, and MCF were reported separately, as each sample behaved differently during compression. On day 1, all samples disintegrated prior to 60% stress but retained MCF. On day 5, all samples required more force to initiate fracture than compression at 60% stress. The force required to break was significantly affected by the ingredients used.

LPP30 remained more adhesive than the control on day 1. An increase in the adhesion of all samples during storage was observed. The greatest change was recorded for LPP20. The smallest change in adhesion was noted in the control sample—50%.

Water activity is an important factor that is directly linked to the shelf life of food products. The water activity of the bonbons was evaluated on days 1 and 5 of their production, and the established results are presented in Table 7.

Table 7. Water activity (a_w) of bonbon formulations.

Bonbon Formulations	Water Activity, a_w	
	Day 1	**Day 5**
Control sample	0.559 ± 0.007 [c]	0.546 ± 0.06 [c]
LPP10	0.503 ± 0.009 [b]	0.496 ± 0.06 [b]
LPP20	0.492 ± 0.003 [a]	0.482 ± 0.003 [b]
LPP30	0.468 ± 0.013 [a]	0.458 ± 0.013 [a]

Different letters in the same column indicate statistically significant differences ($p < 0.05$), according to ANOVA (one-way) and the Tukey test.

A tendency toward reduced water activity during bonbon storage was observed. A higher value for water activity in the control sample may be due to its higher moisture content. Neves [49] reports that fruit bars with added protein, high in carbohydrates, have higher water activity due to the higher moisture content, which decreases during storage. Additionally, the a_w of the bonbons varied depending on the components and the storage period. According to Silva et al. [50], foods with intermediate humidity usually have an

a_w from 0.9 to 0.6, which is low enough to keep the product from microbial spoilage and ensure its stability. Water activity in the range of 0.65 to 0.75 contributes to shorter food life due to intermediate humidity levels. Outside this range, products can be stored for a longer period of time [51]. Similarly, the control sample and formulations LPP20 and LPP30 showed higher reductions in water activity during 5 days of storage.

The currently stated values correspond well to the ones reported for cereal bars, ranging from 0.557 to 0.597 [52]. Other authors [53] also document similar a_w values for sweet-cherry, almond, and honey snack bars (0.467–0.508). This suggests that the developed formulations are favorable in terms of microbial growth inhibition.

The microbial load of the studied formulations is presented in Table 8 and Figure 4. All formulations can be considered safe for consumption, although the control sample is the most contaminated one during storage. This indefinitely hints that the introduced lyophilized peach powder contributes well to several quality parameters including microbial load.

Table 8. Microbial count of bonbon formulations: YM—yeasts and molds; AMM—aerobic mesophilic microorganisms.

Bonbon Formulations	YM, CFU/mL			AMM, CFU/mL		
	Day 1	Day 3	Day 5	Day 1	Day 3	Day 5
Control sample	1500	2100	3600	1050	3400	25,000
LPP10	1150	2200	1450	400	1000	2250
LPP20	500	700	600	100	950	1000
LPP30	1300	1000	500	600	200	1100

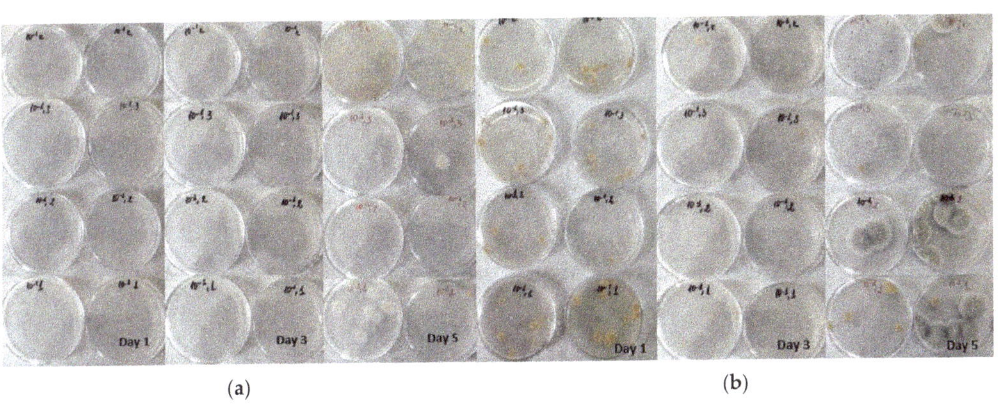

Figure 4. View of microbial growth in bonbon formulations: (a) AMM; (b) YM.

On day one, the highest plate count was noted in the control sample, and the lowest value in LPP20, while the mold count of the formulations varied from 500 to 1500 CFU/mL. The observed differences are most probably due to the difference in major ingredients part of the recipe.

The inclusion of a higher amount of lyophilized peach powder (LPP20 and LPP30) led to a decreased plate count on day 3; in addition, on day 5, a plate count similar to the control on day 1 was observed. This could be due to the small particles of peach powder drawing moisture from the product and leaving less water available for microbial activity [54] and to the antimicrobial properties of the "Evmolpiya" peach itself. Microorganisms are mostly neutrophilic and cannot grow at less than 4.5 pH and 0.8 water activity [35]. Lower water activity, in general, helps to prevent the proliferation of microorganisms [55]. Fruits usually have acidic pH and contribute well in this regard; the pH registered for the "Evmolpiya" variety is 3.65 [56]. Furthermore, the reported water activity for all formulations was below 0.8.

There are various factors that influence mold growth, including water activity, relative humidity, temperature, pH, and storage time. Mold count decreased for all formulations with lyophilized peach powder on day 5. This may be induced by the low a_w, bearing in mind that the survival of mold at lower a_w depends on various factors, i.e., nutrient availability, temperature, and pH [55]. Sugar concentrations also aid in the inhibition of mold growth in the formulations [57].

4. Conclusions

The results of this research revealed the possibility of developing raw vegan bonbons with the addition of lyophilized peach powder.

Considering the content of the bonbons, the reported health claims indicate that they are sources of fiber, with no added sugar, and contain naturally occurring sugars. Color measurements demonstrated similarity in the values. When it comes to texture, the hardness of the bonbon formulations increased progressively for 5 days of storage. A_w decreased, and the microbial load showed that the lyophilized peach powder as an ingredient has a favorable influence on the plate count and mold growth.

The TPC results showed the highest value in the control sample and the lowest value in LPP30. When the TFC is considered, LPP20 was the formulation with the highest values. All formulations containing lyophilized peach powder showed similar AOA results, which may lead to the conclusion that the percentage incorporated does not influence the overall antioxidant activity.

The newly developed bonbon formulations can be used as a quick snack throughout the day or as enrichment to one's daily healthy meal plan. Further research can pave a way for the incorporation of protein-added ingredients in these formulations.

Author Contributions: Conceptualization, A.P. and D.M.; methodology, A.P., D.M. and Z.G.; software, D.M. and Z.G.; validation, A.P., D.M. and Z.G.; formal analysis, A.P., D.M., Z.G. and P.D.; investigation, A.P. and D.M.; resources, A.P. and D.M.; data curation, A.P. and D.M.; writing—original draft preparation, A.P., D.M., Z.G. and P.D.; writing—review and editing, A.P. and D.M.; visualization, A.P.; supervision, A.P. and D.M.; project administration, D.M.; funding acquisition, D.M. All authors have read and agreed to the published version of the manuscript.

Funding: This research was supported by the Bulgarian National Science Fund, project No. КП-06-H37/23. The APC was provided by the same project.

Institutional Review Board Statement: Not applicable.

Informed Consent Statement: Not applicable.

Data Availability Statement: The data presented in this study are available on request from the corresponding author.

Acknowledgments: The authors would like to express their gratitude to Argir Zhivondov from the Fruit Growing Institute, Plovdiv (Bulgaria), and his team for kindly providing the peach variety "Evmolpiya" used to prepare the formulations, object of analysis.

Conflicts of Interest: The authors declare no conflict of interest.

References

1. Rothman, A.J.; Sheeran, P.; Wood, W. Reflective and Automatic Processes in the Initiation and Maintenance of Dietary Change. *Ann. Behav. Med.* **2009**, *38*, 4–17. [CrossRef] [PubMed]
2. González-Monroy, C.; Gómez-Gómez, I.; Olarte-Sánchez, C.M.; Motrico, E. Eating Behaviour Changes during the COVID-19 Pandemic: A Systematic Review of Longitudinal Studies. *Int. J. Environ. Res. Public Health* **2021**, *18*, 11130. [CrossRef] [PubMed]
3. Craig, W.J.; Mangels, A.R.; Fresán, U.; Marsh, K.; Miles, F.L.; Saunders, A.V.; Haddad, E.H.; Heskey, C.E.; Johnston, P.; Larson-Meyer, E.; et al. The Safe and Effective Use of Plant-Based Diets with Guidelines for Health Professionals. *Nutrients* **2021**, *13*, 4144. [CrossRef] [PubMed]

4. World Health Organization European Office for the Prevention and Control of Noncommunicable Diseases. *Plant-Based Diets and Their Impact on Health, Sustainability and the Environment: A Review of the Evidence: WHO European Office for the Prevention and Control of Noncommunicable Diseases*; World Health Organization, Regional Office for Europe: Copenhagen, Denmark, 2021.

5. Kamiński, M.; Skonieczna-Żydecka, K.; Nowak, J.K.; Stachowska, E. Global and local diet popularity rankings, their secular trends, and seasonal variation in Google Trends data. *Nutrition* **2020**, *79–80*, 110759. [CrossRef] [PubMed]

6. Amao, I. Health Benefits of Fruits and Vegetables: Review from Sub-Saharan Africa. In *Vegetables: Importance of Quality Vegetables to Human Health*; IntechOpen: Rijeka, Croatia, 2018; pp. 33–53, Chapter 3; ISBN 978-1-78923-507-4.

7. Slavin, J.L.; Lloyd, B. Health Benefits of Fruits and Vegetables. *Adv. Nutr. Int. Rev. J.* **2012**, *3*, 506–516. [CrossRef] [PubMed]

8. Mead, M.L.; Hill, A.; Carter, S.; Coates, A. The Effect of Nut Consumption on Diet Quality, Cardiometabolic and Gastrointestinal Health in Children: A Systematic Review of Randomized Controlled Trials. *Int. J. Environ. Res. Public Health* **2021**, *18*, 454. [CrossRef]

9. Pradhan, C. Nuts as Dietary Source of Fatty Acids and Micro Nutrients in Human Health. In *Nuts and Nut Products in Human Health and Nutrition*; Peter, N., Ed.; IntechOpen: Rijeka, Croatia, 2021; Chapter 3; ISBN 978-1-78985-511-1.

10. Vadivel, V.; Kunyanga, C.N.; Biesalski, H.K. Health benefits of nut consumption with special reference to body weight control. *Nutrition* **2012**, *28*, 1089–1097. [CrossRef]

11. Mihaylova, D.; Popova, A.; Desseva, I.; Manolov, I.; Petkova, N.; Vrancheva, R.; Peltekov, A.; Slavov, A.; Zhivondov, A. Comprehensive Evaluation of Late Season Peach Varieties (Prunus persica L.): Fruit Nutritional Quality and Phytochemicals. *Molecules* **2021**, *26*, 2818. [CrossRef]

12. Mihaylova, D.; Desseva, I.; Popova, A.; Dincheva, I.; Vrancheva, R.; Lante, A.; Krastanov, A. GC-MS Metabolic Profile and α-Glucosidase-, α-Amylase-, Lipase-, and Acetylcholinesterase-Inhibitory Activities of Eight Peach Varieties. *Molecules* **2021**, *26*, 4183. [CrossRef]

13. Potter, M.; Vlassopoulos, A.; Lehmann, U. Snacking Recommendations Worldwide: A Scoping Review. *Adv. Nutr. Int. Rev. J.* **2018**, *9*, 86–98. [CrossRef]

14. Barnes, T.L.; French, S.A.; Harnack, L.J.; Mitchell, N.R.; Wolfson, J. Snacking Behaviors, Diet Quality, and Body Mass Index in a Community Sample of Working Adults. *J. Acad. Nutr. Diet.* **2015**, *115*, 1117–1123. [CrossRef] [PubMed]

15. Dos Prazeres, I.C.; Domingues, A.F.N.; Campos, A.P.R.; Carvalho, A.V. Elaboration and characterization of snack bars made with ingredients from the Amazon. *Acta Amaz.* **2017**, *47*, 103–110. [CrossRef]

16. Ahmad, S.; Vashney, A.; Srivasta, P. Quality Attributes of Fruit Bar Made from Papaya and Tomato by Incorporating Hydrocolloids. *Int. J. Food Prop.* **2005**, *8*, 89–99. [CrossRef]

17. Abedelmaksoud, T.G.; Smuda, S.S.; Altemimi, A.B.; Mohamed, R.M.; Pratap-Singh, A.; Ali, M.R. Sunroot snack bar: Optimization, characterization, consumer perception, and storage stability assessment. *Food Sci. Nutr.* **2021**, *9*, 4394–4407. [CrossRef]

18. Szydłowska, A.; Zielińska, D.; Łepecka, A.; Trząskowska, M.; Neffe-Skocińska, K.; Kołożyn-Krajewska, D. Development of Functional High-Protein Organic Bars with the Addition of Whey Protein Concentrate and Bioactive Ingredients. *Agriculture* **2020**, *10*, 390. [CrossRef]

19. Orrego, C.; Salgado, N.; Botero, C.A. Developments and Trends in Fruit Bar Production and Characterization. *Crit. Rev. Food Sci. Nutr.* **2013**, *54*, 84–97. [CrossRef]

20. Cunniff, P.; Washington, D. Official methods of analysis of aoac international. *J. AOAC Int.* **1997**, *80*, 127A.

21. Analysis, A.A.M. of AACC Method 44-15.02. Moisture—Air-Oven Methods. Available online: http://methods.aaccnet.org/summaries/44-15-02.aspx. (accessed on 20 February 2022).

22. Habiba, U.; Robin, A.; Hasan, M.; Toma, M.A.; Akhter, D.; Mazumder, A.R. Nutritional, textural, and sensory quality of bars enriched with banana flour and pumpkin seed flour. *Foods Raw Mater.* **2021**, *9*, 282–289. [CrossRef]

23. Kujala, T.S.; Loponen, J.M.; Klika, K.D.; Pihlaja, K. Phenolics and Betacyanins in Red Beetroot (*Beta vulgaris*) Root: Distribution and Effect of Cold Storage on the Content of Total Phenolics and Three Individual Compounds. *J. Agric. Food Chem.* **2000**, *48*, 5338–5342. [CrossRef]

24. Kivrak, I.; Duru, M.E.; Öztürk, M.; Mercan, N.; Harmandar, M.; Topçu, G. Antioxidant, anticholinesterase and antimicrobial constituents from the essential oil and ethanol extract of Salvia potentillifolia. *Food Chem.* **2009**, *116*, 470–479. [CrossRef]

25. Brand-Williams, W.; Cuvelier, M.E.; Berset, C. Use of a free radical method to evaluate antioxidant activity. *LWT Food Sci. Technol.* **1995**, *28*, 25–30. [CrossRef]

26. Mihaylova, D.; Lante, A.; Krastanov, A. Total phenolic content, antioxidant and antimicrobial activity ofHaberlea rhodopensisextracts obtained by pressurized liquid extraction. *Acta Aliment.* **2015**, *44*, 326–332. [CrossRef]

27. Re, R.; Pellegrini, N.; Proteggente, A.; Pannala, A.; Yang, M.; Rice-Evans, C. Antioxidant activity applying an improved ABTS radical cation decolorization assay. *Free Radic. Biol. Med.* **1999**, *26*, 1231–1237. [CrossRef]

28. Benzie, I.F.F.; Strain, J.J.B.T.-M. Ferric reducing/antioxidant power assay: Direct measure of total antioxidant activity of biological fluids and modified version for simultaneous measurement of total antioxidant power and ascorbic acid concentration. In *Oxidants and Antioxidants Part A*; Academic Press: Cambridge, MA, USA, 1999; Volume 299, pp. 15–27; ISBN 0076-6879.

29. Apak, R.; Özyürek, M.; Karademir Çelik, S.; Güçlü, K. CUPRAC Method 2004. *JAFC* **2004**, *52*, 7970–7981. [CrossRef] [PubMed]

30. *ISO 4833-1:2013*; Microbiology of Food and Animal Feeding Stuffs—Horizontal Method for the Enumeration of Microorganisms—Colony-Count Technique at 30 °C. International Organization for Standardization: London, UK, 2013.

31. Assaad, H.I.; Zhou, L.; Carroll, R.J.; Wu, G. Rapid publication-ready MS-Word tables for one-way ANOVA. *SpringerPlus* **2014**, *3*, 474. [CrossRef]
32. Memon, N.N. Nutritional Characteristics (Fatty Acid Profile, Proximate Composition and Dietary Feature) of Selected Nuts Available in Local Market. *Pak. J. Anal. Environ. Chem.* **2019**, *20*, 39–46. [CrossRef]
33. Constantin, O.E. Functional Properties of Snack Bars. In *Functional Foods*; Lagouri, D.I.I.E.-V., Ed.; IntechOpen: Rijeka, Croatia, 2019; pp. 1–14, Chapter 4; ISBN 978-1-83881-150-1.
34. EUR-Lex—32006R1924—EN—EUR-Lex. Available online: https://eur-lex.europa.eu/legal-content/en/ALL/?uri=CELEX% 3A32006R1924 (accessed on 21 February 2022).
35. Jabeen, S.; Huma, N.; Sameen, A.; Zia, M.A. Formulation and characterization of protein-energy bars prepared by using dates, apricots, cheese and whey protein isolate. *Food Sci. Technol.* **2021**, *41*, 197–207. [CrossRef]
36. Arocas, A.; Varela, P.; González-Miret, M.L.; Salvador, A.; Heredia, F.J.; Fiszman, S.M. Differences in Colour Gamut Obtained with Three Synthetic Red Food Colourants Compared with Three Natural Ones: pH and Heat Stability. *Int. J. Food Prop.* **2013**, *16*, 766–777. [CrossRef]
37. Srebernich, S.M.; Gonçalves, G.M.S.; De Cássia Salvucci Celeste Ormenese, R.; Ruffi, C.R.G. Physico-chemical, sensory and nutritional characteristics of cereal bars with addition of acacia gum, inulin and sorbitol. *Food Sci. Technol.* **2016**, *36*, 555–562. [CrossRef]
38. Blumberg, J.B.; Camesano, T.A.; Cassidy, A.; Kris-Etherton, P.; Howell, A.; Manach, C.; Ostertag, L.M.; Sies, H.; Skulas-Ray, A.; Vita, J.A. Cranberries and Their Bioactive Constituents in Human Health. *Adv. Nutr.* **2013**, *4*, 618–632. [CrossRef]
39. Schuster, M.J.; Wang, X.; Hawkins, T.; Painter, J.E. A Comprehensive review of raisins and raisin components and their relationship to human health. *J. Nutr. Health* **2017**, *50*, 203–216. [CrossRef]
40. Średnicka-Tober, D.; Kazimierczak, R.; Ponder, A.; Hallmann, E. Biologically Active Compounds in Selected Organic and Conventionally Produced Dried Fruits. *Foods* **2020**, *9*, 1005. [CrossRef] [PubMed]
41. Dobrinas, S.; Soceanu, A.; Popescu, V.; Popovici, I.C.; Jitariu, D. Relationship between Total Phenolic Content, Antioxidant Capacity, Fe and Cu Content from Tea Plant Samples at Different Brewing Times. *Processes* **2021**, *9*, 1311. [CrossRef]
42. Osman, M.O.; Mahmoud, G.I.; Shoman, S. Correlation Between Total Phenols Content, Antioxidant Power and Cytotoxicity. *Biointerface Res. Appl. Chem.* **2020**, *11*, 10640–10653. [CrossRef]
43. Ly, H.T.; Le, V.K.T.; Le, V.M.; Phan, C.S.; Vo, C.T.; Nguyen, M.L.; Phan, T.A.D. Phytochemical analysis and correlation of total polyphenol content and antioxidant properties of Symplocos cochinchinensis leaves. *Minist. Sci. Technol. Vietnam* **2022**, *64*, 43–48. [CrossRef]
44. Kilicgun, H.; Altıner, D.; Kılıçgün, H. Correlation between antioxidant effect mechanisms and polyphenol content of Rosa canina. *Pharmacogn. Mag.* **2010**, *6*, 238–241. [CrossRef]
45. Szczesniak, A.S. Texture is a sensory property. *Food Qual. Prefer.* **2002**, *13*, 215–225. [CrossRef]
46. Ribeiro, A. Elaboração de uma barra de cereal de quinoa e suas propriedades sensoriais e nutricionais. *Aliment. E Nutr. Araraquara* **2011**, *22*, 63–69.
47. Samakradhamrongthai, R.S.; Jannu, T.; Renaldi, G. Physicochemical properties and sensory evaluation of high energy cereal bar and its consumer acceptability. *Heliyon* **2021**, *7*, e07776. [CrossRef]
48. Rao, Q.; Kamdar, A.K.; Labuza, T.P. Storage Stability of Food Protein Hydrolysates—A Review. *Crit. Rev. Food Sci. Nutr.* **2013**, *56*, 1169–1192. [CrossRef]
49. das Neves, J.S.C. Physical Stability of High Protein Bars during Shelf-Life. 2016. Available online: https://fenix.tecnico.ulisboa.pt/downloadFile/1689244997256129/Physical%20Stability%20of%20High%20Protein%20Bars%20during%20Shelf-life.pdf (accessed on 25 May 2022).
50. Da Silva, E.P.; Siqueira, H.H.; Damiani, C.; Boas, E.V.D.B.V. Physicochemical and sensory characteristics of snack bars added of jerivá flour (*Syagrus romanzoffiana*). *Food Sci. Technol.* **2016**, *36*, 421–425. [CrossRef]
51. Loveday, S.M.; Hindmarsh, J.P.; Creamer, L.K.; Singh, H. Physicochemical changes in intermediate-moisture protein bars made with whey protein or calcium caseinate. *Food Res. Int.* **2010**, *43*, 1321–1328. [CrossRef]
52. Agbaje, R.; Hassan, C.Z.; Arifin, N.; Abdul Rahman, A.; Faujan, N.H. Development and physico-chemical analysis of granola formulated with puffed glutinous rice and selected dried Sunnah foods. *Int. Food Res. J.* **2016**, *23*, 498–506.
53. Pintado, C.M.; Veloso, A.; Maria, Z.; Silveira, A.; Beato, H.; de Andrade, L.P.; Delgado, F. New flavour bars with cherry, almond and honey. *Emir. J. Food Agric.* **2020**, *32*, 857–863. [CrossRef]
54. Kang, J.; Tang, S.; Liu, R.H.; Wiedmann, M.; Boor, K.; Bergholz, T.; Wang, S. Effect of Curing Method and Freeze-Thawing on Subsequent Growth of Listeria monocytogenes on Cold-Smoked Salmon. *J. Food Prot.* **2012**, *75*, 1619–1626. [CrossRef]
55. Rahman, M.S. pH in Food Preservation. In *Handbook of Food Preservation*; Rahman, M.S., Ed.; CRC Press: Boca Raton, FL, USA, 2007; pp. 289–295; ISBN 9780429191084.
56. Mihaylova, D.; Popova, A.; Desseva, I.; Petkova, N.; Stoyanova, M.; Vrancheva, R.; Slavov, A.; Slavchev, A.; Lante, A. Comparative Study of Early- and Mid-Ripening Peach (*Prunus persica* L.) Varieties: Biological Activity, Macro-, and Micro- Nutrient Profile. *Foods* **2021**, *10*, 164. [CrossRef]
57. Mohd-Zaki, Z.; Bastidas-Oyanedel, J.-R.; Lu, Y.; Hoelzle, R.; Pratt, S.; Slater, F.R.; Batstone, D.J. Influence of pH Regulation Mode in Glucose Fermentation on Product Selection and Process Stability. *Microorganisms* **2016**, *4*, 2. [CrossRef]

Article

Lavandula pedunculata Polyphenol-Rich Extracts Obtained by Conventional, MAE and UAE Methods: Exploring the Bioactive Potential and Safety for Use a Medicine Plant as Food and Nutraceutical Ingredient

Ana A. Vilas-Boas [1], Ricardo Goméz-García [1,2], Manuela Machado [1], Catarina Nunes [3], Sónia Ribeiro [3], João Nunes [3], Ana L. S. Oliveira [1] and Manuela Pintado [1,*]

1   CBQF—Centro de Biotecnologia e Química Fina—Laboratório Associado, Escola Superior de Biotecnologia, Universidade Católica Portuguesa, Rua Diogo Botelho 1327, 4169-005 Porto, Portugal; avboas@ucp.pt (A.A.V.-B.); rgarcia@ucp.pt (R.G.-G.); mmachado@ucp.pt (M.M.); asoliveira@ucp.pt (A.L.S.O.)
2   Centro de Investigación e Innovación Científica y Tecnológica—CIICYT, Universidad Autónoma de Coahuila, Saltillo 25280, Coahuila, Mexico
3   Association BLC3—Technology and Innovation Campus, Centre Bio R&D Unit, Senhora da Conceição, 3045-155 Oliveira do Hospital, Portugal; catarina.nunes@blc3.pt (C.N.); sonia.ribeiro@blc3.pt (S.R.); joao.nunes@blc3.pt (J.N.)
*   Correspondence: mpintado@ucp.pt

Citation: Vilas-Boas, A.A.; Goméz-García, R.; Machado, M.; Nunes, C.; Ribeiro, S.; Nunes, J.; Oliveira, A.L.S.; Pintado, M. *Lavandula pedunculata* Polyphenol-Rich Extracts Obtained by Conventional, MAE and UAE Methods: Exploring the Bioactive Potential and Safety for Use a Medicine Plant as Food and Nutraceutical Ingredient. *Foods* 2023, 12, 4462. https://doi.org/10.3390/foods12244462

Academic Editors: Sidonia Martinez and Javier Carballo

Received: 10 November 2023
Revised: 29 November 2023
Accepted: 2 December 2023
Published: 13 December 2023

Abstract: Nowadays, plant-based bioactive compounds (BCs) are a key focus of research, supporting sustainable food production and favored by consumers for their perceived safety and health advantages over synthetic options. *Lavandula pedunculata* (LP) is a Portuguese, native species relevant to the bioeconomy that can be useful as a source of natural BCs, mainly phenolic compounds. This study compared LP polyphenol-rich extracts from conventional maceration extraction (CE), microwave and ultrasound-assisted extraction (MAE and UAE). As a result, rosmarinic acid (58.68–48.27 mg/g DE) and salvianolic acid B (43.19–40.09 mg/g DE) were the most representative phenolic compounds in the LP extracts. The three methods exhibited high antioxidant activity, highlighting the ORAC (1306.0 to 1765.5 mg Trolox equivalents (TE)/g DE) results. In addition, the extracts obtained with MAE and CE showed outstanding growth inhibition for *B. cereus*, *S. aureus*, *E. coli*, *S. enterica* and *P. aeruginosa* (>50%, at 10 mg/mL). The MAE extract showed the lowest IC_{50} (0.98 mg DE/mL) for angiotensin-converting enzyme inhibition and the best results for α-glucosidase and tyrosinase inhibition (at 5 mg/mL, the inhibition was 87 and 73%, respectively). The LP polyphenol-rich extracts were also safe on caco-2 intestinal cells, and no mutagenicity was detected. The UAE had lower efficiency in obtaining LP polyphenol-rich extracts. MAE equaled CE's efficiency, saving time and energy. LP shows potential as a sustainable raw material, allowing diverse extraction methods to safely develop health-promoting food and nutraceutical ingredients.

Keywords: *Lavandula pedunculata*; rosmarinic acid; microwave-assisted extraction; ultrasound-assisted extraction; phenolic compounds; nutraceutical; food additive

1. Introduction

The *Lavandula* genus is prominent in the Mediterranean region, belonging to the botanical family Lamiaceae and encompassing 39 distinct species [1]. The *Lavandula* species possess intriguing economic significance as ornamental plants and for various industrial applications, including pharmaceutical, food, aromatherapy, perfumery, and cosmetics, mainly due to their valuable essential oils (EOs) [1]. *Lavandula pedunculata* (Miller) Cav. (LP) (known in Portugal as "rosmaninho-maior") is common in the Iberian Peninsula [2]. With a widespread distribution, LP is considered the most resistant of all species of the *Lavandula* genus, growing in altitudes up to 1700 m, resisting annual temperature variations, and can reach up to 70 cm tall [3]. However, in Portugal, LP is scarcely used in phytotherapy, nor is it distilled to obtain EOs, due to a low ester content and a high amount of borneol and ketones [2]. It is used only in Portuguese folk medicine as infusions for anxiety and insomnia, the digestive system and as a therapeutic agent with antiseptic action for cleaning wounds [2,3].

Previous works have identified phenolic compounds such as rosmarinic acid, chlorogenic acids, salvianolic acid B, lithospermic acid A, apigenin and luteolin in extracts from the LP subspecies, *lusitanica* [3,4]. However, in comparison to EOs, the non-volatile fraction remains poorly described. Phenolic compounds have been shown to positively affect human health, lowering the risk of diabetes, cancer, cardiovascular disease, and neurological problems [5]. Recent studies have shown that LP extracts have potential through their antioxidant, anti-inflammatory, antiproliferative, antimicrobial, anti-tyrosinase and anticholinesterase activity [6]. In addition, LP extracts have also shown an antihyperglycemic effect using in vitro, ex vivo and in vivo methods. In vitro, LP aqueous extract inhibited pancreatic α-amylase and intestinal α-glucosidase enzymes [7]. Therefore, there is scientific evidence that LP can be helpful as a new potential source of natural BCs, mainly phenolic compounds, which could contribute to valorizing the existing biodiversity and resources of the Portuguese native species of LP. This represents an opportunity in the agricultural sector to promote the sustainable use of this aromatic plant and produce new sources of income for farmers, since the distillery industry does not use this plant species.

Based on a search using Web Of Science (selecting all databases; all of the time span until September 2023, and term: "*Lavandula pedunculata*" in Topic), a total of 44 articles in English were detected. In addition, a careful analysis was conducted to evaluate the articles' titles and abstracts, and it was concluded that only nine research papers addressed the chemical and/or biological characterization of EOs or extracts from LP. Regarding studies about LP polyphenol-rich extracts, the most common extraction was conventional maceration extraction (CE) with non-green solvents (n-hexane, dichloromethane, and methanol) from non-renewable resources. Despite being widely used to extract phenolic compounds due to their strong extraction and dissolving capacities, these solvents and techniques have a negative environmental impact since they require high amounts of solvents and are energy- and time-consuming [8]. Furthermore, their use is associated with both environmental and human health risks. In 2015, the United Nations established the 2030 Agenda for Sustainable Development to overcome these drawbacks, which promotes the wise use of resources and energy. However, there is a pressing need to decrease the usage of organic solvents and improve the energy efficiency of procedures used to extract bioactive compounds (BCs) from plants or foods [9]. Therefore, green chemistry-based extraction methods emerge as a possible alternative to the CE method with organic solvents, given their advantageous attributes, namely decreasing or eliminating hazardous solvents and limiting the cost of solvent waste disposal [10]. Moreover, it allows safety extracts to be obtained that could be used in the food, pharmaceutical and cosmetic industries. Green techniques such as ultrasound-assisted extraction (UAE) and microwave-assisted extraction (MAE), among others, are being used for the recovery of various BCs from plants and food by-products. A recent study from Mansinhos et al. [4] used a combination of (UAE) and deep eutectic solvents to extract phenolic compounds from LP subsp. *lusitanica* Franco. However, a comparative study has not yet been reported between CE and green extractions such as UAE and MAE. Furthermore, to our knowledge, the use of the MAE technique for the phenolic compounds extraction from LP has not yet been evaluated.

The demand for plant-derived BCs and functional ingredients in the food industry has increased due to the current demand for "natural" products, free of harmful chemicals, which are expected to be safe and healthy [11]. The enzymatic browning of fruits and vegetables is one of the food industry's most significant problems [12]. To inactivate the enzymes that cause browning (such as tyrosinase and other polyphenol oxidase enzymes (PPO)), the dipping technique employed frequently uses antioxidant agents and/or browning inhibitors, mainly in the IV gamma products [12,13]. Due to their wealth of phenolic compounds with strong antioxidant activity, LP-based extracts could be a natural option for the food industry to prevent browning. In addition, these natural extracts could also improve the shelf-life, reduce microbial contamination, and replace the use of synthetic food additives.

On the other hand, using natural extracts as food supplements to prevent and/or improve non-communicable diseases or enhance food's nutritional value is on the rise. Natural BCs like flavonoids and phenolic acids have shown the ability to inhibit enzyme activity mainly involved in the mechanism of hypertension and diabetes (α-glucosidase and angiotensin-converting enzyme) [14,15]. Some studies have demonstrated that these natural extracts are similar to certain synthetic drugs in terms of activity; consequently, due to their safety and lower risk of side effects, the consumer's interest is increasing. In the last decade, the market has seen an increase in the availability of natural food additives, functional foods, nutraceuticals and supplements. This industry is expected to grow to around USD 210 billion by 2026 [16]. Europe is gathering clinical evidence to determine the health benefits and toxicity of emerging natural BCs, as the food industry seeks sustainable products with various applications.

Therefore, the main objective of this research was to study the impact of the extraction methodology (MAE, UAE, and CE) on obtaining bioactive polyphenol-rich extracts from LP to select the best technique for producing a novel functional ingredient. The bioactive properties were explored regarding the antioxidant, antimicrobial, antidiabetic, and antihypertensive potential and for tyrosinase inhibition. In addition, cytotoxicity and genotoxicity evaluations were performed to validate the LP extracts' safety.

2. Materials and Methods

2.1. Chemicals

The 1,5-Di-*O*-caffeoylquinic acid, 2-azinobis-3-ethylbenzothiazoline-6-sulphonic acid (ABTS), 2,2'-azo-bis-(2-methylpropionamidine)-dihydrochloride (AAPH), 2,2-diphenyl-1-picrylhydrazyl (DPPH), 2,5-dihydroxybenzoic acid, 3-*O*-caffeoylquinic acid, 4-*O*-caffeoylquinic acid, 4,5-Di-*O*-caffeoylquinic acid, 5-*O*-caffeoylquinic acid, acarbose, angiotensin-I converting enzyme (peptidyl-dipeptidase A, EC 3.4.15.1, 5.1 U/mg), ascorbic acid, caffeic acid, ferulic acid, fluorescein, food grade ethanol 99%, formic acid, gallic acid, rosmarinic acid, salvianolic acid B, sodium carbonate, trifluoroacetic acid, Trolox, α-glucosidase from *S. cerevisiae*, *p*-nitrophenyl-α-D-glucopyranoside, and dimethyl sulfoxide were purchased from Sigma-Aldrich (Sintra, Portugal). Luteolin-7-*O*-glucoside, apigenin-7-*O*-glucoside, and quercetin-3-*O*-glucoside were purchased from Extrasynthese (Genay, France). The tripeptide Abz-Gly-Phe(NO2)-Pro was purchased from Bachem Feinchemikalien (Bubendorf, Switzerland). Tris (tris (hydroxymethyl) aminomethane) was purchased from Fluka (Fluka Gmbh, Germany). Acetonitrile and methanol were purchased from Fischer Scientific (Oeiras, Lisbon, Portugal). Folin–Ciocalteu reagent and potassium persulfate were purchased from Merck (Algés, Lisbon, Portugal). Trifluoroacetic acid was purchased from VWR (Carnaxide, Lisbon, Portugal). Mueller–Hinton broth was purchased from Biokar Diagnostics (Patin, Paris, France).

2.2. Plant Material and Preparation of Extracts

2.2.1. Plant Material

The studied *Lavandula pedunculata* (Mill.) Cav. (LP) consisted of young and old explants of stems, leaves and fruit harvested in 2019 on various locations, such as Gouveia, Fornos de Algodres, Vila Chã and Caldas da Cavaca, in the district of Viseu and Coimbra, Portugal, in March 2019. The raw material (LP) studied in this research work was dried in an oven at 40 °C for 48 h, immediately packed in a vacuum to preserve its properties and kept in a cool and dark place. The sample of LP to be studied encompassed its constituent parts: stems, leaves and fruit. They were reduced to particles of smaller dimensions (6 mm) in a conventional cutting mill (FRITSCH P-15; Idar-Oberstein, Birkenfeld, Germany) before the application of the extraction methodology.

2.2.2. Extraction Methodologies

Conventional Extraction with temperature (CE): The experimental procedure was adapted from Dobros et al. [17]. Briefly, 3 g of dried LP was mixed with 30 mL of extraction solvent (ethanol 50% (v/v)) that was added into a round-bottom flask coupled with a condenser. The assembly was placed in a water bath at 65 °C for 1 h.

Microwave-assisted Extraction (MAE): The methodology was adapted from Leocádio [18]. Briefly, 3 g of LP was weighed into a glass beaker; then, 30 mL of ethanol 50% (v/v) was added. The solution was transferred to the microwave Teflon reactor to perform extraction by applying the Milestone START E microwave program with SK-12 rotor: 500 W of power, temperature increments of 16 °C/min up to 65 °C for 5 min, after reaching 65 °C, maintain for 15 min.

Ultrasound-assisted Extraction (UAE): The experimental procedure was adapted from Leocádio [18]. Briefly, 3 g of an LP plant was placed into a glass flask and 30 mL of ethanol (50% v/v) was added. The solution was sonicated in an ultrasonic bath (Fritsch; model LABORETTE 17) with a power of 120 W and a frequency range between 50 and 60 Hz for 15 min (without exceeding 25 °C).

After each extraction method, the extract was filtered through vacuum filtration, and then the filtrate was evaporated to remove ethanol using a rotary evaporator (40 °C; 175 mbar). The remaining aqueous extract was frozen at −80 °C and dried via freeze-drying (LyoQuest-85; Telstar, Portugal). Finally, the dry extracts were stored in polypropylene flasks and kept in a desiccator during the analysis. The extraction methods were performed in three independent extractions.

The extraction yield was calculated based on the amount of dry plant used to make the dried extracts (Equation (1)):

$$\text{Extractive Yield}(\%) = \frac{\text{Dried extract (g)}}{\text{weight of dried plant (g)}} \times 100 \tag{1}$$

2.3. Total Phenolic Content

The LP extracts' total phenolic content (TPC) was determined using the Folin–Ciocalteu method [19]. Briefly, 80 µL of Folin–Ciocalteu reagent 10% (v/v) was added to 20 µL of extract (previously dissolved in distilled water), followed by 100 µL of sodium carbonate (7.5% (m/v)). The reagents were allowed to react in the dark at room temperature (25 °C) for 1 h. Later, absorbance was measured at 750 nm (Multiskan GO Microplate Spectrophotometer; Thermo Fisher Scientific Inc., Waltham, MA, USA) in a 96-well microplate (Nunc™; Thermo Fisher Scientific Inc., USA). Gallic acid was used as a standard for the calibration curve (0.010–0.125 mg/mL, y = 6.0796x + 0.1314, R^2 = 0.999), and the results were expressed as milligrams equivalent of gallic acid per gram of dry extract (mg GAE/g DE). Three independent analyses were performed in each triplicate extract obtained for each methodology.

2.4. Phenolic Compounds Identification by LC-ESI-QqTOF-HRMS

The polyphenol-rich LP extracts were dissolved in ultrapure water at 10 mg/mL for further analysis by LC-ESI-UHR-QqTOF-MS, according to the method reported by Vilas-Boas et al. [19] Briefly, the separation was performed in a UHPLC UltiMate 3000 Dionex (Thermo Scientific, Waltham, MA, USA), coupled to an ultrahigh-resolution, Qq-time-of-flight (UHR-QqTOF) mass spectrometer with 50,000 full-sensitivity resolution (FSR) (Impact II; Bruker Daltonics, Bremen, Germany). The separation was accomplished with an Acclaim RSLC 120 C18 column (100 mm × 2.1 mm, 2.2 µm) (Dionex, Sunnyvale, CA, USA). The injection volume was 5 µL. The mobile phases consisted of (A) 0.1% aqueous formic acid and (B) acetonitrile with 0.1% of formic acid and the gradient elution conditions were: 0 min, 0% B; 10 min, 21.0% B; 14 min, 27% B; 18.30 min, 58%; 20.0 min, 100%; 24.0 min, 100%; 24.10 min, 0%; 26.0 min, 0% at a flow rate of 0.25 mL/min. Parameters for MS analysis were set using negative ionization mode with spectra acquired over a range from m/z 20 to 1000. The parameters were as follows: capillary voltage, 3.0 kV; drying gas temperature,

200 °C; drying gas flow, 8.0 L/min; nebulizing gas pressure, 2 bar; collision RF, 300 Vpp; transfer time, 120 μs; and prepulse storage, 4 μs. Post-acquisition internal mass calibration used HCOONa clusters, delivered by a syringe pump at the start of each chromatographic analysis. High-resolution mass spectrometry was used to identify the compounds. The elemental composition for the compound was confirmed according to accurate mass and isotope rate calculations designated mSigma (Bruker Daltonics, Billerica, MA, USA). The accurate mass measurement was within 5 mDa of the assigned elemental composition, and mSigma values of <20 provided confirmation. Compounds were identified based on their accurate mass $[M-H]^-$. One independent analysis was performed in each triplicate extract obtained for each methodology.

2.5. Phenolic Compounds Quantification by HPLC–DAD

The quantitative profiling of phenolic compounds in polyphenol-rich LP extracts (previously dissolved in ultrapure water at 20 mg/mL) was performed using a Waters Alliance e2695 separation module system interfaced with a photodiode array UV/Vis detector 2998 (PDA 190–600 nm; Waters, Mildford, MA, USA). The separation occurred in a reversed-phase C18 column (ZORBAX Eclipse XDB-C18, 80A°; 4.6 × 250 mm; 5 μm; Agilent, Santa Clara, CA, USA) at 25 °C. The mobile phase and the gradient elution used were prepared according to Vilas-Boas et al. [19]. The injection volume and the flow rate were 20 μL and 1 mL/min, respectively. Data acquisition and analysis were carried out using Software Empower 3. The detection was performed at 280, 320, 350, and 360 nm, and the phenolic compound identification was performed by comparing the retention time and absorbance spectra with pure standards. The quantification was performed using the calibration curves' interpolation, and the results were expressed as milligrams per gram of dry extract (mg/g DE). Three independent analyses were performed in each triplicate extract (CE, MAE and UAE) obtained for each methodology.

2.6. Antioxidant Activity

2.6.1. ABTS Method

The CE, UAE and MAE extracts described in Section 2.2.2 were evaluated for their antioxidant capacity against the ABTS free radical, as described by Vilas-Boas et al. [19]. The ABTS stock solution was prepared by mixing ABTS aqueous solution (7 mmol/L) with potassium persulfate aqueous solution (2.45 mmol/L) and kept for 16 h in the dark at 25 °C. On the analysis day, the ABTS stock solution was diluted with water to an absorbance of 0.700 ± 0.020 at 734 nm (ABTS working solution). After that, 15 μL of sample (previously dissolved in distilled water) was allowed to react in the darkness at room temperature (25 °C) with 200 μL of ABTS working solution and the absorbance was read at 734 nm 6 min precisely after initial mixing in a 96-well microplate (Nunc™; Thermo Fisher Scientific Inc., USA). A blank was taken with distilled water (A0). The inhibition percentage of the sample was calculated using Equation (2) and compared with the ascorbic acid standard calibration curve (0.0088–0.0880 mg/mL; $r^2 = 0.999$). The assay was realized with the multidetector plate reader Synergy H1 (BioTek Instruments, Winooski, VT, USA) controlled by the Gen5 BioTek software version 3.04.

The results were expressed as milligrams of ascorbic acid equivalent per gram of dry extract (mg AAE/g DE). Three independent analyses were performed in each triplicate extract obtained for each methodology.

2.6.2. DPPH Method

The DPPH assay was carried out according to the procedure described by Vilas-Boas et al. [19]. Briefly, 175 μL of DPPH work solution (60 μM) was allowed to react with 25 μL of extract (previously diluted in distilled water) in a 96-well microplate (Nunc™; Thermo Fisher Scientific Inc., USA) for 30 min in darkness. Then, the absorbance was measured at 515 nm (multidetection plate reader Synergy H1; BioTek Instruments, Winooski, VT, USA), and a blank was taken with distilled water (A0). The inhibition percentage of the sample was calculated using Equation (2), and Trolox was used as a standard to prepare a calibration curve (0.0075–0.0750 mg/mL; r^2 = 0.999). The results were expressed as milligrams of Trolox equivalent per gram of dry extract (mg TE/g DE). Three independent analyses were performed in each triplicate extract obtained for each methodology.

$$\% \text{ Inhibition} = \frac{\text{Abs}_{A0} - \text{Abs}_{sample}}{\text{Abs}_{A0}} \times 100 \tag{2}$$

Abs_{A0} is the absorbance of blank, and Abs_{sample} is the absorbance of the reaction between the sample and the radical after the incubation.

2.6.3. ORAC Method

An oxygen radical absorbance capacity (ORAC) assay was carried out following the methodology described by Coscueta et al. [20]. The reaction was carried out in a 75 mM phosphate buffer (pH 7.4), with a final reaction volume of 200 μL. The sample (20 μL), previously dissolved in distilled water, and fluorescein solution (FL) (120 μL; 70 nM, final concentration in well), was added to the well of the black 96-well microplate (Nunc, Denmark). Each plate experiment also included a calibration curve (1–8 μM of Trolox, final concentration in well) and a blank (FL + AAPH + phosphate buffer). The mixture was preincubated at 37 °C for ten minutes. After that, the AAPH solution (60 μL; 12 mM, final concentration in well) was added rapidly using a multichannel pipet. The fluorescence was recorded at intervals of 1 min for 80 min. A multidetector plate reader (Synergy H1; Biotek, Winooski, VT, USA) with 485 nm excitation and 520 nm emission filters was used. The equipment was controlled by the Gen5 Biotek software version 3.04. The results were expressed as milligrams of Trolox equivalent/g of dry extract (mg TE/g DE). Three independent analyses were performed in each triplicate extract obtained for each methodology.

2.7. Antimicrobial Activity

2.7.1. Bacterial Strains

The Gram-negative bacteria used were *Escherichia coli* (ATCC (American type culture collection) 25922), *Salmonella enterica serovar Enteritidis* (ATCC 13076) and *Pseudomonas aeruginosa* (ATCC 278539) while *Listeria monocytogenes* (NCTC (National Collection of Type Cultures) 10357), *Bacillus cereus* (NCTC 2599) and *Staphylococcus aureus* (ATCC 25923) were the Gram-positive bacteria used. These bacteria were cultured on Mueller–Hinton broth (MHB) medium and incubated at 37 °C overnight. The inoculum test concentration after this period should be 10^8 CFU/mL.

2.7.2. Growth Inhibition Curves

The dry extracts of LP were re-dissolved in MHB at three different concentrations (10, 5 and 2.5% (w/v)) and sterilized by filtration through a 0.22 μm filter (FriLabo—Maia, Portugal). The growth inhibition curves were determined using the broth microdilution assay, following the standards for antimicrobial susceptibility testing provided by the Clinical and Laboratory Standards Institute (CLSL) [21]. At the time of testing, each fresh overnight culture of bacteria (prepared in Section 2.7.1) was adjusted spectrophotometrically (optical density (OD) at 625 nm) between 0.08 and 0.13, representing a concentration of an inoculum of 1×10^8 CFU/mL. Afterwards, each well of a sterile 96-well microplate was filled with a total volume of 200 μL containing approximately 10^6 CFU/mL of test bacteria (1% (v/v) inoculum) and variable concentrations of each extract prepared. A positive control containing only MHB inoculated at 1% (v/v) and two negative controls: (i) containing only sterilized medium and (ii) containing only sterilized extract, were also tested. The OD at 625 nm was assessed for a period of 24 h at 37 °C (1 h intervals) by using a microplate reader (Multiskan GO Microplate Spectrophotometer; Thermo Scientific, Thermo Fisher Scientific Inc., Waltham, MA, USA). The increase in OD was considered a direct consequence of bacterial growth. Each condition was assayed in triplicate. An inhibition percentage [22] was calculated using the following Equation (3):

$$\text{Inhibition Percentage}(\%) = \frac{\text{OD control} - \text{OD bacteria}}{\text{OD control}} \times 100 \tag{3}$$

OD control bacteria and OD bacteria represent the OD (at 625 nm) after 24 h of incubation of the control bacteria in the growth medium and in the presence of the LP extracts, respectively.

2.8. Tyrosinase Inhibition Assay

The capability of LP extracts to inhibit tyrosinase enzyme was assessed with a colorimetric tyrosinase inhibitor screening kit (ab204715, Abcam). Briefly, a mixture of extracts previously diluted in the buffer (20 µL) and tyrosinase solution (50 µL) was incubated in a 96-well microplate at 25 °C for 10 min. Then, 30 µL of substrate solution was added, and the absorbance was continuously read at 510 nm in a microplate reader (Synergy H1; BioTek Instruments, Winooski, VT, USA). Data were recorded every 2 min time interval for 30 min. An inhibition control (IC) was performed with Kojic acid (0.021 mg/mL). A positive control for the enzyme was performed with a mixture of enzyme and substrate in the presence of buffer. The average was made for reading duplicates. Two time points (T1 and T2) were chosen in the linear range of the plot and obtained the corresponding values for absorbance (A1 and A2). The slope was calculated for all samples (S), inhibition control (IC) and enzyme control (EC) by dividing the net ΔA (A2 − A1) values with the time ΔT (T2 − T1). The percentage inhibition was calculated using Equation (4):

$$\text{Tyrosinase inhibition}(\%) = \frac{\text{Slope}_{EC} - \text{Slope}_S}{\text{Slope}_{EC}} \tag{4}$$

2.9. α-Glucosidase Inhibition Assay

The α-glucosidase inhibition was determined using a colorimetric-based method according to Know et al. [23]. Firstly, the dry extracts were dissolved and diluted in the phosphate buffer (pH 6.9; 0.1 M), and the final concentrations tested were between 5.0 and 1.25 mg/mL. A total of 50 µL of extract solution and 100 µL of phosphate buffer (pH 6.9; 0.1 M) containing α-glucosidase (1.0 U/mL) were incubated in 96-well plate at 25 °C during 10 min. After that, 50 µL of a 5 mM *p*-nitrophenyl-α-D-glucopyranoside solution prepared in phosphate buffer (pH 6.9; 0.1 M) was added to each well with a multichannel pipettor and the absorbance was read. The plate was incubated at 25 °C for 5 min, and absorbance readings were recorded at 405 nm with a microplate reader (Synergy H1, BioTek Instruments, Winooski, VT, USA). As a negative control, 50 µL of buffer solution replaced the extracts. Acarbose was used as a positive control at a 5 mg/mL concentration. A blank without the enzyme (100 µL of 0.1 mol/L phosphate buffer instead) was performed for each sample. The α-Glucosidase inhibitory activity was expressed as inhibition (%) and calculated using Equation (5):

$$\alpha\text{-glucosidase inhibition}(\%) = \frac{\Delta \text{Abs}_C - \Delta \text{Abs}_S}{\Delta \text{Abs}_C} \tag{5}$$

where ΔAbs_C is the variation in absorbance of the control, and ΔAbs_S is the variation in the samples' absorbance.

2.10. Angiotensin-Converting Enzyme-I Inhibition Assay (iACE)

The in vitro iACE assay was conducted according to the procedure outlined by Coscueta et al. [24]. A commercial angiotensin-I converting enzyme (EC 3.4.15.1, 5.1 U/mg) was diluted in 5 mL of a glycerol solution in 50% ultra-pure water. The stock solution was then diluted 1:24 with a 150 mM Tris buffer solution, pH 8.3, containing 1 µM of $ZnCl_2$, reaching a 42 mU/mL concentration in the final reaction solution. Then, 40 µL of the working solution of ACE was added to each well of the reaction microplate and then adjusted to 80 μL by adding ultra-pure water for the control and the sample to be analyzed. The enzymatic reaction was initiated by adding 160 μL of a 0.45 mM solution of ABz-Gly-Phe(NO_2)-Pro, dissolved in 150 mM Tris buffer (pH 8.3), containing 1.125 M NaCl, mixing immediately and incubating at 37 °C. The microplate was automatically shaken before the first reading, and the fluorescence generated was measured for 30 min. The measurements were completed with the Synergy H1, (BioTek Instruments, Winooski, VT, USA) plate reader with excitation filters at 350 nm and emission at 420 nm. Nunc^{TM} black 96-well polystyrene microplates (ThermoScientific, Waltham, MA, USA) were used for the assay. Each extract was evaluated in triplicate, and the iACE was expressed as the concentration capable of inhibiting 50% of the ACE activity (IC_{50}). Non-linear modelling was applied to determine the IC_{50}, and the results were expressed as mg/mL to inhibit 50% of the enzymatic activity.

2.11. Cytotoxicity

2.11.1. Cell Line Growth Conditions

Human Caucasian colon carcinoma epithelial cells, Caco-2 (86010202 European Collection of Authenticated Cell Cultures) cells were grown in Dulbecco's modified Eagle's medium (Lonza, Basel, Switzerland) supplemented with 10 % (*v*/*v*) heat-inactivated fetal bovine serum (Biowest, Nuaillé, France), 1% (*v*/*v*) penicillin-streptomycin-fungizone (Lonza, Basel, Switzerland) and 1% (*v*/*v*) non-essential amino acids 100× (Lonza, Basel, Switzerland). All cells were incubated at 37 °C in a humidified atmosphere with 5% CO_2.

2.11.2. Cytotoxicity Assay

The cytotoxicity evaluation was performed according to the ISO 10993-5:2009 standard [25], Cells were grown to 80–90% confluence, detached using TrypLE Express (ThermoScientific, Waltham, MA, USA), and seeded at 1.0×10^5 cells/mL into 96-well tissue cultured plates (Nunclon ThermoScientific, Waltham, MA, USA) and allowed to adhere for 24 h. Afterwards, the media were carefully removed and replaced with media supplemented with LP extracts at 4.8–0.4 mg/mL concentrations. DMSO at 40% in culture media was used as the death control, and plain culture media was used as the growth control. After the incubation time, Presto Blue (ThermoFisher, Waltham, MA, USA) was added to each well and incubated at 37 °C in the darkness for 1 h. After this period, fluorescence (Ex: 560 nm; Em: 590 nm) was measured using a microplate reader (Synergy H1; Biotek Instruments, Winooski, VT, USA). All assays were performed in quintuplicate.

2.12. Mutagenicity Evaluation—AMES Assay

AMES assay was performed in accordance with OECD guideline 471 [26], using the MOLTOX® Salmonella Mutagenicity Assay Kit (reference number: 31-110.2; Trinova Biochem GmbH, Giessen, Germany), according to the manufacturer's instructions. *Salmonella typhimurium* TA 98 was previously inoculated and incubated at 37 °C and held stationary overnight for this assay. After that, the inoculum was set in an orbital (130 rpm at 37 °C) until a density of $1–2 \times 10^9$ bacteria/mL (approximately 1.0–1.4 at OD 650 nm) was reached. Samples were tested at 1 mg/plate, 0.5 mg/plate and 0.125 mg/plate, corresponding to 10, 5 and 1.25 mg/mL, respectively. Two controls were assessed: one with chemical control capable of inducing bacterial mutation Daunomycin and 2-Aminoanthracene) and another with water instead of the extract (solvent control). The assay was placed in a 37 °C incubator for approximately 48 h. The assays were performed with and without metabolic activation (S9). Two plates per three separate experiments were assayed for each concentration tested and for positive and negative controls. According to the kit instructions, a positive result (genotoxic effect) should be at least 2.5 times in a treatment group concerning the corresponding negative control (solvent control).

2.13. Statistical Analysis

The results in Section 3 are presented as the average $\pm$ standard deviation of three independent extractions (n = 3). Shapiro–Wilk test was used to evaluate the normality of data distribution, Levene's test was used for homogeneity of variances and the one-way analysis of variance (ANOVA) was used to examine the significance of the differences between the extracts. For ANOVA, the null hypothesis (H_0) (means are equal) was rejected when the differences between means showed a $p < 0.05$. In this case, multiple comparisons were examined using Tukey's post-hoc test (homogeneity of variance was assumed) at the $p < 0.05$ significance level. The statistical analysis was carried out entirely with SPSS version 23.

3. Results

Three LP polyphenol-rich extracts were prepared from dried LP using three different techniques: CE, UAE and MAE. The aim was to assess the effects of each technique and identify the optimal method for deriving a nutraceutical or health-enhancing food ingredient from a traditionally underexplored Portuguese medicinal plant. While these *Lavandula* species are seldom used for the commercial distillation of EOs, most studies describe their health benefits and applications. However, the potential of its polar extracts rich in phenolic compounds remains relatively unexplored. Therefore, the hydroethanolic polyphenol-rich extracts were further examined for their phenolic compounds and bioactivities using chemical in vitro assays. Additionally, the cytotoxicity and mutagenicity of the polyphenol-rich LP extracts were evaluated to ensure their safe application.

3.1. Extractive Yield

The primary extraction objective was to obtain the desired BCs with high extraction yields while reducing the concentration of unwanted compounds, such as proteins and sugars, which could affect the stability and quality of the final extract. The extraction efficiency may be affected by several factors, the most important of which are the solvent, mass-to-solvent ratio, temperature, and pH [27]. The variables regarding the type of solvent and ratio were fixed because one of the main goals of this research study was to develop extracts by applying distinct techniques and evaluating their impact. For this, an ethanol solution at 50% (v/v) was chosen because it is one of the most frequently used solvents; ethanol is food grade and is safe for the consumer even if not completely removed from the final ingredient; it is environmentally friendly and it can be reused in the process.

The total extraction yield (w/w, %) of the LP polyphenol-rich extracts obtained by different extraction techniques is shown in Table 1. The use of the CE technique resulted in a higher extraction yield ($23.91 \pm 2.00\%$, w/w) ($p < 0.05$), while the remaining green extraction techniques (UAE and MAE) recovery yield was about 16–17% w/w ($p > 0.05$). Nonetheless, it is important to emphasize that achieving a higher extractive yield does not necessarily equate to the extraction of a greater quantity of phenolic compounds, because specific extraction methods or conditions might favor the extraction of other components such as sugars, organic acids, minerals, and proteins. It is possible that the highest yields were caused by the high temperature, as the plant:solvent ratio, solvent type, and moisture content were the same for all extraction procedures. High temperatures (65 °C) were applied in the CE and MAE extractions. It influenced the extraction efficiency since a higher temperature promotes increased solubility and diffusion into the solvent, leading to a higher mass transfer rate [28]. According to research from Lezoul et al. [29], using high temperatures associated with CE helps access a better yield and, consequently, higher TPC. However, this effect cannot be generalized since it strongly depends on the type of compound extracted, the binomial time:temperature used and the heating type. While, statistically, the yield is lower when employing MAE, it is important to note that this technique is a high-temperature extraction method with significantly reduced extraction times, resulting in lower energy consumption than CE. UAE is reported to be a technique that achieves high yields quickly since it is especially good at breaking the cell walls, leading to the release of BCs. However, the lower yield obtained with UAE in this study could be attributed to the fact that high temperatures were not used in this methodology. Nevertheless, the results obtained are in line with those reported by Costa et al. [30], which brought extraction yields of 19.4% for LP extracts using CE (without use of high temperature and ethanol 50% (v/v)).

Table 1. Extractive yields (w/w, %) obtained by the different extraction methods (CE, UAE and MAE) from LP. Values are represented by the average $\pm$ standard deviation. Different letters mean significant differences between extractions ($p < 0.05$).

Matrix	Extraction Method	Extractive Yield (%)
Lavandula pedunculata	CE	23.91 $\pm$ 2.00 [a]
	UAE	16.17 $\pm$ 2.81 [b]
	MAE	17.64 $\pm$ 1.63 [b]

3.2. Total Phenolic Content

Phenolic compounds are widely distributed in plants such as the *Lavandula* species and are known for their antioxidant properties; however, in recent years, they have been linked to other properties such as enzyme inhibition involved in diabetes and hypertension, and fruit and vegetable browning [31]. Table 2 shows the results of the total phenolic content (TPC), which varied between 183.1 $\pm$ 4.9 and 183.7 $\pm$ 17.8 mg GAE/g DE. Despite the absence of statistically significant differences ($p > 0.05$), it is worth noting that methods involving high temperatures, such as CE and MAE, exhibited a higher total phenolic content (TPC), whereas UAE yielded the lowest value. These values follow the extraction yields obtained. As previously mentioned, the temperature seems to have an important impact on the extraction of phenolic compounds. The warmth may cause the plant tissues to soften and the bindings between polysaccharides and/or proteins and phenolic compounds to weaken, which could cause disruption and migration to the solvent [32]. However, because homogenous heating might be achieved more quickly and with less energy usage thanks to the energy transmission mechanism, MAE heating has been found to offer greater benefits than convection. In addition, UAE emerges as an excellent alternative to establishing an environmentally friendly extraction method. It involves physical and chemical phenomena completely different from those applied with MAE and CE with convectional heating. The ultrasound waves induce the cavitation forces, which induce the explosive collapse of bubbles and generate pressure causing plant tissue rupture and improving the release of intracellular compounds into the solvent. However, in this study, the UAE was less efficient than CE in extracting phenolic compounds. On the contrary, a recent study reported that UAE (15 min with 80% ethanol) was more efficient than CE [4]. However, the UAE took place at 50 °C. In addition, Dobros et al. [17] reported a higher yield in TPC (30.25 mg GAE/g dry weight (DW) with UAE (ethanol 50% for 30 min) than CE (ethanol 50% for 30 min) (28.01 mg GAE/g DW). It is important to highlight that both studies reported a TPC of approximately 30 mg GAE/g DE, while our LP extract presents a TPC of 181.4 mg/g DE. To the best of our knowledge, no one has ever published a comparison of these two green extraction methods with CE for extracting phenolic compounds from LP. There are only studies with CE (using polar and non-polar solvents), UAE extraction (with methanol, ethanol 80% (v/v) and water), and UAE in combination with deep eutectic solvents (DES). Based on the results from Baptista et al. [33], the higher value was obtained with a CE with methanol for 24 h (1040.3 $\pm$ 17.8 mg GAE/g), while the CE with 100% water allowed a TPC of about 675 mg GAE/g. Considering the last value, our results, regardless of the extraction method, were about 3.4-fold times less than reported in Baptista et al.'s study. However, our extraction time was lower than in that study. On the other hand, LP extracts obtained with CE using ethanol 80% (v/v) showed a TPC value ranging between 56.1 and 136 mg/g DE [3], while LP extracts obtained with CE (1 h; 200 rpm; 50 °C) showed a TPC value lower than 30 mg/g DE [4].

Table 2. The total phenolic (TPC) and total flavonoid content in LP polyphenol-rich extracts. Values are represented by the average ± standard deviation. Different letters mean significant differences between extraction methods ($p < 0.05$).

Matrix	Extraction Method	TPC (mg GAE/g DE)
	CE	183.7 ± 17.8 [a]
Lavandula pedunculata	UAE	181.4 ± 6.5 [a]
	MAE	183.1 ± 4.9 [a]

Overall, TPC represented about 18% of the total LP polyphenol-rich extract. Therefore, other compounds were extracted, including sugars, minerals, organic acids, fibers and proteins.

3.3. Phenolic Compounds Profile and Quantification

A total of thirty phenolic compounds and seven organic acids were identified using the high-resolution mass spectrometry technique (Table 3). The identification of these compounds led to their distribution into three structurally related classes, i.e., hydroxycinnamic acids (14 compounds), hydroxybenzoic acids (6 compounds) and flavonoids (9 compounds). The identification of all proposed compounds presented a good mass accuracy (<2 mDa), increasing the confidence of the predicted identification. In detail, only five studies were found in the literature that identified phenolic compounds from LP extracts using mass spectrometry without differences (Figure 1). Two of these studies used plant material grown in Portugal [3,4]. The remaining three articles used LP grown in Morocco [7,34] and Turkey [35]. The most abundant compounds found in all the LP extracts were the hydroxycinnamic acids (50%), which agrees with the literature available for LP [3,4] and other *Lavandula* species [6]. Nevertheless, previous studies from Lopes et al. [3] identified only thirteen phenolic compounds, nine hydroxycinnamic acids and four flavonoids. Conversely, a study conducted by Mansinhos et al. [4] discovered twenty-three phenolic compounds in extracts obtained using UAE in conjunction with either 80% (*v/v*) ethanol or deep eutectic solvents. Similar to our findings, over 50% of the compounds were identified as hydroxycinnamic acids. Chlorogenic acids (CGAs) were widely distributed in *Lavandula* species. The main compounds were 3-*O*-caffeoylquinic acid, 5-*O*-caffeoylquinic acid and 4-*O*-caffeoylquinic acid, with a molecular ion [M-H]$^-$ at 353, which was consistent with the molecular formula $C_{16}H_{18}O_9$ and showed the typical fragment of loss as quinic acid (m/z 191). Moreover, another important class of CGAs was found: diCQAs (1,5-Di-*O*-caffeoylquinic acid and 4,5-Di-*O*-caffeoylquinic acid) with a [M-H]$^-$ at 515 and a fragment m/z at 353 representing the caffeoylquinic acid.

Table 3. LC-ESI-UHR-QqTOF-MS data of organic acids and phenolic compounds identified in CE, MAE and UAE in LP polyphenol-rich extracts.

Proposed Name	Molecular Formula	Rt	m/z Measured Mass [M-H]$^-$	MS^2 Fragments (m/z, % Base Peak Intensity)	Error (mDa)
Gluconic Acid	$C_6H_{12}O_7$	1.36	195.0510	75 (100)	1.0
Tartaric acid	$C_4H_6O_6$	1.5	149.0092	72 (100)	0.1
2-Furoic acid	$C_5H_4O_3$	2.2	111.0088	69 (100)	0.5
Succinic acid	$C_4H_6O_4$	2.8	117.0193	73 (100)	−0.3
Malic acid	$C_4H_6O_5$	1.7	133.0143	71 (100)	0.3
Citric acid	$C_6H_8O_7$	2.1	191.0197	87 (100), 111 (39)	0.8
Azelaic acid	$C_9H_{16}O_4$	13.2	187.0975	97 (100), 125 (66)	0.9

Table 3. *Cont.*

Proposed Name	Molecular Formula	Rt	m/z Measured Mass [M-H]$^-$	MS^2 Fragments (m/z, % Base Peak Intensity)	Error (mDa)
3-*O*-caffeoylquinic acid	C$_{16}$H18O$_9$	6.4	353.0878	191 (100), 179 (63), 135 (25),	0.7
5-*O*-caffeoylquinic acid	C$_{16}$H$_{18}$O$_9$	8.2	353.0878	191 (100), 173 (97), 179 (80)	1.1
4-*O*-Caffeoylquinic acid	C$_{16}$H$_{18}$O$_9$	8.9	353.0887	173 (100), 179 (80), 191 (62), 135 (21)	1.0
1,5-Di-*O*-caffeoylquinic acid	C$_{25}$H$_{24}$O$_{12}$	10.1	515.3949	163 (98), 353 (20)	0.8
4,5-Di-*O*-caffeoylquinic acid	C$_{25}$H$_{24}$O$_{12}$	12.9	515.1008	353 (100), 173 (80), 179 (40)	0.9
Caffeic acid	C$_9$H$_8$O$_4$	8.5	179.0351	135 (100), 179 (40)	1.4
Isoferulic acid	C$_{10}$H$_{10}$O$_4$	8.6	193.0506	134 (100)	1.0
Ferulic acid	C$_{10}$H$_{10}$O$_4$	11.8	193.0506	134 (100), 178 (74), 193 (34)	0.7
p-coumaric acid	C$_9$H$_8$O$_3$	10.4	163.0401	119 (100), 163 (20)	−0.6
Lithospermic acid A	C$_{27}$H$_{21}$O$_{12}$	13.9	537.1038	359 (100), 295 (80), 197 (42), 179 (50), 493 (18), 313 (10)	1.8
Rosmarinic acid	C$_{18}$H$_{15}$O$_9$	13.6	359.0772	161 (100), 197 (60), 179 (54)	1.5
Salvianolic acid A	C$_{26}$H$_{22}$O$_{10}$	13.9	493.1141	185 (68), 295 (100)	−2
Sagerinic acid	C$_{36}$H$_{32}$0$_{16}$	13.6	719.1684	161 (100), 359 (80), 197 (20), 179 (11)	1.9
Salvianolic acid B	C$_{36}$H$_{30}$O$_{16}$	14.8	717.1520	537 (50), 519 (40), 339 (8), 321 (100), 197 (6), 179 (27)	1.9
trans-4-Hydroxycinnamate	C$_9$H$_8$O$_3$	9.1	163.0401	119 (100)	0.6
1-*O*-Vanilloyl-beta-D-glucose	C$_{14}$H$_{17}$O$_9$	5.7	329.0878	167 (100)	0.1
Protocatechuic acid	C$_7$H$_6$O$_4$	5.7	153.0193	109 (100)	0.1
2,5-Dihydroxybenzoic acid	C$_7$H$_6$O$_4$	7.3	153.0193	109 (100), 81 (35), 53 (32)	0.3
3,4-Dihydroxybenzaldehyde	C$_7$H$_5$O$_3$	6.9	137.0244	108 (100)	1.0
4-Hydroxybenzoate-*O*-glucoside	C$_{13}$H$_{15}$O$_8$	7.2	299.0772	137 (100)	0.7
Vanillylmandelic acid	C$_9$H$_{10}$O$_5$	5.3	197.0455	72 (100), 123 (55), 135 (60)	0
Apigenin-8-*O*-glucoside	C$_{21}$H$_{20}$O$_{10}$	13.3	431.0984	341 (100), 268 (87), 311 (75)	0.7
Luteolin-7-*O*-glucoside	C$_{21}$H$_{20}$O$_{11}$	11.9	447.0933	447 (20), 285 (100)	1.4
Luteolin	C$_{15}$H$_9$O$_6$	16.5	285.0131	285 (100), 133 (85)	1.1
Quercetin 3-*O*-glucoside	C$_{21}$H$_{20}$O$_{12}$	12.2	463.0882	301 (100)	0.1
Apigenin-7-*O*-glucuronide	C$_{21}$H$_{17}$O$_{11}$	13.4	445.0345	269 (100)	1.2
Luteolin-7-*O*-glucuronide	C$_{21}$H$_{18}$O$_{12}$	12.1	461.9984	285 (100)	0.5
Apigenin-7-*O*-glucoside	C$_{21}$H$_{20}$O$_{10}$	13.5	432.378	268 (100), 431 (20)	0.9
Apigenin	C$_{15}$H$_{10}$O$_5$	17.6	269.0429	269 (100)	1.3
6-Hydroxyluteolin-7-glucoside	C$_{21}$H$_{20}$O$_{12}$	11.7	463.0882	287 (100)	0.4

In addition, the presence of caffeic acid is characteristic of the family *Lamiaceae* and is responsible for its potential bioactivity. Ferulic acid and isoferulic acid were identified in all extracts, which agrees with the report by Mansinhos et al. [4] concerning extracts obtained with UAE. However, this is the only study with LP that reports this compound. This can be explained by the fact that this compound is strongly linked to the cell wall and requires high temperatures or, for example, ultrasonic waves to extract it. Caffeic acid [M-H]$^-$ at 179 was detected in the extracts, which was also reported in all the studies about LP. Within hydroxycinnamic acids, rosmarinic acid (RA) claims to be the most representative of the Lavandula genus. RA is an ester of caffeic acid with an [M-H]$^-$ at 359. Therefore, for its correct identification, a m/z fragment should be at 179 (caffeic acid loss) and 161 (salvianic acid A loss). Other important compounds in the *Lavandula* genus belonging to this class are lithospermic acid A and salvianolic acid (also known as lithospermic acid B). The difference between RA and lithospermic acid A is the RA first lost the caffeic acid or the salvianic acid from [M-H]$^-$, while the lithospermic acid A first lost CO_2 from [M-H]$^-$ m/z 537 (which corresponds with the m/z fragment 493) and then lost caffeic acid (fragment m/z 313) and salvianic acid A (fragment m/z 295). In the case of salvianolic acid B, it successively lost two molecules of caffeic acid or salvianic acid A loss, resulting in the fragment ion at m/z 537, m/z 519, m/z 339 and m/z 321 [36].

Hydroxybenzoic acids are secondary metabolites responsible for signaling in the defense of plants to pathogens, and high levels are related to systemic infections. Furthermore, they have interesting biological properties making them attractive as nutraceuticals or functional food additives [37]. Despite 20% of the phenolic compounds detected in LP polyphenol-rich extracts being classified as hydroxybenzoic acids, none of these individual compounds have been identified explicitly in LP of Portuguese origin. However, extracts obtained with UAE showed gallic and vanillic acid in their composition [4]. On the other hand, Boutahiri et al. [7,34] also identified protocatechuic acid and gentisic acid in methanolic and aqueous extracts from LP grown in Morocco. Therefore, this study represents a groundbreaking discovery as it is the first report of a profusion of compounds belonging to this class being identified in LP, grown in a Portuguese geographical origin.

Flavonoid subclasses observed in the genus *Lavandula* mainly include flavones, flavonols, and anthocyanins. However, in LP extracts obtained using CE, MAE, and UAE, flavones (luteolin and apigenin) and one flavonol (quercetin 3-*O*-glucoside) were predominantly observed. According to the previous literature studies, compounds such as luteolin-7-*O*-glucoside, luteolin-7-*O*-glucuronide, luteolin-*O*-hexosyl-glucuronide are the most representative in LP. In agreement with these studies, hydroethanolic and aqueous extracts obtained with LP showed these three flavones. However, in addition to reporting these compounds, Mansinhos et al. [4] also reported the presence of apigenin-7-*O*-glucoside and apigenin in extracts obtained with UAE, which aligns with the results observed in our extracts. It is worth noting that although apigenin-8-*O*-glucoside and quercetin 3-*O*-glucoside have already been reported in *L. stoechas* and other *Lavandula* species, this is the first time that the existence of these flavonoids have been reported in LP.

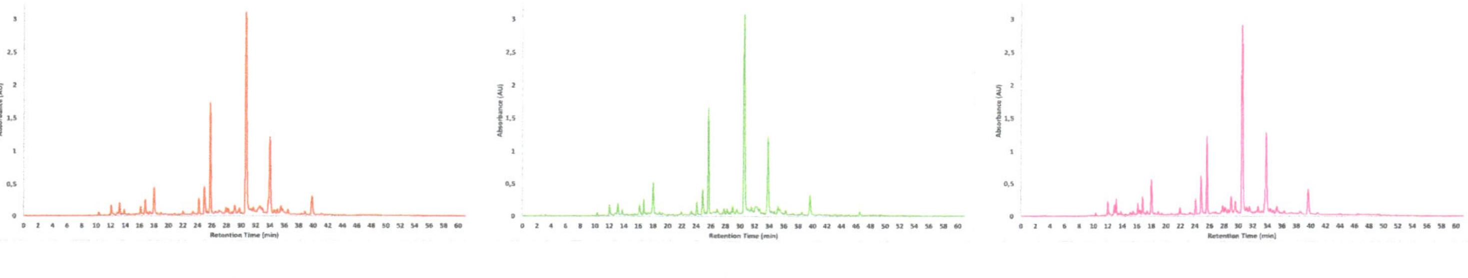

(A) (B) (C)

Figure 1. Chromatograms of phenolic compounds (HPLC-DAD) at 320 nm for CE (**A**), UAE (**B**) and MAE (**C**) extraction. Peaks: 1: 2,5-Dihydroxybenzoic acid; 2: 3-*O*-caffeoylquinic acid; 3: 5-*O*-caffeoylquinic acid; 4: 4-*O*-caffeoylquinic acid; 5: caffeic acid; 6: 1,5-Di-*O*-caffeoylquinic acid; 7: ferulic Acid; 8: luteolin-7-*O*-glucoside; 9: luteolin derivative; 10: 4,5-Di-*O*-caffeoylquinic acid; 11: apigenin-7-*O*-glucoside; 12: apigenin derivative; 13: quercetin-3-*O*-glucoside; 14: rosmarinic Acid; 15: salvianolic acid B: lithospermic acid.

Sixteen phenolic compounds were quantified by HPLC-DAD in comparison with commercial standards (Table 4). The polyphenol-rich LP extracts (MAE, UAE, and CE) presented a similar profile (Figure 1), as previously discovered using mass spectrometry analysis. RA was the most abundant compound found in all extracts, followed by salvianolic acid B. The values range from 58.68 ± 1.42 to 42.27 ± 1.92 mg/g DE and 43.19 ± 1.09 to 40.09 ± 1.61 mg/g DE, respectively. Applying two green extractions showed less RA recovery than the CE ($p < 0.05$). This follows the study of Sik et al. [38], which compares different extraction methods to recovery RA from six Lamiaceae plants. However, these authors reported that the RA yield was lower in MAE when the extraction time was increased from 5 mins to 30 mins using high temperatures (80 °C), possibly due to the lower thermal stability of hydroxycinnamic acid derivatives. Hence, the high temperature disrupts the C-C bonds [39]. This is because the amount of energy converted to heat in the dielectric material is controlled by microwave power, which is correlated with the extraction temperature. Nevertheless, MAE total extraction time was four times less, reducing the energy expenditure. Regardless of salvianolic acid B extraction, the MAE showed the same efficiency as the CE ($p > 0.05$), but the UAE showed lower efficiency. However, this result could be reversed if UAE was optimized using a response surface model for the simultaneous higher yield extraction of salvianolic acid B and RA, which are the main phenolic compounds present in this plant.

Overall, the results obtained with LP extracts are in accordance with a study by Lopes et al. [3], who reported that RA and salvianolic acid B were the major compounds among all thirteen compounds identified in LP. A paper by Costa et al. [30] also reported RA as the most abundant compound in the water:ethanol extract of *L. pedunculata* subsp. lusitanica, but the concentration was smaller in comparison with our results. Moreover, in that study, only six different phenolic compounds were detected (3-*O*-caffeoylquinic acid, 4-*O*-caffeoylquinic acid, 5-*O*-caffeoylquinic acid, RA, luteolin and apigenin). A more recent study by Mansinhos et al. [4] used UAE to improve the recovery of phenolic compounds from LP subsp. lusitanica, and identified twenty-four phenolic compounds, with RA, ferulic acid and salvianolic acid B as the major compounds. Besides RA and salvianolic acid B, chlorogenic acid, caffeic acid, and Luteolin-7-*O*-glucoside were also identified in our LP extract from UAE, but in higher amounts.

The highest content of RA was found in the extract obtained by CET, followed by UAE and MAE ($p < 0.05$). Recent studies of Caleja et al. [40], comparing the same three techniques to extract RA from *Melissa officinalis* L. (Lamiaceae family), showed that UAE proved to be the most effective method than MAE to recover RA. However, Ince et al. [41] demonstrated that MAE was more efficient in extracting RA than CE and UAE. These discrepancies in the results may be because in the studies mentioned, the proportion of the water:ethanol in the extraction solvent was optimized based on response surface models. The same efficiency of UAE was observed for 2,5-dihydroxybenzoic acid and 4,5-*O*-Dicaffeoylquinic acid ($p < 0.05$).

In contrast, CGAs, ferulic acid, salvianolic acid B, lithospermic acid A, and flavonoids were found in greater quantities in the extracts obtained by MAE ($p < 0.05$). UAE showed a lower concentration ($p < 0.05$) of flavonoids; this can be justified because this method, despite applying acoustic cavitation due to the propagation of mechanical waves, did not reach high temperatures, as was the case with CE and MAE. Therefore, the MAE extract showed higher amount of flavonoids (22%) in comparison to CE (15%) and UAE (12%). Conversely, this extract contained fewer phenolic acids than the extracts of UAE (88%) and CE (85%). Regarding caffeic acid and 1,5-Di-*O*-caffeoylquinic acid, the lowest content was found in the CET extract ($p < 0.05$) and no significant differences ($p > 0.05$) were found between UAE and MAE.

RA was the most abundant compound found in the extracts, followed by ferulic acid and salvianolic acid B. These results are following results previously obtained for other *Lavandula* species. Lopes et al. [5] identified thirteen compounds in *L. pedunculata* (Mill.) Cav., being RA (50.9 mg/g DE) and salvianolic acid B (44.3 mg/g DE) the major compounds (in LP grown in Portalegre). However, in same study LP from Bragança showed only 7.5 mg/g DE of RA and 8.7 mg/g DE of salvianolic acid B. Similarly, Costa et al. [30] obtained an LP extract with 6.07 mg/g of RA. However, this extract does not report the presence of salvianolic acid B.

Table 4. Quantification (mg/g DW) via HPLC-DAD of phenolic compounds from LP polyphenol-rich extracts obtained by different extraction techniques. Values are represented by the average ± standard deviation. Different letters in the same row mean significant differences between extraction techniques ($p < 0.05$).

	Phenolic Compound	CET	UAE	MAE
1	2,5-Dihydroxybenzoic acid	1.60 ± 0.06 [a]	1.08 ± 0.09 [b]	0.92 ± 0.03 [c]
2	3-*O*-caffeoylquinic acid	0.48 ± 0.01 [b]	0.43 ± 0.08 [b]	1.46 ± 0.07 [a]
3	5-*O*-caffeoylquinic acid	1.37 ± 0.17 [b]	1.54 ± 0.11 [b]	1.72 ± 0.05 [a]
4	4-*O*-caffeoylquinic acid	1.56 ± 0.06 [b]	0.77 ± 0.07 [c]	1.92 ± 0.06 [a]
5	Caffeic acid	0.31 ± 0.04 [c]	0.92 ± 0.12 [a]	0.66 ± 0.11 [b]
6	1,5-Di-*O*-caffeoylquinic acid	7.13 ± 0.24 [b]	7.86 ± 0.13 [a]	7.83 ± 0.06 [a]
7	Ferulic Acid	0.50 ± 0.09 [b]	0.50 ± 0.07 [b]	1.43 ± 0.06 [a]
8	Luteolin-7-*O*-glucoside	12.79 ± 0.32 [b]	10.82 ± 0.33 [c]	17.56 ± 0.19 [a]
9	Luteolin derivative *	6.04 ± 0.11 [b]	5.12 ± 0.18 [c]	6.53 ± 0.11 [a]
10	4,5-Di-*O*-caffeoylquinic acid	28.52 ± 0.56 [a]	25.25 ± 0.65 [b]	19.28 ± 0.45 [c]
11	Apigenin-7-*O*-glucoside	3.75 ± 0.19 [b]	2.60 ± 0.15 [c]	6.23 ± 0.12 [a]
12	Apigenin derivative +	0.87 ± 0.05 [b]	0.84 ± 0.03 [b]	1.40 ± 0.06 [a]
13	Quercetin-3-*O*-glucoside	2.62 ± 0.18 [c]	2.28 ± 0.07 [b]	4.83 ± 0.10 [a]
14	Rosmarinic acid	58.68 ± 1.42 [a]	52.73 ± 1.86 [b]	48.27 ± 1.92 [c]
15	Salvianolic acid B	42.19 ± 0.71 [ab]	40.09 ± 1.61 [b]	43.19 ± 1.09 [a]
16	Lithospermic acid #	4.64 ± 0.24 [b]	4.48 ± 0.34 [b]	6.30 ± 0.16 [a]
	Total Phenolic Compounds	173.05	157.31	169.53
	Total Phenolic Acids	146.98	135.65	132.98
	Total Flavonoids	26.07	21.66	36.55

* quantified based on luteolin calibration curve; + quantified based on apigenin calibration curve; # quantified based on salvianolic acid B.

On the other hand, luteolin (4.98 mg/g DE) and apigenin (2.74 mg/g DE) were also reported in this LP extract. However, the concentration is lower than obtained in our study with CE and green techniques. In agreement with our results, Costa et al. [30] reported CGAs (0.47 mg/g DE) however, the LP extract obtained with MAE presents a total of CGAs of 5.1 mg/g DE. While Mansinhos et al. [4] reported a few traces of CGA in the UAE extract, in our study, about 2.74 mg/g DE of CGAs are reported. In contrast, the ferulic acid content in this study is six times higher than that obtained in the UAE extract (0.50 ± 0.07 mg/g DE). The achieved variability in phenolic quantification could have been attributed to several factors. The absence of optimization parameters for each extraction method might be the most relevant. Additional factors might include different parameters in the techniques such as power, temperature, or solvent composition, and in addition, the use of other regional LP varieties and gathering conditions.

According to the HPLC-DAD analysis, CET showed the highest content of phenolic compounds (173.05 mg/g DE). Moreover, for the UAE method, the total phenolic compounds by HPLC are lower (157.31 mg/g DE) than MAE (163.53 mg/g DE). The same tendency was observed in the TPC assay, although those differences were not statistically significant ($p < 0.05$) in this assay. In agreement with this, a recent study aiming to improve the extraction of RA from Lamiaceae herbs showed that MAE and CE methods result in similar yields [38]. Based on this, it can be concluded that MAE is only superior to CE in relation to extraction time. In addition, with HPLC analysis, it was possible to identify that although MAE presents similar levels of TPC, its flavonoid content is much higher than CE. Overall, the differences observed between studies could be explained by the geographical area influencing the chemical composition, namely the edaphoclimatic factors. Soil micro-biota, climacteric environment, air humidity, and daily sun exposure are established by the scientific community as primarily responsible for the changes in phenolic compounds observed in plants of the same species.

3.4. Antioxidant Activity

The use of plant-based ingredients and extracts in the food industry has grown due to concerns about the potential adverse effects of synthetic antioxidants. Studies have indicated that natural antioxidants are associated with a lower risk of non-communicable diseases, making them a viable replacement for synthetic antioxidants in food products. *Lavandula* species are recognized for their natural antioxidant properties, which have biological significance and offer economic benefits, mainly when extracted from under-utilized plants like LP. The antioxidant activity was evaluated using three different in vitro assays—ABTS, DPPH and ORAC—and the results are shown in Figure 2. The antioxidant activity measured by ABTS ranged from 156.7 ± 4.3 to 166.5 ± 7.2 mg AAE/g DE, and no statistically significant differences were found between the CE and MAE extraction methods ($p > 0.05$). The DPPH values ranged from 115.7 ± 12.0 to 170.5 ± 12.1 mg TE/g DE, with the lowest value belonging to UAE and the highest to CET and MAE ($p < 0.05$). The radical scavenging effects measured by the ORAC assay ranged from 1306.0 ± 159.8 to 1765.5 ± 214.3, mg TE/g DE, with the lowest value belonging to MAE and the highest to CET and UAE ($p < 0.05$). As shown, the values obtained by the ORAC method were 10-fold higher than those obtained by ABTS and DPPH methods. These differences probably result from the different mechanisms, single electron transfer and hydrogen atom transfer, involved during DPPH/ABTS and ORAC assays. Additionally, DPPH was dissolved in methanol, while ABTS was performed under an aqueous solution; therefore, the DPPH method results in the aprotic reaction being better adapted for determining the effects of less-polar phenolic compounds such as flavonoids. This explains why the extract obtained with MAE presents higher values than the ones obtained with CE and UAE in the DPPH method. This extract has more flavonoids (Section 3.2). In addition, the values of DPPH revealed an excellent correlation ($r^2 = 0.898$) with the flavonoid content.

These discrepancies between the obtained values show why combining more than one method is important to determine the in vitro antioxidant capacity. LP polyphenol-rich extracts obtained by CET showed the highest antioxidant capacity on ORAC assay. These values are consistent with the total phenolic content from HPLC-DAD analysis, which indicates that phenolic compounds are important contributors to the antioxidant properties in LP. Compared to MAE, the high antiradical activity of this extract could be partially attributed to the presence, in the highest amounts, of RA, a compound known for its high redox potential. Among all bioactivities, the antioxidant activity is the most studied. Studies from Baptista et al. [33] have shown that when tested at 50 µg/mL, the methanolic extract from LP showed DPPH radical scavenging inhibition value of about 60%., while the extract obtained with water at the same concentration showed an inhibition percentage of approximately 15%. Curiously, the aqueous extracts only exhibited significant DPPH radical scavenging activity values when tested at 400 µg/mL ($68.9 \pm 2.9\%$). Ferreira et al. [42] tested the ethanolic extracts and decoctions of LP using the DPPH method and the samples revealed antioxidant activity. Baptista et al. [33] also studied the antioxidant potential of different extracts from L. pedunculata, and all samples displayed a high DPPH scavenging activity. Moreover, Costa et al. [30] assessed the radical scavenging effects of the essential oil and polar extracts from *L. pedunculata* subsp. *lusitanica* by ABTS and ORAC assays. After the infusion method, the CE with ethanol 50% (*v*/*v*) without temperature was the most efficient at neutralizing ABTS and peroxyl radicals, while the essential oil was the weakest. Finally, a more recent study with LP extracts from UAE [4], also reported good results in the ABTS, DPPH and ORAC assays. Nevertheless, the results obtained in our study were more promising in terms of antioxidant potential since in ABTS and DPPH, the values range from 20 to 70 mgTE/g DE and in ORAC from 50 to 275 mgTE/g DE.

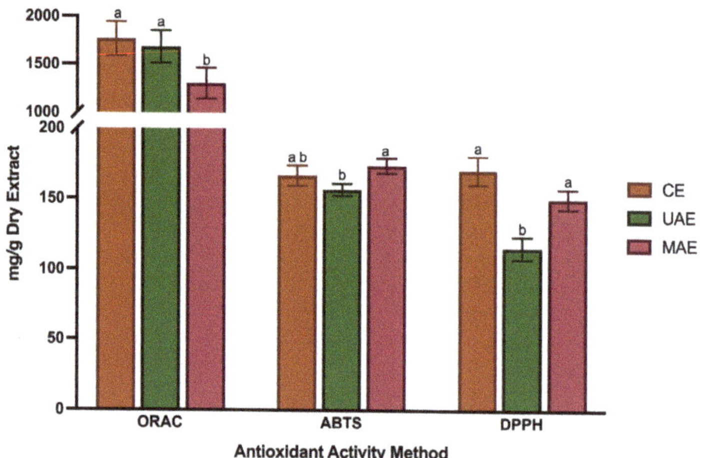

Figure 2. Antioxidant activity of LP extracts by ORAC method (expressed as mg TE/g DE), ABTS method (expressed as mg AAE/g DE) and DPPH method (expressed as mg TE/g DE). Values are represented by the average ± standard deviation. Different letters (a,b) mean significant differences between extraction methods ($p < 0.05$).

3.5. Antimicrobial Activity

The antibacterial activity of LP polyphenol-rich extracts was tested against a panel of six bacteria, including three Gram-positive bacteria (*S. aureus*, *B. cereus*, and *L. monocytogenes*) and three Gram-negative bacteria (*E. coli*, *S. enterica*, and *P. aeruginosa*), specifically selected basis on their importance to public health. Figure 3 presents the growth inhibition curves obtained for each LP extract at 10 mg/mL (highest concentration tested) and Table 5 shows the growth inhibition (%) after 24 h exposure to LP polyphenol-rich extracts. Overall, the LP extract presents good bacterial growth inhibition at 10 mg/mL regardless of the extraction method. However, no bactericide concentration was detected in our study. Conversely, the work of Lopes et al. [3] demonstrated that the MIC/MBC with LP hydroethanolic extracts for *B. cereus*, *S. aureus*, *L. monocytogenes*, *E. coli* and *P. aeruginosa* was 0.075/015, 0.45/0.6, 0.45/0.6, 0.2/0.3 and 0.3/0.45 mg/mL for LP growth in Portugal, respectively. While the study indicates significantly greater antimicrobial activity than our LP extracts, the discrepancy in phenolic compound content suggests that the variance in antimicrobial results could be attributed to the analytical method.

As shown, the least resistant microorganisms were *B. cereus*, *P. aeruginosa* and *L. monocytogenes* since they were inhibited by all extracts regardless of the concentration used. However, the greatest inhibition occurs at the highest concentration (10 mg/mL) with MAE for *B. cereus* ($p < 0.05$), while for *L. monocytogenes*, no statistical differences were found between the CE and MAE extract. In addition, regardless of the extraction method, at 10 mg/mL, all extracts demonstrated the same inhibition capacity for *P. aeruginosa* ($p > 0.05$). The extract obtained with MAE showed higher inhibition for both Gram-positive bacteria (*B. cereus* and *L. monocytogenes*) and all Gram-negative bacteria. Meanwhile, the LP extract obtained with UAE showed a lower inhibition, which agrees with the lower TPC present in this extract and antioxidant activity. Phenolic compounds and other natural agents such as

EOs showed different antimicrobial mechanisms of action; however, the mechanism that is mainly reported is interference on the cell membrane including changes in their structure and function, which may affect electron chain transport, enzyme activity, nutrient uptake and synthesis of nucleic acids and proteins [43].

On the other hand, the most resistant bacteria was *E. coli* since only 10 mg/mL could inhibit 40.5, 39.4 and 59.0% of the growth with CE, UAE and MAE extracts, respectively. Gram-positive bacteria are generally more susceptible because the cell wall has a thick peptidoglycan layer. In contrast, Gram-negative bacteria have a thin peptidoglycan layer and an extra layer, an outer membrane, that consists of phospholipids and lipopolysaccharides [44]. Despite *P. aeruginosa* showing outstanding inhibition with LP extracts, the results obtained in this study agree with those findings in the literature. However, experiments with pure phenolic compounds showed that phenolic acids, such as RA, chlorogenic acid, caffeic acid and some flavonoids (quercetin, luteolin-7-*O*-glucoside and apigenin-7-*O*-glucoside), may be effective against both Gram-negative and Gram-positive bacteria [45]. The presence of caffeic acids in our extracts may explain why *P. aeruginosa* was also inhibited with lower extracts concentration. It is known that a high content of hydroxycinnamic acid and flavonoids is correlated with antioxidant and antimicrobial activities [45,46]. According to our findings from the HPLC-DAD analysis, the substantial presence of identified phenolic compounds, such as RA, chlorogenic acid and caffeic acid in *Lavandula* extracts obtained with the MAE and CE methods, could be the main reason for the antibacterial capacity of these extracts. RA is a caffeic acid ester found in different plants of the Lamiaceae family with antioxidant, anti-inflammatory, and antibacterial effects. Chlorogenic and caffeic acids have also been reported as compounds with significant antioxidant and antimicrobial activity [3,45,47]. However, both CE and MAE extracts presented similar antibacterial activity despite the differences in the average content of each phenolic identified in both extracts. This may happen because of the additive and synergistic effects of phytochemicals in the extracts [45]. Overall, the production of LP extracts with MAE has several utilities and applications in different fields, such as food preservation to replace synthetic additives and to extend the shelf-life of perishable foodstuffs.

Table 5. Growth inhibition (%) for each pathogenic microorganism after 24 h exposure to LP polyphenol-rich extracts (CE, UAE, MAE) at different concentrations. Values are represented by the average ± standard deviation. Different letters (a–f) mean significant differences between the concentration tested ($p < 0.05$).

Microorganism		10 mg/mL			5 mg/mL			2.5 mg/mL		
		CE	UAE	MAE	CE	UAE	MAE	CE	UAE	MAE
Gram +	*Bacillus cereus*	68.6 ± 2.2 [b]	64.0 ± 1.3 [c]	75.8 ± 1.6 [a]	63.6 ± 2.4 [c]	48.8 ± 0.8 [e]	69.0 ± 1.4 [b]	56.5 ± 1.3 [d]	20.5 ± 1.4 [f]	50.2 ± 1.3 [e]
	Listeria monocytogenes	47.1 ± 2.0 [a]	43.7 ± 1.0 [b]	47.1 ± 1.5 [a]	42.5 ± 1.8 [b]	38.1 ± 1.2 [c]	41.0 ± 2.9 [bc]	38.0 ± 2.0 [c]	28.1 ± 1.3 [d]	30.5 ± 0.9 [d]
	Staphylococcus aureus	65.6 ± 2.4 [a]	40.5 ± 2.1 [d]	57.0 ± 2.3 [b]	48.4 ± 1.1 [c]	32.6 ± 1.0 [d]	43.5 ± 2.1 [e]	ni	ni	ni
Gram −	*Escherichia coli*	40.5 ± 1.3 [b]	39.4 ± 2.2 [b]	59.0 ± 1.6 [a]	ni	ni	ni	ni	ni	ni
	Salmonella enterica	65.0 ± 1.5 [a]	61.3 ± 3.0 [ab]	64.6 ± 1.2 [a]	57.5 ± 1.4 [b]	50.3 ± 0.8 [c]	48.2 ± 1.2 [d]	ni	ni	ni
	Pseudomonas aeruginosa	60.7 ± 2.0 [a]	57.1 ± 1.7 [ab]	57.6 ± 1.8 [a]	55.6 ± 1.7 [bc]	44.0 ± 1.2 [d]	52.5 ± 1.4 [c]	45.9 ± 1.2 [d]	28.3 ± 0.7 [f]	40.6 ± 1.4 [e]

ni, no inhibition growth.

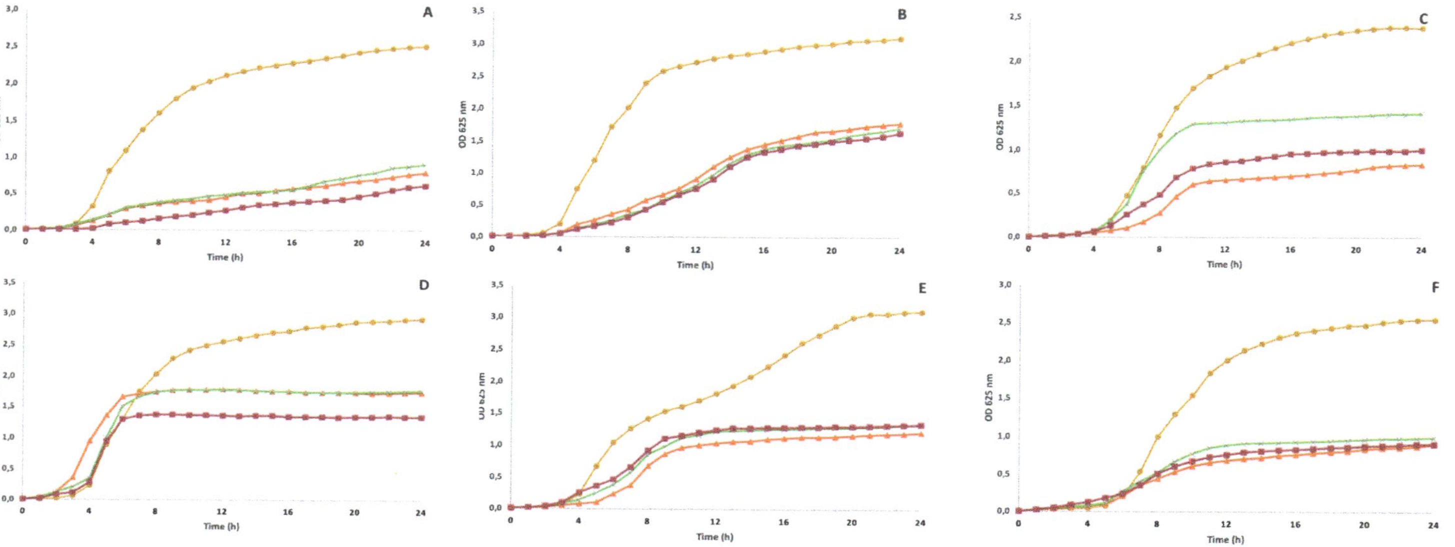

Figure 3. Growth inhibition curves by optical densities (at 625 nm) of *Bacillus cereus* (**A**), *Listeria monocytogens* (**B**), *Staphylococcus aureus* (**C**), *Escherichia coli* (**D**), *Pseudomonas aeruginosa* (**E**) and *Salmonella enterica* (**F**) when incubated with different LP polyphenol-rich extracts extracted based on CE (▲, in orange), UAE (✖, in green) and MAE (■, in pink) at 10 mg/mL and positive control (●, in yellow).

3.6. Tyrosinase Inhibition

Tyrosinase is copper-containing enzyme, also known as PPO, found in mammals, bacteria, fungi and plants, which controls the production of melanin and catalyzes the hydroxylation of monophenols to o-diphenols and their subsequent oxidation to *o*-quinones [48]. Tyrosinase has been implicated in skin diseases and esthetic characteristics such as freckles, melasma, age spots, and in Parkinson's and Huntington's diseases [4]. In addition, this enzyme is responsible for enzymatic browning reactions in fruits and vegetables [37]. Therefore, in the food industry, tyrosinase inhibition is key, as the enzymatic browning contributes to undesirable oxidation in fruit and vegetables, which causes unwelcome effects on their safety, organoleptic properties, and nutrition [49]. Phenolic compounds can interfere with the production of browning catalyzed by tyrosine via (i) reducing the products of the reaction (quinones) and (ii) inhibiting the tyrosinase activity. Therefore, many tyrosinase inhibitors based on polyphenol-rich extracts have been discussed, but only a few have possessed enough potency and safety for food industry application [48]. Hence, pursuing novel, safe, and efficient tyrosinase inhibitors has emerged as an appealing objective. To the best of our knowledge, only two research studies were found concerning the capability of LP polar extracts (obtained with UAE and with CE [4,35]) to inhibit tyrosinase activity. Therefore, this study reports for the first time tyrosinase inhibition with LP extracts obtained with MAE. Figure 4 displays the results for the tyrosinase inhibitory potential of LP extracts. Kojic acid and arbutin are a natural tyrosinase inhibitor clinically used to cure human hyperpigmentation and prevent browning in food products [50]. Thus, this study used kojic acid as a positive control, at 0.02 mg/mL showed a capacity to inhibit tyrosinase of 93.4%. LP polyphenol-rich extracts (5–1.25 mg/mL) could inhibit tyrosinase between 73.01% and 21.33%. However, the inhibition is not directly dependent on extract concentration. The CE and MAE showed better enzyme inhibition at 2.5 mg/mL ($p < 0.05$). However, despite lower inhibition (30.13%) at 1.25 mg/mL, the MAE extract demonstrated greater tyrosinase inhibition capacity than the other extracts ($p < 0.05$).

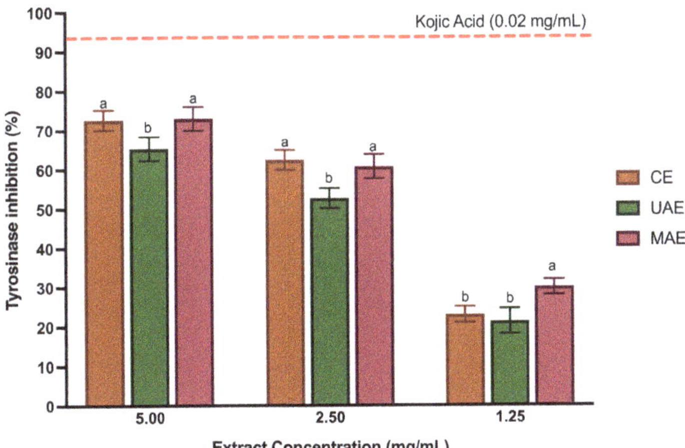

Figure 4. Tyrosinase inhibition (%) for the LP polyphenol-rich extracts at different concentrations (5, 2.5 and 1.25 mg/mL). Values are represented by the average ± standard deviation. Different letters (a,b) mean significant differences between extraction techniques for the same tested concentration ($p < 0.05$).

The big difference between the extracts obtained with MAE and CE is in the number of flavonoids. The structure of flavonoids is compatible with the competitive inhibition of tyrosinase, and detailed studies have shown that some flavonoids such as luteolin, apigenin and quercetin are quite potent inhibitors, some of them with a lower IC_{50} than kojic acid [51]. Otherwise, regardless of tested concentration, the UAE extract showed the lowest % inhibition ($p < 0.05$). This is explained by the fact that this extract generally has fewer phenolic compounds. In agreement, Mansinhos et al. [4] also detected less enzyme inhibition with the LP extract obtained by UAE with ethanol 80% (*v/v*) than in the extracts obtained with DES. In addition, Zengin et al. investigated the capacity of water extracts from LP as tyrosinase inhibitors and showed a good inhibitory capacity, which agrees with our results.

Isolated compounds in high amounts in LP extracts showed a good ability to inhibit tyrosinase. In a prior study, Kang et al. [52] demonstrated the inhibitory potential of RA against mushroom tyrosinase, achieving an IC_{50} value of 16.8 μM. Notably, they observed that the bioactivity of RA was comparable to that of kojic acid, with IC_{50} values of 22.4 μM. Regarding hydroxycinnamic acids, they were shown to have a dual role, functioning either as tyrosinase inhibitors or as enzyme substrates. Furthermore, studies showed the activity of diCGAs were twice as effective as those of CQAs [53]. Likewise, some natural extracts rich in ferulic acid also have strong anti-tyrosinase activity [54]. Furthermore, combining various phenolic compounds effectively inhibits tyrosinase activity rather than employing individual compounds.

As previously mentioned, enzymatic browning represents a significant challenge in the food industry, especially for IV gamma products. To prevent this, natural food additives based on LP polyphenol-rich extract could replace synthetic compounds such as ascorbic acid and sulfite-containing compounds.

3.7. α-Glucosidase and ACE Inhibition

As per the World Health Organization (WHO), metabolic syndrome, characterized by elements like obesity, insulin resistance, and hypertension, stands as a significant health threat in the modern era. Among these, hypertension and type 2 diabetes are considered leading risk factors. Newly released data indicate that approximately 425 million individuals globally are currently living with diabetes, with a potential increase of 48% projected by 2045 [55]. The population affected by hypertension is estimated to be 1 billion, contributing to an annual mortality of around 9.4 million people worldwide [56]. Numerous studies have supported the pharmacological properties of plants in managing diabetes and hypertension, presenting them as a viable alternative in healthcare [57]. The antihypertensive potential of LP polyphenol-rich extracts remains unexplored, regardless of the extraction method used. Previous research has extensively examined ACE inhibitory properties in *Lavandula* genus compounds like EOs [58]. ACE is a metalloproteinase that regulates blood pressure by converting angiotensin I into angiotensin II, a potent vasoconstrictor. In addition, ACE can degrade bradykinin (BK), a potent endogenous peptide vasodilator. ACE inhibition enhances the effects of vasodilator bradykinin (BK) and reduces angiotensin II formation, explaining the benefits of ACE inhibition [59]. The traditional antihypertensive drugs such as captopril and lisinopril showed several side effects [60]. Hence, there is growing interest in discovering safe and efficient ACE inhibitory compounds from plants. The IC_{50} results of CE, MAE and UAE extract were 1.06, 0.98 and 1.17 mg extract/mL, respectively, as illustrated in Table 6. The results can also be expressed based on the TPC, thus demonstrating that extracts with the highest content of phenolic compounds have the lowest IC_{50} (MAE and CE). The specific mechanisms responsible for this enzyme inhibition remain unknown. Still, some of the cardiovascular effects produced by RA, the main phenolic acid present in the LP extracts, include ACE inhibition and/or modulation. For this reason, some reports associated this compound with significant reductions in arterial pressure [59].

Table 6. Antihypertensive activity based on iACE assay. Values are represented by the average ± standard deviation. Different letters (a,b) mean significant differences between extraction techniques for the same tested concentration ($p < 0.05$).

Matrix	Extraction Method	iACE	
		IC_{50} (mg Phenolic Compounds/mL)	IC_{50} (mg Extract/mL)
Lavandula pedunculata	CE	0.19 ± 0.01 [ab]	1.06 ± 0.05 [b]
	UAE	0.21 ± 0.01 [a]	1.17 ± 0.04 [a]
	MAE	0.18 ± 0.02 [b]	0.98 ± 0.05 [b]

In addition, the bradykinin dose–response curves showed that RA promoted a systolic blood pressure reduction similar to captopril. Salvianolic acid B also presents a huge quantity in the extracts, impacting ACE inhibition. Ye et al. [61], using molecular docking methods, found that salvianolic acid B affected ACE and renin to relax blood vessels and regulate hypertension. Other phenolic compounds can inhibit ACE activity and reduce blood pressure. Indeed, using in vivo assays with spontaneous hypertensive rats, blood pressure lowering effects have been confirmed for quercetin and ferulic acid [62]. The phenolic acids and flavonoids are reported to inhibit ACE activity via interaction with the zinc ion on the active site. In addition, the significance of hydroxyl and methoxy groups for zinc metalloproteinase inhibition is also reported [15]. The presence of specific functional groups, such as hydroxyl seems to increase the potency to inhibit ACE as these can act as hydrogen bond acceptor or donor. Conversely, the presence of methoxy groups negatively influences ACE inhibitory activity. Regarding the flavonoids, some studies reported the importance of the hydroxy group on position 7 in the structure of flavonoids for inhibiting ACE enzyme activity [15]. Therefore, the presence of caffeic acid, luteolin-7-*O*-glucoside and apigenin-7-*O*-glucoside could be important for the significant antihypertensive activity observed in the extracts.

However, ACE activity inhibition is not solely attributed to phenolic compounds. This study did not assess triterpenoids and small peptides, recognized for their ACE inhibitor capabilities [63]. To our knowledge, this is the first time that iACE activity in different LP polyphenol-rich extracts was investigated. Therefore, the extracts developed from this medicinal plant serve as an natural alternative as a nutraceutical promoting the cardiovascular health of the population.

Over the past decade, there has been an increase in the number of publications studying the effect of polyphenols on diabetes. All classes of phenolic compounds have shown antidiabetic potential [64]. The research suggests that the primary mechanism of action is associated with inhibiting glucose absorption by inhibiting digestive enzymes such as α-glucosidase and α-amylase [7,65,66]. In the small intestinal epithelial cell, α-glucosidase is an oligosaccharide hydrolase responsible for breaking down oligosaccharides and disaccharides into monosaccharides. Inhibiting the activity of α-glucosidase can slow down carbohydrate absorption, effectively regulating postprandial hyperglycemia and aiding in managing type 2 diabetes [37]. Some in vitro and in vivo studies have reported antidiabetic activity through enzyme inhibition in LP extracts and other *Lavandula* species, obtained through CE with water and methanol [7,35,67]. However, the bioactivity of LP extracts related to α-glucosidase inhibition or other mechanisms has never been studied with LP extracts obtained through green techniques such as UAE and MAE. In addition, to the best of our knowledge, it is the first time that LP of Portuguese geographical origin has been tested for α-glucosidase inhibition. Figure 5 displays the results of the α-glucosidase inhibition by the LP extracts at three different concentrations (5, 2.5 and 1.25 mg/mL) and acarbose, a positive control. Acarbose is the most frequently used drug for diabetes and, in most studies, it has been shown to increase the life expectancy in type 2 diabetes patients. In some cases, a variety of severe side effects such as abdominal pain, diarrhea, flatulence, and skin problems can also occur [68]. The three polyphenol-rich extracts, independent of concentration, showed higher α-glucosidase inhibition. However, as the extract concentration decreases, the % inhibition decreases, even if not directly proportional. At 5 mg/mL, the LP polyphenol-rich extracts and the commercial drug (acarbose) showed a similar activity; however, the MAE extracts presented a higher activity than the remaining extracts ($p < 0.05$). Furthermore, as the tested concentration decreased, this extract and the CE extract remained the most significantly active ($p < 0.05$) in inhibiting the enzyme. These results agree that these two extracts display a higher content of phenolic compounds than the extract obtained with UAE. In particular, the MAE extract showed a higher amount of flavonoids (Section 3.3), which have been reported by Ali et al. [69] using molecular docking techniques to have a significant role in the α-glucosidase inhibition. Despite differences in extraction techniques and solvents, our results agree with studies reported with extracts with LP from Morocco and Turkey. The LP extracts obtained in this studies (CE with methanol) showed considerable α-glucosidase inhibition in vitro [35] and acute and chronic oral administration of LP aqueous extract reduced the peak of the glucose concentration [7]. Several important hydroxycinnamic acids present in high concentration in LP extracts, such as ferulic acid [70], chlorogenic acid [71,72], caffeic acid [73] and RA [74,75] have shown considerable hypoglycemic activity via in vitro and in vivo experiments. The potential interaction mechanism between these BCs and α-glucosidase reported in the literature describes the formation of hydrogen bonds or hydrophobic forces in the enzyme's active site [70]. In particular, the in vitro studies of Kubínova et al. [74] showed that IC_{50} of RA was four-times lower than acarbose. In addition, this phenolic acid has been demonstrated to regulate glucose homeostasis, restoring the blood glucose level and regulating adiponectin and leptin levels in diabetic rats [76,77]. Also, the salvianolic acid had a potent α-glucosidase inhibitory activity, whilst the interaction mechanisms remain unclear [68]. Other important flavonoids such as apigenin and luteolin present in LP extracts are reported for their capacity to inhibit α-glucosidase competitively, forming complexes where the main forces driving the interaction were hydrophobic and hydrogen bonding [78].In recent studies, gentisic acid showed a moderate antidiabetic activity by inhibiting α-glucosidase [79].

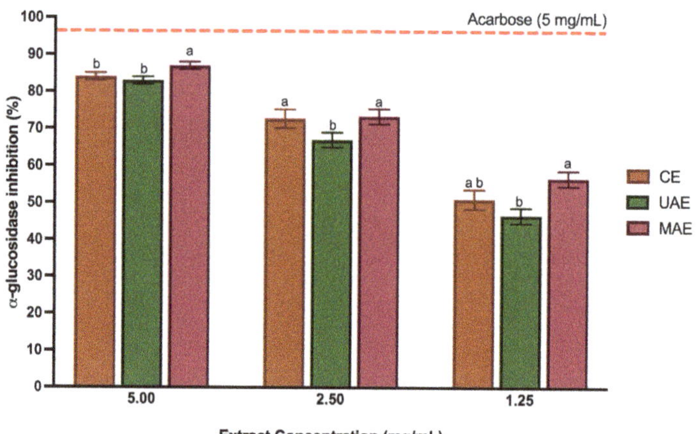

Figure 5. Antidiabetic activity (α-glucosidase inhibition assay) of LP polyphenol-rich extracts. Values are represented by the average $\pm$ standard deviation. Different letters (a,b mean significant differences between extraction techniques for the same tested concentration ($p < 0.05$).

Although the values of α-glucosidase are inferior to those obtained for acarbose, the results show high potential. In addition, the LP polyphenol-rich extracts were obtained from a natural, unexplored source with green techniques and solvents, allowing the safe incorporation into food products. These extracts may be a sustainable alternative to regulate type 2 diabetes.

3.8. Cytotoxicity and Mutagenicity

The human epithelial cell line caco-2 monolayers have been widely used for mimicking intestinal conditions, providing a reliable model for investigating the absorption of BCs after oral administration in humans [80]. They are routinely used to determine the mechanism of action of BCs or extracts to move towards preclinical animal studies and human clinical trials for ensuring the safety of the final food product [81]. As shown in Figure 6, LP extracts did not exert any inhibitory activity upon caco-2 cellular metabolism at all tested concentrations. This is in line with the standard ISO 10993-5:2009 [25], which considers as cytotoxic when the cell viability is lower than 70%. However, the MAE extract at 4.8 mg/mL had a value of 28.67 $\pm$ 4.16%, which is very close to the limit established as toxic.

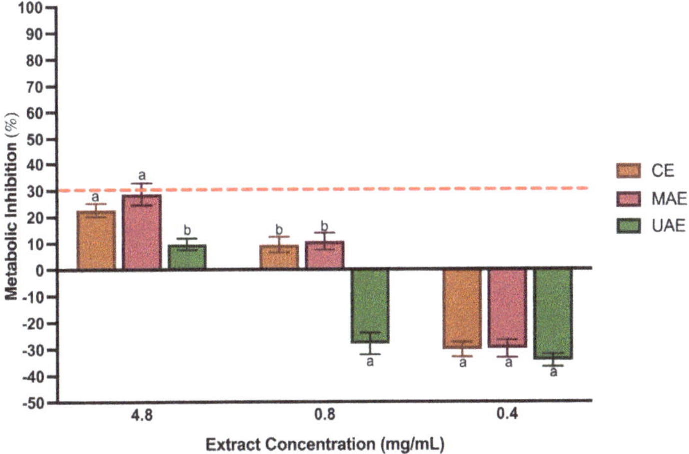

Figure 6. Cytotoxicity of LP polyphenol-rich extract at different concentrations. The dotted line represents the 30% cytotoxicity limit defined by the ISO 10993-5 [25]. Values are the mean $\pm$ standard deviation. Different letters mean significant differences ($p < 0.05$).

In contrast, the UAE extract at the same concentration showed about 3-fold less metabolic inhibition ($p < 0.05$). Interestingly, when considering the impact of the lowest LP concentration upon caco-2 metabolism, an apparent metabolic stimulation (metabolism inhibition negative values) represents a metabolic activity higher than the growth control. Considering this result, it is recommended to quantify the DNA using Pico Green Assay, as performed by Rodrigues et al. [82], to assess the actual effect of LP extracts on metabolic activity.

As far as we know, no cytotoxicity studies on caco-2 cells exist in the literature with LP extracts. However, Costa et al. [83] studied the cytotoxicity of *Lavandula viridis* L'Hér (*Lavandula* genus) extract at 0.5 mg/mL obtained under CE with water:ethanol 50% (*v/v*) and showed that the viability of the caco-2 cell line was not affected by the extracts. The same study also proved no metabolic inhibition for RA (at concentrations of 0.125 and 0.250 mg/mL), the main phenolic compound in *Lavandula viridis* L'Hér. [75].

Evaluating genotoxic properties is an essential step of the safety assessment of substances intended to be used as pharmaceuticals, nutraceuticals, or food additives. The Ames test was standardized in the 1975s by Ames et al. [84] to evaluate the mutagenic capabilities of chemicals. Recently, the Ames test has become a commonly employed method for identifying both the mutagenic and antimutagenic properties of medicinal plants and extracts derived from plants. The Ames test uses *Salmonella typhimurium* strains that are auxotrophic for histidine to detect gene mutations, specifically as base pair substitutions and frameshift mutations. Moreover, it is emphasized how crucial it is to conduct the Ames test both in the absence and presence of the exogenous metabolic system (S9). The addition of the S9 fraction is essential to replicate the metabolism of the test substance as it would occur in mammals. Consequently, tests lacking these two conditions jeopardize the accuracy of findings and compromise the safety assessment of the extracts under evaluation.

In the genotoxic profile of the polyphenol-rich LP extracts (Table 7), no harmful effects were found, as none of the tested concentrations exerted any genotoxic effects, either with and or without metabolic activation, against the tested strain, according to parameters defined by the manufacturer of the Ames test kit and the OECD guidelines. Briefly, to conclude the genotoxicity, the values should be at least 2.5 times higher than those obtained with the solvent control (water in this study). So, with S9, the values should be lower than 28.75; without S9 activation, the values should be lower than 41.25. To our knowledge, this paper reported for the first time the genotoxicity of LP extracts whether obtained using the CE methodology or via green technologies such as MAE and UAE.

Table 7. Results obtained for LP polyphenol-rich extracts AMES genotoxicity assay against *S. typhymurium* TA98 with and without the metabolic activation (S9). All values represent the average number of positive (revertant) wells $\pm$ standard deviation. Different letters (a,b) mean significant differences between only the LP polyphenol-rich extracts ($p < 0.05$).

Sample Test	TA 98	
	With S9	**Without S9**
Solvent Control (water)	11.51 ± 0.50	16.50 ± 1.50
Positive Control	303.00 ± 12.50	481.50 ± 3.50
CE	18.00 ± 2.00 [a]	27.50 ± 0.50 [a]
MAE	13.50 ± 3.00 [ab]	20.00 ± 1.00 [b]
UAE	10.00 ± 2.50 [b]	18.50 ± 2.00 [b]

The absence of cytotoxicity and mutagenicity in the extracts suggests that the present components did not induce cellular or DNA damage. Therefore, they can be considered safe at the concentrations tested. Among the extraction methods employed, UAE emerged as the safest choice due to its slightly lower phenolic compound content.

4. Conclusions

To achieve a change encouraging the conservation and valorization of Portuguese flora, this research article highlights important insights concerning evidence of the bioactivity of LP polyphenol-rich extracts obtained with different extraction methodologies (CE, MAE, and UAE) such as antioxidant, antimicrobial, anti-hypertensive, antidiabetic and anti-browning effects, as well as proving the safe use for human consumption through cytotoxicity and mutagenicity assays.

The findings demonstrate that MAE is a superior method in terms of the time and energy saved compared to UAE; however, the extraction yields and phenolic compounds obtained in the extracts from *LP* are very similar to those obtained through CE. MAE enabled the extraction of a comparable amount of TPC to CE but in a shorter extraction time, reducing energy consumption. Both polyphenol-rich extracts exhibited higher quantities of RA, salvianolic acid, 4,5-Di-*O*-caffeoylquinic acid and luteolin-7-*O*-glucoside; therefore, demonstrated increased antioxidant capacity evaluated by in vitro methods. In addition, the extracts showed great growth inhibition for *B. cereus*, *S. aureus*, *S. enterica* and *P. aeruginosa* (>50%) at very low concentrations (10 mg/mL). As we understand it, the significant positive antihypertensive and antidiabetic activity in LP extracts are mainly obtained with MAE; this extract showed potent antihypertensive activity (IC_{50} = 0.98 mg DE/mL), and antidiabetic activity (87% at 5 mg/mL) was reported for the first time. Moreover, the same extract also showed significant tyrosinase inhibition (73% at 5 mg/mL). Ultimately, the extracts demonstrated no cytotoxic or genotoxic effects, irrespective of the extraction method, affirming their safe application within the food industry, particularly as antioxidant and antimicrobial food additives. Additionally, this underscores the substantial potential to utilize LP extracts in the production of nutraceutical products, offering antihypertensive and antidiabetic effects.

Overall, the results and evidence will be the beginning of a possible increase in the exploitation and valorization of LP, which may trigger its major production and industrial processing to develop and optimize sustainable food-based products. Explored Portuguese plants bring value to the country while minimizing environmental impact by using other important raw materials to produce natural functional ingredients, promoting the bioeconomy.

Author Contributions: Conceptualization, A.A.V.-B. and M.P.; methodology, investigation, validation, A.A.V.-B., C.N., S.R., M.M., R.G.-G. and A.L.S.O.; formal analysis, A.A.V.-B., C.N., S.R. and A.L.S.O.; writing—original draft preparation, A.A.V.-B.; writing—review and editing, R.G.-G. and A.L.S.O.; supervision, M.P. and J.N.; project administration, J.N. and M.P. All authors have read and agreed to the published version of the manuscript.

Funding: This work received financial support from the Ministry of Agriculture and Rural Development, co-financed by the European Agricultural Fund for Rural Development (EAFRD), through the partnership agreement Portugal PDR-2020, under the project Nature Bioactive Food: Desenvolvimento de produtos e ingredientes alimentares bioativos através de recursos agrícolas endógenos portugueses para uma alimentação saudável (Ref. PDR2020-101-031531). In addition, this work was also supported by the Transfer Empreende project (Norte-02-0651-FEDER-000081), supported by Norte Portugal Regional Operational Programme (NORTE 2020), under the Portugal 2020 Partnership Agreement, through the European Regional Development Fund (ERDF).

Data Availability Statement: All data generated or analyzed during this study are included in this published article.

Acknowledgments: The authors would like to thank the Nature Bioactive Food project (PDR2020-101-031531), the Transfer Empreende project (Norte-02-0651-FEDER-000081), and the scientific collaboration of the Escola Superior de Biotecnologia–Universidade Católica Portuguesa through CBQF under the FCT project UIDB/50016/2020.

Conflicts of Interest: The authors declare no conflict of interest.

References

1. Habán, M.; Korczyk-Szabó, J.; Čerteková, S.; Ražná, K. Lavandula Species, Their Bioactive Phytochemicals, and Their Biosynthetic Regulation. *Int. J. Mol. Sci.* **2023**, *24*, 8831. [CrossRef]
2. da Cunha, A.P.; Nogueira, M.T.; Roque, O.R. *Plantas Aromáticas e Óleos Essenciais: Composição e Aplicações*; Fundação Calouste Gulbenkian: Lisbon, Portugal, 2012; ISBN 972311450X.
3. Lopes, C.L.; Pereira, E.; Soković, M.; Carvalho, A.M.; Barata, A.M.; Lopes, V.; Rocha, F.; Calhelha, R.C.; Barros, L.; Ferreira, I.C.F.R. Phenolic Composition and Bioactivity of *Lavandula pedunculata* (Mill.) Cav. Samples from Different Geographical Origin. *Molecules* **2018**, *23*, 1037. [CrossRef]
4. Mansinhos, I.; Gonçalves, S.; Rodríguez-Solana, R.; Luis Ordóñez-Díaz, J.; Manuel Moreno-Rojas, J.; Romano, A.; Barba, F.J. Ultrasonic-Assisted Extraction and Natural Deep Eutectic Solvents Combination: A Green Strategy to Improve the Recovery of Phenolic Compounds from *Lavandula pedunculata* Subsp. Lusitanica (Chaytor) Franco. *Antioxidants* **2021**, *10*, 582. [CrossRef]
5. Arruda, H.S.; Neri-Numa, I.A.; Kido, L.A.; Júnior, M.R.M.; Pastore, G.M. Recent Advances and Possibilities for the Use of Plant Phenolic Compounds to Manage Ageing-Related Diseases. *J. Funct. Foods* **2020**, *75*, 104203. [CrossRef]
6. Domingues, J.; Delgado, F.; Gonçalves, J.C.; Zuzarte, M.; Duarte, A.P. Mediterranean Lavenders from Section Stoechas: An Undervalued Source of Secondary Metabolites with Pharmacological Potential. *Metabolites* **2023**, *13*, 337. [CrossRef] [PubMed]
7. Boutahiri, S.; Bouhrim, M.; Abidi, C.; Mechchate, H.; Alqahtani, A.S.; Noman, O.M.; Elombo, F.K.; Gressier, B.; Sahpaz, S.; Bnouham, M.; et al. Antihyperglycemic Effect of *Lavandula pedunculata*: In Vivo, in Vitro and Ex Vivo Approaches. *Pharmaceutics* **2021**, *13*, 2019. [CrossRef] [PubMed]
8. Alara, O.R.; Abdurahman, N.H.; Ukaegbu, C.I. Extraction of Phenolic Compounds: A Review. *Curr. Res. Food Sci.* **2021**, *4*, 200–214. [CrossRef] [PubMed]
9. Panzella, L.; Moccia, F.; Nasti, R.; Marzorati, S.; Verotta, L.; Napolitano, A. Bioactive Phenolic Compounds from Agri-Food Wastes: An Update on Green and Sustainable Extraction Methodologies. *Front. Nutr.* **2020**, *7*, 60. [CrossRef]

10. Castiello, C.; Junghanns, P.; Mergel, A.; Jacob, C.; Ducho, C.; Valente, S.; Rotili, D.; Fioravanti, R.; Zwergel, C.; Mai, A. GreenMedChem: The Challenge in the next Decade toward Eco-Friendly Compounds and Processes in Drug Design. *Green Chem.* **2023**, *25*, 2109–2169. [CrossRef]

11. Asioli, D.; Aschemann-Witzel, J.; Caputo, V.; Vecchio, R.; Annunziata, A.; Næs, T.; Varela, P. Making Sense of the "Clean Label" Trends: A Review of Consumer Food Choice Behavior and Discussion of Industry Implications. *Food Res. Int.* **2017**, *99*, 58–71. [CrossRef]

12. Dias, C.; Fonseca, A.M.A.; Amaro, A.L.; Vilas-Boas, A.A.; Oliveira, A.; Santos, S.A.O.; Silvestre, A.J.D.; Rocha, S.M.; Isidoro, N.; Pintado, M. Natural-Based Antioxidant Extracts as Potential Mitigators of Fruit Browning. *Antioxidants* **2020**, *9*, 715. [CrossRef] [PubMed]

13. De Corato, U. Improving the Shelf-Life and Quality of Fresh and Minimally-Processed Fruits and Vegetables for a Modern Food Industry: A Comprehensive Critical Review from the Traditional Technologies into the Most Promising Advancements. *Crit. Rev. Food Sci. Nutr.* **2020**, *60*, 940–975. [CrossRef] [PubMed]

14. Ali Asgar, M.D. Anti-Diabetic Potential of Phenolic Compounds: A Review. *Int. J. Food Prop.* **2013**, *16*, 91–103. [CrossRef]

15. Al Shukor, N.; Van Camp, J.; Gonzales, G.B.; Staljanssens, D.; Struijs, K.; Zotti, M.J.; Raes, K.; Smagghe, G. Angiotensin-Converting Enzyme Inhibitory Effects by Plant Phenolic Compounds: A Study of Structure Activity Relationships. *J. Agric. Food Chem.* **2013**, *61*, 11832–11839. [CrossRef]

16. Choudhary, P.; Khade, M.; Savant, S.; Musale, A.; Chelliah, M.S.; Dasgupta, S. Empowering Blue Economy: From Underrated Ecosystem to Sustainable Industry. *J. Environ. Manag.* **2021**, *291*, 112697. [CrossRef]

17. Dobros, N.; Zawada, K.; Paradowska, K. Phytochemical Profile and Antioxidant Activity of *Lavandula angustifolia* and *Lavandula × intermedia* Cultivars Extracted with Different Methods. *Antioxidants* **2022**, *11*, 711. [CrossRef] [PubMed]

18. Leocádio, J.C.S. A Influência Do Método de Extração Na Bioatividade e Perfil Metabólico de Extratos de Rosmaninho. Master's Thesis, Universidade de Coimbra, Coimbra, Portugal, 2018.

19. Vilas-Boas, A.A.; Campos, D.A.; Nunes, C.; Ribeiro, S.; Nunes, J.; Oliveira, A.; Pintado, M. Polyphenol Extraction by Different Techniques for Valorisation of Non-Compliant Portuguese Sweet Cherries towards a Novel Antioxidant Extract. *Sustainability* **2020**, *12*, 5556. [CrossRef]

20. Coscueta, E.R.; Reis, C.A.; Pintado, M. Phenylethyl Isothiocyanate Extracted from Watercress By-Products with Aqueous Micellar Systems: Development and Optimisation. *Antioxidants* **2020**, *9*, 698. [CrossRef]

21. CLSI. *Methods for Dilution Antimicrobial Susceptibility Test for Bacteria That GrowAerobically*, 11th ed.; CLSI standard M07; CLSI: Wayne, NJ, USA, 2015.

22. Ribeiro, T.B.; Oliveira, A.; Coelho, M.; Veiga, M.; Costa, E.M.; Silva, S.; Nunes, J.; Vicente, A.A.; Pintado, M. Are Olive Pomace Powders a Safe Source of Bioactives and Nutrients? *J. Sci. Food Agric.* **2021**, *101*, 1963–1978. [CrossRef]

23. Kwon, Y.-I.I.; Vattem, D.A.; Shetty, K. Evaluation of Clonal Herbs of Lamiaceae Species for Management of Diabetes and Hypertension. *Asia Pac. J. Clin. Nutr.* **2006**, *15*, 107.

24. Coscueta, E.R.; Campos, D.A.; Osório, H.; Nerli, B.B.; Pintado, M. Enzymatic Soy Protein Hydrolysis: A Tool for Biofunctional Food Ingredient Production. *Food Chem. X* **2019**, *1*, 100006. [CrossRef] [PubMed]

25. Wallin, R.F.; Arscott, E.F. A Practical Guide to ISO 10993-5: Cytotoxicity. *Med. Device Diagn. Ind.* **1998**, *20*, 96–98.

26. OECD 471: Bacterial Reverse Mutation Test. *OECD Guidel. Test. Chem. Sect.* **1997**, *4*, 1–11.

27. Drăghici-Popa, A.-M.; Boscornea, A.C.; Brezoiu, A.-M.; Tomas, Ș.T.; Pârvulescu, O.C.; Stan, R. Effects of Extraction Process Factors on the Composition and Antioxidant Activity of Blackthorn (*Prunus spinosa* L.) Fruit Extracts. *Antioxidants* **2023**, *12*, 1897. [CrossRef] [PubMed]

28. Vo, T.P.; Nguyen, N.T.U.; Le, V.H.; Phan, T.H.; Nguyen, T.H.Y.; Nguyen, D.Q. Optimizing Ultrasonic-Assisted and Microwave-Assisted Extraction Processes to Recover Phenolics and Flavonoids from Passion Fruit Peels. *ACS Omega* **2023**, *8*, 33870–33882. [CrossRef] [PubMed]

29. Lezoul, N.E.H.; Belkadi, M.; Habibi, F.; Guillén, F. Extraction Processes with Several Solvents on Total Bioactive Compounds in Different Organs of Three Medicinal Plants. *Molecules* **2020**, *25*, 4672. [CrossRef]

30. Costa, P.; Gonçalves, S.; Valentão, P.; Andrade, P.B.; Almeida, C.; Nogueira, J.M.F.; Romano, A. Metabolic Profile and Biological Activities of *Lavandula pedunculata* Subsp. *lusitanica* (Chaytor) Franco: Studies on the Essential Oil and Polar Extracts. *Food Chem.* **2013**, *141*, 2501–2506. [CrossRef]

31. Gonçalves, S.; Romano, A. Inhibitory Properties of Phenolic Compounds against Enzymes Linked with Human Diseases. *Phenolic Compd.-Biol. Act.* **2017**, *40*, 100–120.

32. Vilas-Boas, A.A.; Oliveira, A.; Ribeiro, T.B.; Ribeiro, S.; Nunes, C.; Gómez-García, R.; Nunes, J.; Pintado, M. Impact of Extraction Process in Non-Compliant 'Bravo de Esmolfe' Apples towards the Development of Natural Antioxidant Extracts. *Appl. Sci.* **2021**, *11*, 5916. [CrossRef]

33. Baptista, R.; Madureira, A.M.; Jorge, R.; Adão, R.; Duarte, A.; Duarte, N.; Lopes, M.M.; Teixeira, G. Antioxidant and Antimycotic Activities of Two Native Lavandula Species from Portugal. *Evid.-Based Complement. Altern. Med.* **2015**, *2015*, 570521. [CrossRef]

34. Boutahiri, S.; Eto, B.; Bouhrim, M.; Mechchate, H.; Saleh, A.; Al Kamaly, O.; Drioiche, A.; Remok, F.; Samaillie, J.; Neut, C.; et al. *Lavandula pedunculata* (Mill.) Cav. Aqueous Extract Antibacterial Activity Improved by the Addition of *Salvia rosmarinus* Spenn., *Salvia lavandulifolia* Vahl and *Origanum compactum* Benth. *Life* **2022**, *12*, 328. [CrossRef] [PubMed]

35. Zengin, G.; Yagi, S.; Selvi, S.; Cziáky, Z.; Jeko, J.; Sinan, K.I.; Topcu, A.A.; Erci, F.; Boczkaj, G. Elucidation of Chemical Compounds in Different Extracts of Two Lavandula Taxa and Their Biological Potentials: Walking with Versatile Agents on the Road from Nature to Functional Applications. *Ind. Crops Prod.* **2023**, *204*, 117366. [CrossRef]

36. Zeng, G.; Xiao, H.; Liu, J.; Liang, X. Identification of Phenolic Constituents in Radix Salvia Miltiorrhizae by Liquid Chromatography/Electrospray Ionization Mass Spectrometry. *Rapid Commun. Mass Spectrom.* **2006**, *20*, 499–506. [CrossRef] [PubMed]

37. Oliveira, A.L.S.; Carvalho, M.J.; Oliveira, D.L.; Costa, E.; Pintado, M.; Madureira, A.R. Sugarcane Straw Polyphenols as Potential Food and Nutraceutical Ingredient. *Foods* **2022**, *11*, 4025. [CrossRef]

38. Sik, B.; Hanczné, E.L.; Kapcsándi, V.; Ajtony, Z. Conventional and Nonconventional Extraction Techniques for Optimal Extraction Processes of Rosmarinic Acid from Six Lamiaceae Plants as Determined by HPLC-DAD Measurement. *J. Pharm. Biomed. Anal.* **2020**, *184*, 113173. [CrossRef]

39. Kim, D.-S.; Kim, M.-B.; Lim, S.-B. Enhancement of Phenolic Production and Antioxidant Activity from Buckwheat Leaves by Subcritical Water Extraction. *Prev. Nutr. Food Sci.* **2017**, *22*, 345. [CrossRef]

40. Caleja, C.; Barros, L.; Prieto, M.A.; Barreiro, M.F.; Oliveira, M.B.P.P.; Ferreira, I.C.F.R. Extraction of Rosmarinic Acid from Melissa Officinalis L. by Heat-, Microwave-and Ultrasound-Assisted Extraction Techniques: A Comparative Study through Response Surface Analysis. *Sep. Purif. Technol.* **2017**, *186*, 297–308. [CrossRef]

41. Ince, A.E.; Şahin, S.; Şümnü, S.G. Extraction of Phenolic Compounds from Melissa Using Microwave and Ultrasound. *Turk. J. Agric. For.* **2013**, *37*, 69–75. [CrossRef]

42. Ferreira, A.; Proença, C.; Serralheiro, M.L.M.; Araújo, M.E.M. The in Vitro Screening for Acetylcholinesterase Inhibition and Antioxidant Activity of Medicinal Plants from Portugal. *J. Ethnopharmacol.* **2006**, *108*, 31–37. [CrossRef]

43. Gómez-Garcýa, R.; Campos, D.A.; Vilas-Boas, A.; Madureira, A.R.; Pintado, M. Natural Antimicrobials from Vegetable By-Products: Extraction, Bioactivity, and Stability. *Food Microbiol. Biotechnol.* **2020**, *2020*, 249–286.

44. Jackson, M.; Stevens, C.M.; Zhang, L.; Zgurskaya, H.I.; Niederweis, M. Transporters Involved in the Biogenesis and Functionalization of the Mycobacterial Cell Envelope. *Chem. Rev.* **2020**, *121*, 5124–5157. [CrossRef] [PubMed]

45. Zenão, S.; Aires, A.; Dias, C.; Saavedra, M.J.; Fernandes, C. Antibacterial Potential of Urtica Dioica and Lavandula Angustifolia Extracts against Methicillin Resistant Staphylococcus Aureus Isolated from Diabetic Foot Ulcers. *J. Herb. Med.* **2017**, *10*, 53–58. [CrossRef]

46. Jurić, T.; Mićić, N.; Potkonjak, A.; Milanov, D.; Dodić, J.; Trivunović, Z.; Popović, B.M. The Evaluation of Phenolic Content, in Vitro Antioxidant and Antibacterial Activity of Mentha Piperita Extracts Obtained by Natural Deep Eutectic Solvents. *Food Chem.* **2021**, *362*, 130226. [CrossRef] [PubMed]

47. Nunes, R.; Pasko, P.; Tyszka-Czochara, M.; Szewczyk, A.; Szlosarczyk, M.; Carvalho, I.S. Antibacterial, Antioxidant and Anti-Proliferative Properties and Zinc Content of Five South Portugal Herbs. *Pharm. Biol.* **2017**, *55*, 114–123. [CrossRef] [PubMed]

48. Zolghadri, S.; Bahrami, A.; Hassan Khan, M.T.; Munoz-Munoz, J.; Garcia-Molina, F.; Garcia-Canovas, F.; Saboury, A.A. A Comprehensive Review on Tyrosinase Inhibitors. *J. Enzyme Inhib. Med. Chem.* **2019**, *34*, 279–309. [CrossRef]

49. Peng, Z.; Wang, G.; He, Y.; Wang, J.J.; Zhao, Y. Tyrosinase Inhibitory Mechanism and Anti-Browning Properties of Novel Kojic Acid Derivatives Bearing Aromatic Aldehyde Moiety. *Curr. Res. Food Sci.* **2023**, *6*, 100421. [CrossRef]

50. Mukherjee, P.K.; Biswas, R.; Sharma, A.; Banerjee, S.; Biswas, S.; Katiyar, C.K. Validation of Medicinal Herbs for Anti-Tyrosinase Potential. *J. Herb. Med.* **2018**, *14*, 1–16. [CrossRef]

51. Fan, M.; Ding, H.; Zhang, G.; Hu, X.; Gong, D. Relationships of Dietary Flavonoid Structure with Its Tyrosinase Inhibitory Activity and Affinity. *LWT* **2019**, *107*, 25–34. [CrossRef]

52. Kang, H.S.; Kim, H.R.; Byun, D.S.; Park, H.J.; Choi, J.S. Rosmarinic Acid as a Tyrosinase Inhibitors from Salvia Miltiorrhiza. *Nat. Prod. Sci.* **2004**, *10*, 80–84.

53. Iwai, K.; Kishimoto, N.; Kakino, Y.; Mochida, K.; Fujita, T. In Vitro Antioxidative Effects and Tyrosinase Inhibitory Activities of Seven Hydroxycinnamoyl Derivatives in Green Coffee Beans. *J. Agric. Food Chem.* **2004**, *52*, 4893–4898. [CrossRef]

54. Yu, Q.; Fan, L. Understanding the Combined Effect and Inhibition Mechanism of 4-Hydroxycinnamic Acid and Ferulic Acid as Tyrosinase Inhibitors. *Food Chem.* **2021**, *352*, 129369. [CrossRef]

55. Saeedi, P.; Petersohn, I.; Salpea, P.; Malanda, B.; Karuranga, S.; Unwin, N.; Colagiuri, S.; Guariguata, L.; Motala, A.A.; Ogurtsova, K. Global and Regional Diabetes Prevalence Estimates for 2019 and Projections for 2030 and 2045: Results from the International Diabetes Federation Diabetes Atlas. *Diabetes Res. Clin. Pract.* **2019**, *157*, 107843. [CrossRef] [PubMed]

56. Berek, P.A.L.; Irawati, D.; Hamid, A.Y.S. Hypertension: A Global Health Crisis. *Ann. Clin. Hypertens.* **2021**, *5*, 8–11.

57. Chukwuma, C.I.; Matsabisa, M.G.; Ibrahim, M.A.; Erukainure, O.L.; Chabalala, M.H.; Islam, M.S. Medicinal Plants with Concomitant Anti-Diabetic and Anti-Hypertensive Effects as Potential Sources of Dual Acting Therapies against Diabetes and Hypertension: A Review. *J. Ethnopharmacol.* **2019**, *235*, 329–360. [CrossRef] [PubMed]

58. Biltekin, S.N.; Karadağ, A.E.; Demirci, B.; Demirci, F. ACE2 and LOX Enzyme Inhibitions of Different Lavender Essential Oils and Major Components Linalool and Camphor. *ACS Omega* **2022**, *7*, 36561–36566. [CrossRef] [PubMed]

59. Ferreira, L.G.; Evora, P.R.B.; Capellini, V.K.; Albuquerque, A.A.; Carvalho, M.T.M.; da Silva Gomes, R.A.; Parolini, M.T.; Celotto, A.C. Effect of Rosmarinic Acid on the Arterial Blood Pressure in Normotensive and Hypertensive Rats: Role of ACE. *Phytomedicine* **2018**, *38*, 158–165. [CrossRef] [PubMed]

60. Fernandez Cunha, M.; Coscueta, E.R.; Brassesco, M.E.; Marques, R.; Neto, J.; Almada, F.; Gonçalves, D.; Pintado, M. Exploring Bioactivities and Peptide Content of Body Mucus from the Lusitanian Toadfish *Halobatrachus didactylus*. *Molecules* **2023**, *28*, 6458. [CrossRef]

61. Ye, L.; He, Y.; Ye, H.; Liu, X.; Yang, L.; Cao, Z.; Tang, K. Pathway-Pathway Network-Based Study of the Therapeutic Mechanisms by Which Salvianolic Acid B Regulates Cardiovascular Diseases. *Chin. Sci. Bull.* **2012**, *57*, 1672–1679. [CrossRef]

62. Häckl, L.P.N.; Cuttle, G.; Sanches Dovichi, S.; Lima-Landman, M.T.; Nicolau, M. Inhibition of Angiotensin-Converting Enzyme by Quercetin Alters the Vascular Response to Bradykinin and Angiotensin I. *Pharmacology* **2002**, *65*, 182–186. [CrossRef]

63. Ribeiro, T.B.; Oliveira, A.; Campos, D.; Nunes, J.; Vicente, A.A.; Pintado, M. Simulated Digestion of an Olive Pomace Water-Soluble Ingredient: Relationship between the Bioaccessibility of Compounds and Their Potential Health Benefits. *Food Funct.* **2020**, *11*, 2238–2254. [CrossRef]

64. Voss, G.B.; Oliveira, A.L.S.; da Cruz Alexandre, E.M.; Pintado, M.E. Importance of Polyphenols: Consumption and Human Health. In *Technologies to Recover Polyphenols from AgroFood By-Products and Wastes*; Elsevier: Amsterdam, The Netherlands, 2022; pp. 1–23.

65. Williamson, G. Possible Effects of Dietary Polyphenols on Sugar Absorption and Digestion. *Mol. Nutr. Food Res.* **2013**, *57*, 48–57. [CrossRef]

66. Vasquez-Ramos, C.S.; Garcia-Moreno, M.G.; García-García, D.; Martínez-Medina, G.A.; Niño-Herrera, S.A.; Luna-García, H.; Paciós-Michelena, S.; Ilyina, A.; Segura-Ceniceros, E.P.; Chávez-González, M.L. Natural Extracts and Compounds as Inhibitors of Amylase for Diabetes Treatment and Prevention. *Funct. Foods Nutraceuticals Hum. Health Adv. Nat. Wellness Dis. Prev.* **2021**, *2021*, 69.

67. Mustafa, S.B.; Akram, M.; Muhammad Asif, H.; Qayyum, I.; Hashmi, A.M.; Munir, N.; Khan, F.S.; Riaz, M.; Ahmad, S. Antihyperglycemic Activity of Hydroalcoholic Extracts of Selective Medicinal Plants Curcuma Longa, Lavandula Stoechas, Aegle Marmelos, and Glycyrrhiza Glabra and Their Polyherbal Preparation in Alloxan-Induced Diabetic Mice. *Dose-Response* **2019**, *17*, 1559325819852503. [CrossRef] [PubMed]

68. Tang, H.; Ma, F.; Zhao, D.; Xue, Z. Exploring the Effect of Salvianolic Acid C on α-Glucosidase: Inhibition Kinetics, Interaction Mechanism and Molecular Modelling Methods. *Process Biochem.* **2019**, *78*, 178–188. [CrossRef]

69. Ali, A.; Cottrell, J.J.; Dunshea, F.R. Antioxidant, Alpha-Glucosidase Inhibition Activities, In Silico Molecular Docking and Pharmacokinetics Study of Phenolic Compounds from Native Australian Fruits and Spices. *Antioxidants* **2023**, *12*, 254. [CrossRef]

70. Son, M.J.; Rico, C.W.; Nam, S.H.; Kang, M.Y. Effect of Oryzanol and Ferulic Acid on the Glucose Metabolism of Mice Fed with a High-fat Diet. *J. Food Sci.* **2011**, *76*, H7–H10. [CrossRef]

71. Aleixandre, A.; Gil, J.V.; Sineiro, J.; Rosell, C.M. Understanding Phenolic Acids Inhibition of α-Amylase and α-Glucosidase and Influence of Reaction Conditions. *Food Chem.* **2022**, *372*, 131231. [CrossRef]

72. Mei, X.; Zhou, L.; Zhang, T.; Lu, B.; Sheng, Y.; Ji, L. Chlorogenic Acid Attenuates Diabetic Retinopathy by Reducing VEGF Expression and Inhibiting VEGF-Mediated Retinal Neoangiogenesis. *Vascul. Pharmacol.* **2018**, *101*, 29–37. [CrossRef] [PubMed]

73. Xu, W.; Luo, Q.; Wen, X.; Xiao, M.; Mei, Q. Antioxidant and Anti-Diabetic Effects of Caffeic Acid in a Rat Model of Diabetes. *Trop. J. Pharm. Res.* **2020**, *19*, 1227–1232. [CrossRef]

74. Kubínova, R.; Pořízková, R.; Navrátilová, A.; Farsa, O.; Hanáková, Z.; Bačinská, A.; Čížek, A.; Valentová, M. Antimicrobial and Enzyme Inhibitory Activities of the Constituents of *Plectranthus madagascariensis* (Pers.) Benth. *J. Enzyme Inhib. Med. Chem.* **2014**, *29*, 749–752. [CrossRef]

75. Govindaraj, J.; Sorimuthu Pillai, S. Rosmarinic Acid Modulates the Antioxidant Status and Protects Pancreatic Tissues from Glucolipotoxicity Mediated Oxidative Stress in High-Fat Diet: Streptozotocin-Induced Diabetic Rats. *Mol. Cell Biochem.* **2015**, *404*, 143–159. [CrossRef] [PubMed]

76. Jayanthy, G.; Subramanian, S. RA Abrogates Hepatic Gluconeogenesis and Insulin Resistance by Enhancing IRS-1 and AMPK Signalling in Experimental Type 2 Diabetes. *RSC Adv.* **2015**, *5*, 44053–44067. [CrossRef]

77. Runtuwene, J.; Cheng, K.-C.; Asakawa, A.; Amitani, H.; Amitani, M.; Morinaga, A.; Takimoto, Y.; Kairupan, B.H.R.; Inui, A. Rosmarinic Acid Ameliorates Hyperglycemia and Insulin Sensitivity in Diabetic Rats, Potentially by Modulating the Expression of PEPCK and GLUT4. *Drug Des. Dev. Ther.* **2016**, *10*, 2193–2202.

78. Zeng, L.; Zhang, G.; Lin, S.; Gong, D. Inhibitory Mechanism of Apigenin on α-Glucosidase and Synergy Analysis of Flavonoids. *J. Agric. Food Chem.* **2016**, *64*, 6939–6949. [CrossRef] [PubMed]

79. Mechchate, H.; Es-Safi, I.; Mohamed Al Kamaly, O.; Bousta, D. Insight into Gentisic Acid Antidiabetic Potential Using in Vitro and in Silico Approaches. *Molecules* **2021**, *26*, 1932. [CrossRef] [PubMed]

80. Awortwe, C.; Fasinu, P.S.; Rosenkranz, B. Application of Caco-2 Cell Line in Herb-Drug Interaction Studies: Current Approaches and Challenges. *J. Pharm. Pharm. Sci.* **2014**, *17*, 1. [CrossRef] [PubMed]

81. Iftikhar, M.; Iftikhar, A.; Zhang, H.; Gong, L.; Wang, J. Transport, Metabolism and Remedial Potential of Functional Food Extracts (FFEs) in Caco-2 Cells Monolayer: A Review. *Food Res. Int.* **2020**, *136*, 109240. [CrossRef]

82. Rodrigues, C.V.; Sousa, R.O.; Carvalho, A.C.; Alves, A.L.; Marques, C.F.; Cerqueira, M.T.; Reis, R.L.; Silva, T.H. Potential of Atlantic Codfish (*Gadus morhua*) Skin Collagen for Skincare Biomaterials. *Molecules* **2023**, *28*, 3394. [CrossRef]

83. Costa, P.; Grevenstuk, T.; Rosa da Costa, A.M.; Gonçalves, S.; Romano, A. Antioxidant and Anti-Cholinesterase Activities of Lavandula Viridis L'Hér Extracts after in Vitro Gastrointestinal Digestion. *Ind. Crops Prod.* **2014**, *55*, 83–89. [CrossRef]

84. Ames, B.N.; McCann, J.; Yamasaki, E. Methods for Detecting Carcinogens and Mutagens with the Salmonella/Mammalian-Microsome Mutagenicity Test. *Mutat. Res./Environ. Mutagen. Relat. Subj.* **1975**, *31*, 347–363. [CrossRef]

MDPI

St. Alban-Anlage 66

4052 Basel

Switzerland

www.mdpi.com

Foods Editorial Office

E-mail: foods@mdpi.com

www.mdpi.com/journal/foods